Reform, Revolution and Direct Action amongst British Miners

Historical Materialism Book Series

The Historical Materialism Book Series is a major publishing initiative of the radical left. The capitalist crisis of the twenty-first century has been met by a resurgence of interest in critical Marxist theory. At the same time, the publishing institutions committed to Marxism have contracted markedly since the high point of the 1970s. The Historical Materialism Book Series is dedicated to addressing this situation by making available important works of Marxist theory. The aim of the series is to publish important theoretical contributions as the basis for vigorous intellectual debate and exchange on the left.

The peer-reviewed series publishes original monographs, translated texts, and reprints of classics across the bounds of academic disciplinary agendas and across the divisions of the left. The series is particularly concerned to encourage the internationalization of Marxist debate and aims to translate significant studies from beyond the English-speaking world.

For a full list of titles in the Historical Materialism Book Series
available in paperback from Haymarket Books, visit:
https://www.haymarketbooks.org/series_collections/1-historical-materialism

Reform, Revolution and Direct Action amongst British Miners

The Struggle for the Charter in 1919

Martyn Ives

Haymarket Books
Chicago, IL

First published in 2016 by Brill Academic Publishers, The Netherlands
© 2016 Koninklijke Brill NV, Leiden, The Netherlands

Published in paperback in 2017 by
Haymarket Books
P.O. Box 180165
Chicago, IL 60618
773-583-7884
www.haymarketbooks.org

ISBN: 978-1-60846-819-5

Trade distribution:
In the US, Consortium Book Sales, www.cbsd.com
In Canada, Publishers Group Canada, www.pgcbooks.ca
In the UK, Turnaround Publisher Services, www.turnaround-uk.com
All other countries, Ingram Publisher Services International, ips_intlsales@
ingramcontent.com

Cover design by Jamie Kerry of Belle Étoile Studios and Ragina Johnson.

This book was published with the generous support of Lannan Foundation
and the Wallace Action Fund.

Library of Congress Cataloging-in-Publication data is available.

Contents

Preface and Acknowledgements

This book fills a gap in the existing literature on the politics of the Miners' Federation of Great Britain (MFGB) during the period of the Sankey Commission, and offers an alternative analysis of the critical year 1919. It examines previously unknown or un-researched mass mobilisations of miners across a range of coalfields in the early months of 1919, and in the period following the publication of the Reports of the Sankey Commission. Extensive use of local union records and local newspapers, amongst other sources, produces detailed accounts of rank and file actions in the districts that were barely contained by the official leadership and sometimes took on an insurrectionary quality. It also provides an analysis of the influential groups of militants – amongst their numbers syndicalists, revolutionaries and Labour left wingers – who led the local movements, posing a threat to the MFGB Executive Committee's policy of conciliation and compromise over the Miners' Charter. The official Yorkshire Miners' Association's strike of the summer provides a comparative study to the overwhelmingly unofficial strikes of the first quarter of the year. This 'history from below' is integrated into a wider analysis of national politics and industrial politics. In particular, the investigation of radical local movements is linked to an examination of the MFGB leadership, in the context of a discussion about the general tendencies of trade union bureaucracies to seek accommodation with capital and clamp down on rank and file militancy. Particular attention is paid to the decision of the MFGB EC to co-operate with the Sankey Commission and accept the Interim Report, and to the considerations that moulded official MFGB strategy between June and August 1919, from the publication of the Final Report to Lloyd George's definitive rejection of nationalisation. Consideration and space is also given to intrigues at Westminster, and the politics of the Coalition government, which, in the end, successfully navigated its way through the post-war crisis.

Political debates inside the MFGB are located within the context of the impassioned discussions that were taking place within the wider labour movement as to whether social advance would come through parliament, or through the direct action of the working class and their trade unions. The book provides a critique of 'direct action' and syndicalism, and an analysis of its strengths and weaknesses in 1919. It argues that whilst 1919 was not Britain's lost revolution, the situation held revolutionary potential and in the midst of the mass strikes and riots Labour's gradualism could seem at times almost irrelevant. Its eventual hegemony in the 1920s and beyond should not be seen as inevitable; certainly it was not at the time. The consideration of the politics of the miners in this pivotal year raises significant questions about dominant assumptions and images within labour historiography. In particular, the argument is constructed around a belief that reasonable counterfactuals can illuminate history, that suppressed alternatives are worthy of investigation, and that the actual pattern of development should not simply be characterised as 'normal'.

My thanks go to the staff and trade unionists at the various NUM Area Offices who made their unions' records available to me, and who helped me to find what I was looking for. Particular thanks go to Phil Thompson at Barnsley, and Julie Fletcher at Leigh. Thanks to Nicky Stonelake and Hywell Francis at the South Wales Miners' Library, and to all those who painstakingly collected and catalogued the South Wales Coalfield Archive. Thanks also to the staff at the Mining Museum and Library at Pendleton, and to all at the Working Class Movement Library in Salford, Alan Kahan in particular. I am indebted to the Librarians, Archivists and staff at the John Rylands University Library, the Manchester Central Reference Library, the Scottish Records Office, the National Library of Scotland, the National Library of Wales, the Scottish Records Office, the Glamorgan Records Office, the Wigan Records Office, the Durham Records Office, the Public Records Office at Kew, the British Library of Political and Economic Science, and at the International Instituut voor Sociale Geschiedenis in Amsterdam. Special thanks to the staff at the Colindale Newspaper Library for the hundreds of volumes they wheeled my way. I would like to thank the Government Department for the contribution from the Chester Fund. Also thanks to Karen Hall and Joyce Gardner, for getting me to fill in the right forms at the right time. I remain grateful to Anne Robertson, without whose support this book could not have been written. My thanks go to David Howell for his knowledge and guidance, and to Australia, for being so far away that Paul Blackledge had time to read it on a long haul flight home and recommend it for publication.

List of Abbreviations

ASLE&F	Amalgamated Society of Locomotivemen, Enginemen and Firemen
BSP	British Socialist Party
CIC	Coal Industry Commission (1919)
CLC	Central Labour College
CP	Communist Party
CWC	Clyde Workers' Committee
DLB	Dictionary of Labour Biography
DMA	Derbyshire Miners' Association
FK&CMA	Fife, Kinross, and Clackmannan Miners Association
FMRC	Fife Miners Reform Committee
ILP	Independent Labour Party
ISEL	Industrial Syndicalist Education League
IWW	Industrial Workers of the World
IUC	Industrial Unrest Committee
JDC	Joint Demobilisation Committee
JSDC	Joint Standing Disputes Committee
L&CMF	Lancashire and Cheshire Miners' Federation
LMRC	Lanarkshire Miners' Reform Committee
LMU	Lanarkshire Miners' Union
MAGB	Miners' Association of Great Britain
MFGB	Miners' Federation of Great Britain
MFGB EC	Miners' Federation of Great Britain, Executive Committee
MM	Minority Movement
MRC	Miners' Reform Committee
MSWCOA	Monmouthshire and South Wales Coal Owners' Association
NER	North-Eastern Railway
NIC	National Industrial Conference
NLS	National Library of Scotland
NMA	Nottinghamshire Miners' Association
NUPPO	National Union of Police and Prison Officers
NUR	National Union of Railwaymen
NUSMW	National Union of Scottish Mine Workers
PLP	Parliamentary Labour Party
PRO	Public Records Office
RILU	Red International of Labour Unions
SLP	Socialist Labour Party
SWMF	South Wales Miners' Federation

SWML   South Wales Miners' Library
SWSS   South Wales Socialist Society
SYCOA   South Yorkshire Coal Owners' Association
TMNS   The Miners' Next Step
TUC   Trade Union Congress
URC   Unofficial Reform Committee
WEWNC   War Emergency Workers' National Committee
WYCOA   West Yorkshire Coal Owners' Association
YMA   Yorkshire Miners' Association

Introduction

But the diversity of possible itineraries does demonstrate that eventual results cannot be predicted at the outset. Each step proceeds for cause, but no finale can be specified at the start, and none would ever occur a second time in the same way, because any pathway proceeds through thousands of improbable stages. Alter any early event, ever so slightly and without apparent importance at the time, and evolution cascades into a radically different channel.[1]

∴

While the general strike of 1926 is often thought of as the highpoint of industrial struggle in interwar Britain, in many respects it was only an echo of the more insurgent militancy of 1919–21. Indeed, the defeat suffered by the miners in the aftermath of the general strike arguably did little more than compound that which had followed Black Friday five years earlier. Then, on 15 April 1921, the famed Triple Alliance had failed as rail and transport leaders refused to call for strike action in support of the miners. Black Friday and the general strike were important in shaping interwar politics in Britain not least because they cemented the idea that amelioration of the working class could come only through the election of a Labour government rather than via 'self-activity', the collective action of workers and trade unions at workplace, factory or pit. For this reason alone it is understandable that historians of this period have tended to focus on, and give primacy to, the industrial defeats of the 1920s.[2] Not so easy to comprehend is the relative absence of interest in, or study of, the mass militancy of the years that *preceded* Black Friday and the general strike, and the lack of attention paid to the political alternatives to social democracy which accompanied it.[3]

---

1 Gould 1998, p. 51.
2 For an early account see Page Arnot 1953; more recent accounts include A. Campbell et al. 1996; McIlroy et al. 2009; Barron 2009.
3 In 1979, Chris Wrigley published a monograph which invited historians to pay more attention to the 'critical year' of 1919, but it has remained curiously under-researched (Wrigley 1979). John Saville also highlighted the importance of 1919, but he did so on the basis of existing literature, with no additional research. Ralph Miliband's *Parliamentary Socialism*, a study

This book is an attempt to redress the historiographical imbalance; by shift-
ing attention back to the year 1919, the high-water mark of the neo-syndicalist
idea of 'direct action', it hopes to provide a new perspective on the post-war
years, and to excavate labour's challenge from beneath the burial mounds of
1921 and 1926. Pull focus back to this pivotal year and a different view of post-
war labour politics presents itself, one in which Labour's eventual hegemony in
the working class movement loses its air of inevitability. This is nowhere more
visible than in the MFGB, which was at the heart of both the Labour party and
the direct action movement in the post-war period. As such, the miners' history
at this juncture deserves a very close reading.

While agreeing with Leslie Morton and George Tate that 'the year 1919 was
a critical one indeed for the ruling class', I cannot accept their claim that the
position of the ruling class 'was saved only by the absence of a determined
centralised leadership to coordinate the separate streams of turbulent popular revolt'.[4] This overly simplistic analysis of the situation leaves too much out,
not to mention the historical implausibility of a British Lenin leading a British
Bolshevik party in 1919. This is far from the last word on the matter, however.
1919 was witness to a massive upturn in industrial militancy, a militancy characterised by the fact that workers were, according to Ralph Miliband, 'ready
for bold leadership on political as well as industrial issues' but whose potential
union leaders were more concerned to control than exploit.[5] And if the miners
were hoodwinked and their militancy hamstrung by the political cunning of
Lloyd George's Sankey Commission, then the Welsh Wizard could not have succeeded without the complicity of trade union officials. It was they who sold
Sankey to their members, against the prescient warnings of activists, including
revolutionaries, to their left.[6]

While it may be implausible to imagine a British Lenin in 1919, there were,
as we shall see, real British revolutionaries who played significant roles in what

---

sympathetic to the idea of direct action as a radical alternative, places insufficient emphasis
on the importance of 1919. His focus on the leadership within the labour movement limits his
ability to say much about direct action in 1919 once the miners accept the Sankey Commission. More attention is paid to the Council of Action and the Russian issue in 1920, by which
time direct action has been largely stripped of the insurgent qualities which it had displayed
in the earlier year (Saville 1988; Miliband 1964); Campbell, Fishman and Howell's study of the
miners between 1910 and 1947 makes only three mentions of the Sankey Commission, and
these do not reference any particular militancy in 1919 (Campbell et al. 1996).

4   Morton and Tate 1979, p. 279.
5   Miliband, quoted in Fox 1985, p. 291.
6   Foster 2004, p. 27; Campbell 2000, p. 169.

was a genuine mass struggle. And it is far from implausible to imagine that the competition for hegemony within the labour movement between revolutionaries on the one hand, and reformists centred around the union leaderships on the other, could have panned out quite differently than it eventually did.

Research for this book was undertaken with a belief that history is a process, which at certain critical moments has a potentiality to unfold in a variety of radically different ways. In this sense Stephen Jay Gould's 'diversity of possible itineraries' can be applied to historical as well as evolutionary processes. This approach places me in the seemingly difficult position of taking seriously some of the claims normally associated with counterfactual history. According to Tory historian Andrew Roberts, Marxists can have no truck with counterfactual history because 'Marxism requires humans to operate according to a pre-determined dialectical materialism, and not by the caprices of accident or serendipity'.[7] Drawing on Niall Ferguson's discussion of the importance of counterfactuals to the study of history, Roberts cites the examples of 'Marxists' such as E.P. Thompson, Eric Hobsbawm and E.H. Carr (sic) as amongst the most prominent critics of counterfactual history. In the text by Ferguson to which Roberts refers, the evidence that Marxists do indeed scorn counterfactuals as, in E.P. Thompson words, 'unhistorical shit' seems unassailable.[8] However, if we turn to the essay by Thompson from which this phrase was pruned, it is immediately apparent that it is best understood as a passing comment, taken out of context, within a broader defence of the historian's craft against the consequences of (Althusserian) structuralism. Indeed, whereas Thompson supplies no reasons to justify his criticisms of counterfactuals, he does explicitly defend 'history as process, as open-ended and indeterminate eventuation – but not for that reason devoid of rational logic or of determining pressures ... of alternative possibilities'.[9] Now it seems reasonable to suppose that if history does include 'alternative possibilities', then realistic counterfactuals will potentially be a useful tool through which they might be explored.

Ferguson also refers to Hobsbawm's criticisms of counterfactuals, though he does not cite Hobsbawm's claim that 'history is what happened, not what might have happened'.[10] But even here, things aren't quite as simple as they appear. For despite the fact that Hobsbawm does, by contrast with Thompson, provide justification for his criticisms of counterfactual history, these reasons

---

7   Roberts 2004, pp. 2–3.
8   Ferguson 1997, p. 5; Thompson 1978, p. 108.
9   Thompson 1978, pp. 83–4, 103.
10   Ferguson 1997, p. 5; Hobsbawm 1984, p. 8; cf. 1994, p. 42.

are not quite so absolute as Ferguson would have it. Hobsbawm actually suggests that counterfactuals can be of interest methodologically insofar as they may help in the clarification of hypotheses to explain really possible alternatives.[11] Interestingly, the substantive target of Hobsbawm's criticisms was not counterfactual history as such, but rather a tendency he located amongst the 'radicalised generations of students and (in due course) junior professors in the 1960s' to mine history for 'inspiring examples of struggles'.[12] Hobsbawm's claim is that these historians tended rather to romanticise the history of the labour movement by painting it a good deal more radical than it actually was. The problem here is the implication that while those historians under criticism were writing from ideologically motivated perspectives, he was simply telling it like it was. But this argument is open to the devastating criticisms that have been made of empiricism by, especially, Marxists. Clearly, Hobsbawm's own perspective was at least as ideological as the historians he criticised. Specifically, the formative influence on Hobsbawm's Marxism was the 'Popular Front' politics embraced by the Comintern in the mid-1930s as Stalin transformed the Communist parties into what were in essence reformist organisations with a view to forging alliances with Britain and France against Germany.[13] Hobsbawm was a consistent enough Communist to recognise that this perspective assumed that the working class was no longer a revolutionary force and that, therefore, the kind of politics associated with the Bolshevik Party were not relevant to modern conditions. It is this perspective that informed his fairly consistent tendency to dismiss, as Ian Birchall and Norah Carlin argue, 'any concept of the self-activity of the working class as a revolutionary force'.[14]

Unfortunately, it is just this sort of approach that has condemned the revolutionaries and direct action syndicalists of 1919 to the 'enormous condescension of posterity'. The problem with condescension of this sort is that it tends, as E.P. Thompson wrote of Perry Anderson's analysis of English history, to 'flatten history'. 'Awkward facts are not mentioned; awkward decades (e.g. 1920–40) are simply elided'.[15] Militant critics of counterfactual history such as E.H. Carr reinforce this tendency to dismiss the movements of history's losers by their refusal to take seriously the possibility that (sometimes) these losers could have won, or at the very least they could have reshaped the overall outcome of

11   Hobsbawm 1984, p. 8.
12   Hobsbawm 1984, p. 6.
13   Hallas 1985, pp. 142–8.
14   Birchall and Carlin 1983, p. 104; See also Blackledge 2012, pp. 22–4.
15   Thompson 1978, p. 275.

struggles.[16] Carr infamously dismissed counterfactuals as 'parlour games' that have nothing 'to do with history'.[17] His reasons for doing so were, however, far from convincing. He suggested that whereas historians of the Wars of the Roses or the Norman Conquest have been allowed to get on with the job of explaining what happened in the periods they research, because the debates from the Russian Revolution were still very much alive, historians of that period (such as himself) found themselves constantly under attack from one quarter or another.[18]

While Carr was right about the attacks made on serious historians of the Russian Revolution, this does not justify his rejection of counterfactuals. For any theory of history that wants to avoid teleology must be open to the question 'what if?' More to the point, once it is recognised that Marx's determinism has nothing in common with political or historical fatalism, it is apparent that historical materialism becomes open to engagement with counterfactuals. And Marx was not a fatalist. His determinism, as is apparent in his *Theses on Feuerbach*, was aimed not at reducing humans to ciphers of impersonal forces, but rather at grasping the nature of real *human* agency. In the first of these theses Marx famously criticised the fatalistic implications of existing materialist theories of agency and conversely the fundamental limitations of the idealist (moralist) alternative:

> [t]he chief defect of all hitherto existing materialism ... is that the thing, reality, sensuousness, is conceived only in the form of the object or of con-templation, but not as sensuous human activity, practice, not subjectively. Hence, in contradistinction to materialism, the active side was developed abstractly by idealism – which, of course, does not know real, sensuous activity as such.[19]

Later on in the same text, Marx writes that 'the coincidence of the changing of circumstances and of human activity or self-changing can be conceived and rationally understood only as *revolutionary practice*'.[20] This perspective was renewed by Lenin through his critique of the fatalism of Second International

---

16 Ferguson and Roberts are right to associate Carr with the rejection of counterfactual history, though it is a sign of their intellectual shallowness that they label him a 'Marxist'.
17 Carr 2001, p. 91.
18 Ibid.
19 Marx 1975, p. 421.
20 Marx 1975, p. 422.

Marxism. In notes taken from a close reading of the *Science of Logic*, Lenin expressed his break with Second International dualism thus:

> The activity of man, who has made an objective picture of the world for himself, changes external actuality, abolishes its determinates (= alters some sides or other, qualities, of it), thus removes from it the features of semblance, externality and nullity, and makes it as being in and for itself (= objectively true).[21]

Commenting on these notebooks, Stathis Kouvelakis points out that it is 'particularly significant' that Lenin ended the section on 'philosophical materialism' with a reference to the notion of 'revolutionary practical activity'. For Lenin understood that subjective practical activity lay at the centre of the 'objective' world, and consequently insisted that social scientific laws should not be 'fetishised' as things distinct from conscious human activity but instead be recognised as necessarily 'narrow, incomplete, [and] approximate' attempts to frame political intervention.[22] Consequently, whereas Second International theorists interpreted Hegel's claim that to act freely meant to act in accordance with necessity in a reductive manner, for Lenin, as Richard Day argues, 'man's consciousness not only reflects the objective world but creates it'.[23]

If it is true that human activity helps create the social world, surely realistic counterfactuals are an essential counterpart to the idea that history, from time to time, opens up the possibility of radical social transformations that must be fought for. Indeed, Trotsky's analysis of the pivotal role of Lenin in 1917 is an example of exactly such an approach. Imagining Lenin's absence from Russia in 1917, Trotsky argues that while other revolutionaries arrived at similar political conclusions to Lenin, political organisation was needed to win the revolution and none but Lenin had the stature within the Bolshevik Party to revise the Party's programme at the speed necessary to seize the moment in October. Without Lenin, therefore, he concludes there would have been no socialist revolution.[24] This example suggests that the claim that counterfactuals necessarily degenerate into parlour games may be true of the essays collected by Roberts and Ferguson, but it is not necessarily so. Ian Kershaw is surely right that 'historians implicitly operate with short-range counterfactuals in terms of

21   Lenin 1961, *Collected Works*, vol. 38, pp. 217–18, emphasis in original.
22   Kouvelakis 2007, pp. 174, 186.
23   Day, quoted in Anderson 1995, p. 113; cf. Callinicos 2007, pp. 23–8.
24   MacIntyre 2008, pp. 267–76.

alternatives to immediate important occurrences or developments'.[25] And so long as we follow Ralph Darlington's advice about using counterfactuals with care (alternative scenarios cannot be plucked from thin air), we avoid doing violence to the historical record by dismissing real alternatives that were actually debated at the time just because these arguments did not carry the day amongst contemporary actors.[26] To understand history as it actually happened, we therefore need to investigate feasible alternative outcomes with a view to formulating informed conjectures as to why these were not realised. Similarly, the explanation of why certain movements, mobilisations or schools of thought were successful in the past involves an explanation of why their competitors were not, particularly if that competition was keenly felt. Unless space is allowed for history's lost causes, it is all too easy for history to degenerate into teleology.

Those historians who dismiss the idea of a revolutionary potential in 1919 have tended to do so on the basis of a one-sided account according to which the events as they transpired were (more or less) the only feasible outcome of the historical process. Thus, for instance, Chris Wrigley argues that of the three challenges posed by labour to the British state in the period from 1917 to 1920, although the 'constitutional' and 'industrial' were of significant moment, the 'revolutionary challenge' was 'in reality of no great dimension' even if it did provide a cause for 'anxiety' within the government.[27]

Though this claim involves an important element of truth, by skirting over the revolutionary elements of the struggle it tends too quickly to assume the continued hegemony of reformism within the British labour movement. This book, by contrast, is an argument against foregone conclusions. It is an exercise in what Christopher Hill calls the 'worm's eye view' of history, and my hope is that this perspective opens the door, as Hill suggests it should, to a deeper understanding of the past.[28] It is inspired by a belief that in 1919 there was a conjuncture of factors which rendered the situation profoundly unpredictable, and pregnant with alternative possible outcomes. To contemporaries who felt themselves to be standing at the threshold of a new society, the defeats of the future were barely conceivable. If anything, it was the old order that seemed to be crumbling as the revolutionary process unfolded across eastern and central Europe. In Britain, the *Herald* welcomed the New Year with unbounded optimism:

25   Kershaw 2007, p. 6.
26   Darlington 2006, pp. 76–88; cf. Hawthorn 1991, p. 17.
27   Wrigley 1993, p. 263; cf. 2000, p. 182.
28   Hill 1975, pp. 14, 18; cf. Blackledge 2005, p. 219.

It is the simple and dazzling fact that 1919 can be, and must be, the beginning of a veritable New Order, a radiant point of time to which our children's children decisively marking workers' passing out of slavery to freedom, their progression from death in life, to intensive cultivation on all planes of the Commonwealth.[29]

Lansbury wrote these words in the same week that the December 1918 general election had produced a coalition government 'of capitalists and landlords' led by Lloyd George.[30] Small grounds for optimism, one would have thought. This election result alone, some might say, weighs decisively against there being any revolutionary potential in 1919. Yet Lansbury's positivity was rooted in the belief that the election reflected an ephemeral outpouring of triumphant patriotism; beneath the jingoism he detected a working class that was in no mood to be constrained by constitutional or parliamentary timetables. If the reconstruction of society summed up in the promise of 'homes fit for heroes' was not rapidly delivered from above, then it would be undertaken from below, by direct action. Revolt was in the air, blown across to Britain from Russia and Central Europe. There it mingled with the domestic unrest to produce an atmosphere of insurgency and generate an intense discussion as to whether 1919 would become Britain's 1917. Of course, in the event, there was no British revolution; apart from anything else, its rulers were afforded a degree of flexibility by victory in the war, a flexibility which the Kaiser, for one, was denied.

Yet it is not enough to simply dismiss the issue of revolution out of hand. Perhaps with hindsight Basil Thompson's reports to the Cabinet can appear hysterical, but he had no monopoly on panic. Care should be taken that hindsight does not lead to the distortion of certain historical realities. The eventual shape of the postwar order in Europe and in Britain was impossible to predict in 1919, especially in the first half of the year, and the concept of revolution was very much part of the political discourse. Beatrice Webb pondered the question again and again in her diaries. As 1918 drew to a close, she wrote:

The Bolsheviks grin at us from a ruined Russia, and their creed, like the plague of influenza, seems to be spreading westwards from one country to another ... Will western civilization flare up in the flames of an anarchic

---

29   *Herald*, 4 January 1919. During the war, Lansbury was only able to sustain his newspaper on a weekly basis, and it did not revert to daily publication until 30 March 1919.

30   *Herald*, 4 January 1919.

revolution? Individuals brood over these questions and wonder what will have happened this time next year.[31]

Webb welcomed the armistice with considerable misgivings:

> Peace! Thrones are everywhere crashing and the men of property every-where secretly trembling ... How soon will the tide of revolution catch up with the tide of victory? That is a question which is exercising Whitehall and Buckingham Palace, and which is causing anxiety even among the more thoughtful democrats. Will it be six months or a year?[32]

In July 1919, Edward Grey confessed his belief that 'we shall see a world revolution comparable with the break up of the Roman Empire', while Arthur Henderson identified 'the great tide of revolutionary feeling [which] is rising in every country' as 'the outstanding fact of world politics at the present time'.[33] A compilation of all such writings and speeches in 1919 would fill several volumes. Indeed, there is scarcely a newspaper or journal that does not raise revolution as a possibility, and the issue was discussed hotly – and often – at trade union meetings and in government. Sometimes the spectre was raised as a red scare by guardians of constitutionality, but often the fears – or hopes – were genuine. While this book is not an attempt to salvage a lost British revolution, it does argue that any realistic account of Britain in the aftermath of the war should be alive to the fact that 'revolution' was an important and influential currency in 1919. Contemporaries believed revolution was possible, and the language of revolution helped shape the events of that year.

According to John Dunn, a revolution is:

> [a] political struggle of great intensity, initiated by political crises within particular historical societies and resolved ... by the creation of a political capacity to confront the historical problems of these societies in ways that their pre-revolutionary regimes had proved wholly incapable of doing.[34]

He follows Lenin in defining a revolutionary situation as one in which the ruling class is unable to continue ruling in the old way, while the working class refuses

---

31   Cole 1952, p. 134.
32   Cole 1952, p. 136.
33   Cole 1952, p. 164; Henderson 1918, pp. 74–5.
34   Dunn 1989, p. xvi.

to continue to be ruled in that way.[35] Similarly, Charles Tilly follows Trotsky to argue that in a revolutionary situation three proximate causes converge:

> the appearance of contenders ... advancing exclusive competing claims to control the state; commitment to those claims by a significant segment of the citizenry; [and] incapacity or unwillingness of rulers to suppress the alternative ... claims.[36]

Clearly these conditions were not met in Britain in 1919. But it would be wrong to assume a crudely objectivist account of the strengths of the British state against the fragility of, for instance, Tsarism. In part at least, the resilience of the British state was a consequence of Lloyd George's political skills on the one side and the acquiescence of the leadership of the labour movement on the other. That these actors were themselves in no small part products of a prior historical process should not detract from this fact. For his part, Lloyd George sat atop a powerful social base of property-owning and other anti-revolutionary social layers,[37] while the (divided) left vied for leadership of an increasingly militant labour movement. What overly objectivist accounts forget is that the very essence of the social polarisation between these groups was its dynamism. Attitudes and opinions were in a state of flux. On both sides of this polarisation there were respective struggles for hegemony by developing tendencies around emerging strategies. Within the ruling class there were disputes between those backing Churchill, who wanted a more aggressive approach to labour, against Lloyd George's strategy of buying time through concessions whilst sowing divisions in his opponents' ranks. That these disputes did not turn into open divisions can, in part, be explained by the failure of labour to exploit them.

The key leadership positions within the labour movement were taken by trade union officials in general and miners' leader Robert Smillie in particular. Though Smillie had a reputation as one of the most left-wing trade union leaders of his day, he played a key role in selling the Sankey Coal Commission to the miners against the objections of revolutionaries such as Will Lawther who recognised the commission for what it was: a delaying tactic used by the government to ride the storm.[38] Activists like Lawther provided an alternative to the MFGB leadership inside the coalfields. In between these two poles, the

---

35   Dunn 1989, p. 14; Lenin 1993, pp. 104–5.
36   Tilly 1993, p. 10.
37   Wrigley 1993, pp. 280–4.
38   Hinton 1983, p. 111.

mass of the miners were swayed one way and then the other. Even Smillie had to be persuaded to accept Sankey against his gut instincts by fellow ILP er Sidney Webb.[39]

To assume the hegemony of reformism within the workers' movement surely begs the question of how reformist activists overcame the challenge from those committed to the philosophy of direct action. In relation to Russia, Dunn explains the triumph of the Bolsheviks in 1917 as a function of Lenin's 'voluntaristic' break with the Marxist conception of revolution as, as it were, a model of 'plucking the ripe fruits'. Though Dunn's approach involves a caricatured interpretation of Lenin's politics, its great strength flows from his focus on the question of human agency. And if Lenin's ability to make a decisive intervention in Russia in 1917 can be understood against the backdrop of his roots in the Russian revolutionary tradition, it is equally as important to historicise the weaknesses of the British left.

Britain's pre-eminence amongst world powers in the nineteenth century had shielded her from the political turmoil experienced elsewhere. This coloured the class struggle in Britain in the period after Chartism; though such struggles were widespread, they tended to be fragmented and sectional in character. This informed the emergence of a very moderate leadership within the trade union movement.[40] Conversely, the abstract nature of British Marxism is perhaps best understood as the flipside to the non-revolutionary nature of the British working class. Because it was so sectional, the class struggle militated against the emergence of revolutionary political class consciousness within the working class. Indeed, working class politics, especially after the defeats of New Unionism, became increasingly dominated by reformist tendencies. Though this development was rooted in a specifically British context, it mirrored similar processes across Europe at the turn of the last century.[41]

Whereas Marx, in *The German Ideology*, had claimed that 'even a minority of workers who combine and go on strike very soon find themselves compelled to act in a revolutionary way', he subsequently recognised, as he wrote in *Wages, Price and Profit*, that since the 1850s trade unions had increasingly grown to struggle 'against the effects of the existing system, instead of simultaneously trying to change it'.[42] Unfortunately, despite this and similar *ad hoc* insights, Marx never articulated a coherent model of labour reformism.[43]

---

39   Cliff and Gluckstein 1996, p. 84.
40   Burgess 1980, p. 27.
41   Schorske 1983, pp. 1–28.
42   Cliff and Gluckstein 1986, pp. 21–2.
43   Johnson 1980, p. 73; Fernbach 1974, pp. 59, 63; Molyneux, 1986, p. 31.

This gap in Marx's politics is partly explicable as a corollary of the relatively underdeveloped nature of institutionalised reformism before his death. However, as this institutional form grew in strength around the turn of the last century, it became a pressing concern for important students of trade unionism such as Sidney and Beatrice Webb, Robert Michels and Rosa Luxemburg.[44] What all of these commentators recognised in their different ways was the *generally* conservative character of the layer of professional negotiators that emerged within trade unions to organise the terms and conditions of the sale of labour power.

In a recent reassessment of theories of bureaucratic conservatism within trade unions, Ralph Darlington and Martin Upchurch have suggested that this phenomenon is underpinned by, first, their role as negotiators over the conditions of sale of labour power, second, their intermediary role between capital and labour, third, their relatively high pay by contrast with those they represent, and, fourth, their attachment to social-democratic political parties. These factors not only differentiate the conditions of life of the bureaucracy from the ordinary trade union members, but more importantly they tie them to the wage contract itself.[45]

It is important to state that for revolutionaries in 1919, as ever, the problem was not confined to the limitations of the trade union bureaucracy. It was also the limitations of trade unionism itself. Whilst trade unionism unites workers to challenge capitalism, it ultimately seeks accommodation with it. Furthermore, as with the bureaucracy, for most of the time workers too will tend to naturalise the capitalist social relations in which they live, and will often stand to the right of or share the general perspective of their trade union officials – this is the rational core of the claim that trade union bureaucrats reflect a deeper conservatism rooted in the nature of trade unionism itself.[46]

Though this argument reflects an important truth, it takes no account of the tendency, under certain conditions, for trade union struggle to overcome sectionalism, unite the working class, and thereby transcend narrow economism to present a challenge to capitalism as a whole. As a corollary of this, it understates the gap that tends to open between full-time officers and the rank and file, especially in periods of heightened class struggle. As Rosa Luxemburg

---

44   See Blackledge 2013, pp. 32–6; Luxemburg 1986; Michels 1962; Webb and Webb 1907; Schorske, 1983, p. 127; Burgess 1980, p. 27.

45   Darlington and Upchurch 2011, p. 80; cf. Cliff and Gluckstein 1986, p. 290.

46   See Hyman 2011, and for a critique, see Darlington and Upchurch 2011. It was all said before in an earlier exchange between Richard Hyman and Duncan Hallas: see Hyman 1980 and Hallas 1980.

argued, when workers engage in mass collective and self-directed struggles, they can begin to move beyond the limitations of this reformist framework.[47] Trade union officers, by contrast, typically play a much more conservative role: their ties to the union machine and through it to the state mean that they tend to limit these struggles within capitalist parameters.[48] This conservatism is particularly debilitating in periods of austerity. Whereas during economic booms it is easy to see that capitalism can afford better terms and conditions for workers, once boom turns to bust 'realistic' assessments of what the system can afford have been used to justify the logic of cuts. The bureaucracy therefore tends to play a contradictory role. Though it emerged to represent the interests of workers against capitalists, because it does so from within a capitalist perspective it all too often acts as a conservative influence within the working class.[49]

Consequently, trade union bureaucrats tend to come into conflict with rank and file members especially if and when the members engage in generalised industrial conflicts that spill over into political confrontations with capitalist states.[50] The problem for revolutionaries in 1919 was that this conflict was made more difficult because the reformist bureaucracies had had a long time to establish their hegemony within the labour movement. Challenging this hegemony was always going to be an uphill struggle, even in a period of mass milit-

---

47   Luxemburg 1986.
48   Callinicos 1995, pp. 23–6.
49   Darlington and Upchurch 2011, p. 84.
50   Clearly the concept of a 'rank and file' is highly contentious. It describes a diverse group whose only defining characteristic according to many writing both from the point of view of traditional positivist historiography and more recent forms of research influenced by the linguistic turn is its very heterogeneity. Though this criticism of the idea of a rank and file is not without substance, as with similar criticisms of the related concept of class it is essentially superficial. Like the concept of class, the concept of 'rank and file' relates not primarily to states of consciousness but rather to objective material interests rooted in the relations of production. As Darlington and Upchurch write, '[n]otwithstanding such differentiation, it is the exploitative social relations at the heart of capitalist society to which the mass of rank-and-file union members are subject that provides the material basis for collective workers' struggles which distinguish them from FTOs. It is this that makes the idea of the "rank-and-file" a term not devoid of analytical use, even if it encompasses an internally differentiated layer of members' (Darlington and Upchurch 2011, p. 88; Darlington 2014, p. 66). On the power of Marx's objective concept of class, see de. Ste Croix 1983; 1984. For a more popular presentation, see Blackledge 2011. Under capitalism, Marx's concept of class relates to the widely dismissed labour theory of value. For a demolition of the myth that Marx's theory of value is inconsistent, see Kliman 2007, and for the power of Marx's theory, see Weeks 1981.

ancy. But an uphill struggle does not mean a predetermined defeat. There *was* a real struggle, and the extent to which revolutionaries challenged the hegemony of reformism in 1919 can only be discerned through a detailed reconstruction of the events of that year.

Unfortunately, the challenge from the far left was weakened by the way that it tended to reproduce the capitalist division between politics and economics in its own ranks through the division between industrial syndicalists and political revolutionaries (albeit with numerous individuals playing roles in both camps). If the sectarian and abstract nature of much of British Marxism reflected, in part, its divorce from industrial conflict,[51] the anti-political character of syndicalism reflected the real but limited nature of industrial struggle.[52]

Despite these weaknesses, 1919 was characterised in part through hopes and fears for revolution on the part of revolutionaries and counter-revolutionaries alike. That these hopes and fears did not materialise should not be allowed to blind us to the fact that they were a part of the historical narrative. For unless the issue of revolution in 1919 is taken as seriously as it was by those who participated in key events, then historians are likely to miss much that is significant about that tumultuous year. This is true, as I have suggested, of Chris Wrigley's important work on the subject. By counterposing industrial, constitutional, and revolutionary dimensions of the workers' movement, he tends to underestimate the revolutionary potential of the period. For by conceiving of the left in these terms it is difficult to view the British revolutionaries as anything other than a tiny minority within the working class. This allows Wrigley to argue that 'there does not appear to have been a potentially revolutionary situation'.[53] If this argument follows from his tendency to reify the distinctions between industrial, constitutional, and revolutionary struggles, it is susceptible to critique from the perspective of Rosa Luxemburg's conception of mass strike: the way that such strikes allow for a politicisation of those involved such as to blur differences between politics and economics, industrial and revolutionary struggles.[54] The problem with Wrigley's approach is that by fixing the differences between industrial, constitutional, and revolutionary challenges from labour, he does not allow for their cross-pollination. Chanie Rosenburg is, by contrast, surely right to suggest that the situation in Britain in 1919 was 'pregnant with revolutionary potential'.[55] Though Rosenburg pushes the argument

51   Hinton 1983, pp. 89–93; Hallas 1985, p. 44.
52   Darlington 2008, pp. 233–5.
53   Wrigley 1993, p. 262.
54   Luxemburg 1986, pp. 46, 80.
55   Rosenburg 1987, p. 29.

too far, so long as we limit ourselves to recognising that the potential for revolutionary conflict was much greater than that which transpired, we are better able to grasp the dynamic of class struggle in 1919 and potential lessons for today.

It seems to me that by setting fire to the straw person of the lost British revolution (usually using Basil Thompson's reports to fan the flames), historians have served to obscure important radical mobilisations of workers in 1919, and the political debates to which they gave rise. K.O. Morgan's *Consensus and Disunity* is a good case in point. He dismisses Thompson's reports, and the radical historians of the 1960s who took them seriously, as groundless.[56] The upshot is that the militant workers who were the subjects of those reports hardly feature in his account of the period. Almost automatically, the focus for research becomes the higher echelons of society, with little detail about what was going on down below. When the working class does make an appearance, it tends to be in the shape of its more 'reasonable' political representatives and trade union officers, who often do not accurately reflect the opinions and values of those they claim to speak for.

My intention is not to devalue the worth of such scholarship; the politics of the Coalition government form an integral part of our understanding of the immediate postwar period, and anyone seeking to study it must acknowledge the debt they owe to the research that has been done in this area.[57] The point is rather that there is a gap in the historiography; political and industrial histories which focus on national leaders and institutions can only tell us so much about what was happening in society. To complement these it is necessary to begin to dig deeper, to investigate strata which have so far remained hidden.

These are the general views and considerations which prompted me to look at the immediate postwar period and 1919 in particular, and to direct my research towards the industrial unrest and the political ideas which infused it. The intense debate which took place within the labour movement as to whether socialism would come about through parliament or through the direct industrial action of workers provides the thematic context for an alternative analysis of the politics of the Miners Federation of Great Britain (MFGB) in the period of the Sankey Commission.

The miners seemed the obvious lens through which to view the crisis of 1919; one million strong, and widely considered to be in the van of the 'direct action movement', they were at the heart of the debate over the direction in which labour should go. The object of study is their postwar programme for hours,

56   Morgan 1979, pp. 52–3.
57   Apart from Morgan, see Wrigley 1990; Middlemas 1979; Armitage 1969.

wages, and nationalisation with joint control, and the developments that led to the historic compromise of the Sankey Commission. My methodological emphasis is upon developments at a regional or district level; these are then integrated into the national picture. Similarly, within the miners' union, this study maintains a dual focus on the leadership – both at district and national levels – and on the rank and file. In the interests of '[g]iving proper voice to the rank-and-file worker, whose history, attitudes and actions are typically subordinate and sparse in the archival record',[58] I have researched many thousands of editions of local newspapers, which in this period carried highly detailed reports of strikes and mass meetings, and in which the authentic voices of rank and file miners can be heard. In the process I have unearthed what had previously been a largely unknown strike movement that involved hundreds of thousands of miners, was led by socialists, syndicalists and revolutionaries against the officials, and which affected the majority of the coalfields in 1919. This has allowed a reinterpretation of the critical decisions which were taken at the top by government ministers and miners' leaders, and a re-evaluation of the Sankey Commission itself. Its eventual success as a compromise which marked a turning point in the class struggle in 1919 was not a foregone conclusion; there were considerable forces pitted against it, and the entire conciliation process was a great deal more precarious than has hitherto been assumed.

The book begins by investigating the origins of the Miners' Charter, and situating it within its political context. Particular emphasis is given to the importance of the South Wales miners' strike of 1915, which radically changed the industrial environment within which the MFGB operated. The actual campaign for the Miners' Charter can be divided into two stages; the first began in earnest on 14 January 1919 at the Southport Conference of the Miners' Federation of Great Britain (MFGB), and ended with the acceptance of the Interim Sankey Report. The second stage ended with Lloyd George's definitive rejection of nationalisation in August 1919. The book reflects this chronological division in its two parts. The core of both parts consists of studies of four areas of the British coalfield, Fife and Lanarkshire, Nottinghamshire, South Wales and Yorkshire. In Chapter Eight, other coalfields that were affected by the unrest of the summer make appearances, in support of a fresh analysis of the nationalisation controversy. The areas studied were selected first and foremost because they were the ones where direct action was most widespread. Fortunately, they

---

58   McIlroy and Campbell 2009, 'Introduction to the Paperback Edition: Debating the 1926 Mining Lockout', p. xxi, in *Industrial Politics and the 1926 Mining Lockout: The Struggle for Dignity*, edited by John McIlroy, Alan Campbell and Keith Gildart, Cardiff: University of Wales Press.

represent a wide variety of coalfields and district organisations, differing in size, geographical and geological formation, and in their political and industrial traditions.

This selection has involved two omissions, however. Firstly, the strike movements in North and South Staffordshire and in Kent have been excluded, to allow for a level of detail in the area studies that would otherwise have been impossible. Secondly, and more importantly, the North-East areas of Durham and Northumberland have been left out. The reason for this is simply one of a lack of sources; both areas remained relatively quiet throughout the year, and so neither local newspapers nor union records tell us much about what was going on in the lodges and collieries in the crucial months. As areas which were (almost) unaffected by the unrest, their inclusion would have provided a useful basis for comparison with the more militant centres. The gap which they leave is regrettable and frustrating, but not decisive, thanks to the variety of experience which the selected areas represent. In particular, the account of the official strike in the Yorkshire coalfield in July/August provides an interesting contrast to the unofficial strikes in the other areas.

In the concluding chapter, I analyse the decline of the direct action phenomenon in mining, and suggest various ways this impacted upon the politics of the labour movement as a whole. Some of these relate to the themes outlined in the opening section of this introduction. In particular, the emphasis is on the integration of dissent within the Labour Party, and the fact that this integration involved the suppression or marginalisation of other movements. The iconography of a ladder of progress from Keir Hardie to the 1945 Labour government is rejected. Labour's gradualism and electoralism did become hegemonic, but this need not have been so. There were other possibilities. Leave them out, and history is incomplete.

Martyn Ives
Paul Blackledge

PART 1

CHAPTER 1

Political Alternatives in the Labour Movement in 1919

We are either constitutionalists or we are not constitutionalists. If we are constitutionalists, if we believe in the efficacy of the political weapon (and we do, or why do we have a Labour party?), then it is both unwise and undemocratic because we fail to get a majority at the polls to turn around and demand that we should substitute industrial action.[1]

The trade union weapon is the only weapon to contend with the new government.[2]

∴

1   The Impact of the War on the Labour Movement and the Working Class

The First World War dramatically altered the position of the working class in British society. Both on the Western Front, and at home, it was workers who bore the brunt of the war effort. This simple fact meant that long before the war had ended the question of labour and its place in society became, in a way it had never quite been before, the key question in domestic politics. The 'war of production', as government ministers often referred to it, had highlighted the importance of the working class in the British economy and society; it had been, according to W. Basil Worsfold, a contemporary writer on social politics, 'a revelation of the value of the manual workers to the state'.[3] This elevation of the status of the working class had been accompanied, and reinforced, by a tremendous growth in trade unionism. From approximately 4 million in 1914, the number of organised workers had climbed to 6.5 million in 1918, and would

---

1 J. McGurk, L&CMF official, Labour Party Annual Conference Report 1919, p. 44.
2 S.O. Davies, Dowlais SWMF Agent, speaking in Merthyr, New Year's Day 1919, *Merthyr Pioneer* 4 January 1919.
3 Cited in Marwick 1974, p. 7.

continue upwards to almost 8 million by the end of 1919, a figure corresponding to a union density of almost 50 percent.[4] This growth gave a massive injection of power to the trade union movement, which demanded, and received, a new approach to industrial relations from employers and government.[5] One index was the huge extension of national pay bargaining, incorporating an extra 4.5 million workers, a figure representing at least one-third of the total employed workforce.[6] Another was the incorporation of labour leaders into the bureaucracy of the wartime state. To ensure the co-operation of the workers in continuous production, trade union officials were involved in a myriad of committees, consultations and enquiries in influential offices off the corridors of power. For the same reason, Lloyd George in December 1916 summonsed Labour Party leaders to a meeting, offered them the nationalisation (in some form) of mines, railways and shipping, and gave four of them seats in the Coalition. A grateful and surprised J.R. Clynes commented, 'Labour has been curiously elevated by the demands of war'.[7]

The accession of labour leaders to positions of power and responsibility gave a much needed boost to the Labour Party's image as a potential party of government. This huge leap forward in political credibility was further encouraged by wartime developments in the state's relationship to industry and society. The demands of war found laissez-faire orthodoxy, already being questioned prior to 1914, utterly wanting. Albeit in a piecemeal and haphazard manner, the state came to intervene in almost every sphere of national life. By the war's end, the government had total responsibility for all land and sea transport, and controlled production in mines and quarries. It directly owned over 250 National Factories, and supervised operations in 20,000 more. Food and raw materials were purchased and rationed by the state, which also decided what crops should be grown and how the land should be utilised. There were also import/export controls, the regulation of capital and securities, and controls over mobility, working conditions and wages of labour.[8]

The creation of a state controlled economy in Britain presented the Labour Party with a tremendous opportunity; wartime state capitalism provided a blueprint for postwar socialism. The ILP's newspaper, *Labour Leader*, quoted enthusiastically a speech by Churchill in which he said:

---

4 HMSO 1971, p. 395, Table 196.
5 The implications of this have been debated by, amongst others, Reid 1986; Reid 1985; McLean 1983; Hinton 1973.
6 Clegg 1985, pp. 163–8.
7 Cronin 1984, p. 21.
8 Armitage 1969, pp. 1–3; Tawney 1943, pp. 2–7.

I have not been quite convinced by my experience at the Ministry of Munitions that socialism is possible, but I have been very nearly convinced. I am bound to say I consider, on the whole, the achievements of the Ministry of Munitions constitute the greatest argument for state socialism that has ever been produced.[9]

Nonetheless, the controlled economy had been established without any accompanying ideological conversion and government policy remained rapid decontrol as soon as the war was over. 'Labour and the New Social Order' was the Labour Party's bid to nail its ideological colours to the mast of wartime collectivism. The positive aspects of the war experience, so the argument went, could be harnessed to provide progressive social change without the dangers of social upheaval.[10]

Bernard Waites' study *A Class Society at War* emphasised that the war, whilst in general strengthening labour's position in society, produced different and contradictory conclusions as to how that power should be exercised, based on the different wartime experiences of leaders and led within the labour movement.[11] For Arthur Henderson, the war, by enhancing the role of trade unions in ensuring social stability, had conciliatory consequences for capital and labour. Reconstruction would come about, he believed, because of 'the new democratic consciousness' the war had produced in all classes. 'During the war', he wrote, 'we have learnt the meaning of co-operation for common ends. The lesson holds good for the politics of tomorrow'.[12]

However, for many industrial militants the war produced entirely different conclusions. By 1917, 'war weariness' was chipping away at the patriotism which had kept workers' grievances in check, and whereas in 1916 only 276,000 workers were involved in a strike, in 1917 the figure jumped to 872,000.[13] The reports of the Industrial Unrest Commissioners showed that important groups of workers had anything but a positive view of government intervention. Nor were they taken by the new 'democratic consciousness' of the times. Rather, the reports revealed that workers were very often highly concerned that the Munitions of War Act, the Military Service Act and the Defence of the Realm Act, with their restrictions on the liberty and mobility of workers, were the advanced draft

---

9   *Labour Leader*, 30 January 1919.
10   Tawney 1943, p. 2; Winter 1974, pp. 6, 58, 221.
11   Waites 1987, pp. 41–2.
12   Henderson 1918, pp. 20, 77.
13   Department of Employment, British Labour Statistics 1971, p. 396, Table 197.

of a 'servile state'. Furthermore, there was widespread anger at the 'profiteering, exploiting and plundering' in which employers were indulging.[14] Taken together, the Reports leave no doubt that there was considerable anger at both government and employers about the unequal burden of sacrifice the working class was being forced to bear. 'The Co-operative News', not prone to political outbursts, wrote of:

> Conscription at a shilling a day for the private soldier, excess profits for the manufacturer and dealer. Suppression for the workers' papers, license for the Harmondsworth Press; tall prices for producers, queues for consumers; Imprisonment for strikers, autocracy for a War Cabinet of three.[15]

Below the ranks of the Labour and trade union leaders then, the war very often did not encourage consensus politics. As Hinton has commented:

> Important sections of the workers experienced wartime state intervention not as liberating, but as repressive. They saw in the wartime state not an actual or potential ally, but an agency used by the employers to reinforce their class power.[16]

The association of labour leaders with the state was causing serious tensions within many unions by 1917, a development once again highlighted in the district reports of the Industrial Unrest Commissioners. John Hill, in his Presidential address to the TUC in September 1917, drew attention to their findings that: 'Trade union officials are distrusted ... In their [the members'] opinion we had either fallen asleep at our posts, or we have sold their birthright for a mess of pottage'.[17]

As a consequence, much of the industrial unrest that occurred from 1917 onwards was unofficial in its nature, directed not only against the government and employers, but also against the trade union leaders. It was in this context that rank and file committees were established or extended in several major industries, most notably in engineering, the railways and mining. (As we shall

---

14   Commission of Enquiry into Industrial Unrest, No. 3 Division, Report of the Commissioners for the Yorkshire and East Midlands Area 1917, PRO CAB 24/23 Cd.8664, pp. 50–2. Summary of the Reports of the Commission, 1917–18 (PRO CAB 24/23 Cd. 9085) *passim*.
15   *Co-operative News*, 26 May 1917, cited in Waites 1987, p. 70.
16   Hinton 1983, p. 104.
17   Clegg 1985, p. 174.

see, Robert Smillie's refusal to sign the Treasury Agreement did not inoculate the MFGB against this development). With union executives just 'so many bureaus for the interpretation of government orders, mere annexes of the Ministry of Munitions', these rank and file committees gained considerable influence.[18] Drawing on the syndicalism of the prewar labour unrest, they sought to place an entirely different interpretation on the working class experience of the war from that of the Labour Party. They were largely responsible for popularising the notions of 'direct action' and 'workers' control', political concepts which reflected both the potential industrial power of the trade unions and the suspicion of state control. With men like Henderson and Webb in mind, the Scottish Workers' Committees wrote:

> To obtain the support of the worker, this middle class State Capitalism element adopts the language of socialism and dupes the workers with catch phrases of popular rebellion. They stake their all on the capital of Parliament not, be it observed, to destroy that political engine of class rule, but to use it 'in the interests of Labour.' But it cannot be used in the interests of Labour. 'He owns my life who owns the means whereby I live', and until the worker owns the means of life he will be a slave. Direct Action alone can give him control of the means of life and the entire economic forces of the country.[19]

The war had been, therefore, a watershed in the development of both the parliamentary and extra-parliamentary lefts, with each able to point to encouraging developments for their own particular programme or strategy. In different ways, it seemed that the war had transformed their respective projects from idealistic dreams to realisable goals. But what of the mass of the working class population? The development of political programmes is easier to trace than developments in general class attitudes, but significant changes are nonetheless discernible.

In the first place, there seems to have been a heightened sense of self-awareness within the working class. In part, this was a consequence of structural changes, such as the narrowing of the gap between skilled, semi-skilled and unskilled.[20] However, the war had also caused subjective changes in the way workers viewed themselves in relation to other classes in society. 'Rob Roy',

---

18   *Herald*, 22 February 1919.

19   Campbell and Gallacher 1919, p. 2.

20   These changes are outlined in Cronin 1984, p. 22, and Waites 1987, pp. 24–7.

writing in the Glasgow socialist paper 'Forward', observed elements of a new postwar psychological outlook:

> For one thing they have a better conceit of themselves that has been knocked into their heads from high quarters. (This was 'an engineers' war', a 'war won by women' etc). For another, they have a good deal less conceit of at least large sections of the ... high quarters. The part played by landlords and profiteers in winning the war or exploiting those left at home needs no special revelation.[21]

Both Waites and Cronin have studied contemporary accounts of what was described as 'the awakening' of the working class, and they too noted this newfound confidence of workers in their own abilities and power, and an accompanying weakening of deference towards the rich and powerful.[22] In this context, it was a short step from 'self-awareness' to 'self-assertiveness'.[23] War had, in Cronin's words, 'awakened the worker to his tremendous power' and the armistice found a 'working people optimistic and quite confident of their own potential'.[24] Waites has detected a strong shift during the years in 'the language and reference of class', in which capital was depicted as 'immoral and predatory', and in which class conflict became the dominant motif.[25] He cites a government report which identified 'Three ideas now dominating the mass'. These were: (1) The more the working class produce, the more the idle rich will waste; (2) A man cannot work his best in a system run for private profit; (3) Capitalists were not in a position to lecture labour on restricting production. Whilst Waites rejects any suggestion that the British working class reached a revolutionary level of consciousness in 1919, he makes the reasonable suggestion that sentiments such as these represented 'elements of an at least proto-revolutionary consciousness'.[26] The *Times* detected similar elements in the industrial unrest of the first few months of 1919:

> This unrest and restlessness of workers really means that an industrial population has made up its mind that for the future, no matter what the national need may be, it will not be disciplined, managed, or controlled

---

21   *Forward*, 1 February 1919.
22   Cronin 1984, pp. 22–31, Waites 1987, *passim*.
23   Waites 1987, p. 45.
24   Cronin 1984, pp. 26, 30.
25   Waites 1987, pp. 68, 61–2.
26   Waites 1987, p. 221.

by an authority, a private employer or a State department, which it does not choose by its own unfettered will to recognise as fit and suitable and whose dictates it does not consider sound and reasonable.[27]

Greatly raised expectations were a corollary of the elevated status and power of the trade unions at the war's end. Jimmy Thomas told the National Industrial Conference convened by Lloyd George in February that: 'The workers have set their faces towards some order of society which will improve their lives and conditions in accord with the new valuation they set upon themselves'.[28] Cronin has described the explosion of aspirations at the war's end as 'a novel and expanded sense of what was due to workers and what they were morally justified in demanding'.[29]

Contemporary observers described the raised horizons of workers as embracing notions of liberty and emancipation. The *Herald* noted that 'the idea of freedom is stirring in the land'.[30] George Dewar, in an article entitled *The Great Home Problem of 1919* which examined the mood of workers in the north of England, wrote:

> There is an extraordinary ferment throughout the packed industrial districts, which is not exactly like any previous phenomenon of the same character. The people believe they are nearing the portals of a new world which the war and events in Eastern Europe have forced ajar for them.[31]

Again, although it is not possible to talk of a revolutionary mood comparable to Russia, Germany or Hungary, the atmosphere in Europe did not leave British workers unaffected. Concerned politicians often located the domestic unrest in the context of wider European upheavals. Lloyd George, for example, told the Peace Conference that Europe was 'filled with the spirit of revolution'. He continued: 'The whole existing order in its political, social and economic aspects is questioned by the masses of the population from one end of Europe to the other'.[32]

Moving on to the concrete demands of workers in the immediate postwar period, one is struck first of all by the volume of calls for shorter hours and

---

27 *Times*, 29 April 1919.

28 Cited in Williams 1921, p. 122.

29 Cronin 1979, p. 114.

30 *Herald*, 22 February 1919.

31 Dewar 1919, p. 206.

32 Arnot 1975, p. 160.

higher wages. At first glance this might suggest an economism at odds with
the mood of idealism described above. However, a closer look at the nature
of these demands shows that they were consistent with, and indeed an integral part of, the desire for thoroughgoing change in society. In the first place,
they were the demands of an entire class, not just sections of it, and were justified as compensation for the years of overtime, inflation and profiteering.
Increased leisure time, and the resources with which to enjoy it, was a com-
monly held aim. Crucially, these demands were a political response to demobilisation. Shorter hours in particular, described by Robert Williams as 'The
Genesis of the Book of Emancipation',[33] were designed to prevent the employers from capitalising on the return of four million soldiers to create a pool of
unemployed labour and undermine the new found strength of the trade union
movement.[34] The demand for shorter hours was essentially the cutting edge of
the struggle between capital and labour over the shape of post-war reconstruction in Britain. This is what Robert Williams meant when he wrote:

> The demobilisation of the Armed Forces, the introduction of a shorter
> working week, and the general and comprehensive solution of all indus-
> trial and economic problems arising out of the end of the war are essen-
> tially political matters, not necessarily Parliamentary matters, but matters
> of political import.[35]

There was a widespread recognition that the militancy over hours and wages
was just one aspect of a higher and broader collective working class aspiration.
Lloyd George said that: 'All signs go to show that the striving is as much for social
and political changes as for increases in wages'.[36] In his survey of contemporary
writings on the subject, Cronin finds that:

> Workers were interested in more than wages, that their list of grievances
> had been extended and linked to a thoroughgoing critique of society, and
> their attitude to work itself.[37]

---

33 *Merthyr Pioneer*, 3 January 1919. The *Pioneer* was a socialist newspaper with which Hardie
    had been involved. Whilst it was not owned by the ILP, it had close links with it.
34 See Clegg 1985, p. 211 for the general agitation for shorter hours from the late summer of
    1918.
35 *Herald*, 8 February 1919. The issues surrounding demobilisation are dealt with in more
    depth in subsequent chapters on the campaign for the Miners' Charter.
36 Cited in Murphy 1972, p. 172.
37 Cronin 1984, p. 24.

The last point is a reference to the demand for workers' control, which was often linked to the demand for nationalisation, and was the upshot of wartime fears about a servile state. During the war, the guild socialist G.D.H. Cole wrote that the 'State and the municipality as employers have turned out not to differ essentially from the private capitalist', and that therefore democratic control of the workplace had to be an integral part of the demand for nationalisation. By 1919, he could write that 'as far as the Labour movement is concerned the internal battle for the idea of workers' control has been fought and won'.[38] Ramsay MacDonald concurred that: 'At the end of the war, the control of the workshop by labour is as important as the control of Parliament by Labour'.[39] The Labour side's report to the National Industrial Conference emphasised:

> The growing determination of labour to challenge the whole existing structure of capitalist industry ... Labour is too strong to remain within the bounds of the old industrial system and its unsatisfied demand for the reorganisation of industry on democratic lines is not only the most important but also a constantly growing source of unrest.[40]

Quite what was meant by workers' control varied considerably between workers' committees, guild socialists, advocates of Whitleyism, or state socialists, as we shall see when we look at the campaign for this aspect of the Miners' Charter, but the fact that they all embraced it in some form or other indicates its popularity in the labour movement. George Dewar reckoned that 'a great mass of the workers' were animated by the demand.[41]

The experience of war had profound and radicalising consequences for millions of workers, who, with the armistice, demanded a fundamental change to their position in society, commensurate with the sacrifices they had endured, and their recently discovered indispensability to economy and society. Some of these aspirations were idealistic and formless expressions of the changed postwar psychology. Uncovering the mood and atmosphere of 1919, and rescuing it from obscuration by the later defeat of 1921, is of fundamental importance if we are to understand the full significance of this tumultuous year. It is also important to attempt to collate and assess the concrete demands raised by

---

38   Cited in Pribicevic 1959, p. 160.
39   *Forward*, 1 February 1919.
40   Cited in Williams 1921, p. 119.
41   Dewar 1919, pp. 207–8. See Chapters Five and Seven for more on workers' control.

the trade union movement. Six weeks into the new year, 'The Herald' published a ten point programme which it believed summed up the programme
of the working class: (1) No conscription; (2) Discharge not demobilisation;
(3) Full maintenance for all unemployed; (4) A 40 hour week with no loss of
earnings; (5) Nationalisation, with workers' control, of mines, rail and transport; (6) Full recognition of all trade unions, including those in the armed
forces and the police; (7) Fulfilment of all government pledges for restora-
tion of trade union rights; (8) Amnesty to all political and military prisoners;
(9) No further use of the military in industrial disputes; (10) Hands off Russia.[42]

2 The General Election of December 1918 and the Direct Action Controversy

Aspirations were one thing. The question was, how to realise them? In essence
there were two strategies on offer. On the one hand, there was the revamped
state socialist project of Webb and Henderson's new model Labour Party. On
the other was the industrial muscle of the working class summed up in the term
'direct action'.

Direct action, both as a descriptive term and a political concept, had its
modern roots in the syndicalist influenced labour unrest of 1910–14. Lenin
explained the rise of syndicalism in Western Europe as 'a direct and inevitable
result of opportunism, reformism and parliamentary cretinism',[43] and it was
indeed sentiments like these that drove Tom Mann to set up the Industrial
Syndicalist Education League (ISEL) in 1910. 'Labour MPS', he wrote in the first
issue of the *Industrial Syndicalist*,

> seem to have constituted themselves apologists for existing society, show-
> ing a degree of studied respect for bourgeois conditions, and a toleration
> of bourgeois methods, that destroys the probability of their doing any real
> work of a revolutionary character.[44]

Mann turned instead to the trade union movement, advocating the amalgamation of sectional unions into industrial ones, and the adoption of an aggressive

---

42   *Herald*, 8 February 1919.
43   Lenin 1962, *Collected Works*, vol. 13, p. 166.
44   *Industrial Syndicalist*, July 1910.

direct action policy.[45] He rejected the SLP's strategy of building new revolutionary trade unions, choosing instead to work within the existing organisations, and, in the context of the Great Unrest, his syndicalist ideas gained considerable influence.[46]

The lack of enthusiasm for parliamentary politics inside the working class in these years before the war was such that G.D.H. Cole could write in 1913: 'To attack the Parliamentary Labour Party nowadays may look rather like flogging a dead horse'.[47] As we have seen, in many respects the war experience had done little to discourage such views. As one ILP member in South Wales observed, it had undermined confidence in gradualism and had

> made the average man of the working class very anxious to have a quick solution of the difficulties of the time. As a consequence the socialist movement had a growing disposition to believe in cataclysm. The war had imported military terms into the nomenclature of the class struggle.[48]

Ramsey MacDonald similarly noted, in January 1919, that 'the ruling classes have taught the appeal to force, and the workers are appealing to industrial force'.[49] Again, George Dewar writing in 1919 reminded his readers:

> Always bear this in mind: the workers during the last four years have secured various benefits or concessions not through Parliament, they have secured them through outside pressure and demonstration. And they know it.[50]

This was, as we shall see, nowhere truer than in the coalfields. Leo Chiozza-Money, soon to represent the miners on the Sankey Commission, lamented that:

> The economic history of the war simply teems with illustrations of the unfortunate fact that only through strikes ... has any measure of justice been obtained by those engaged in making fortunes for the profiteers.[51]

---

45   *Industrial Syndicalist*, September and December 1910.
46   Holton 1976, *passim*.
47   Winter 1974, p. 103.
48   *Merthyr Pioneer*, 6 March 1920.
49   *Labour Leader*, 30 January 1919.
50   Dewar 1919, pp. 201–2.
51   *Labour Leader*, 9 January 1919.

The Labour Party's ascendant trajectory in the 1920s and beyond is clearer in hindsight than it was to contemporaries who stood at the end of the syndicalist decade. Complacency on this point can cause misleading historical oversight. Jay Winter, for example, in his otherwise useful study of the evolution of socialist thought within the Labour Party, overlooks the robust challenge of the extra-parliamentary left. Yet at one point, Winter himself quotes the Fabian Clifford Sharp to convey a sense of the parental concern which some had for its political infant. In the wake of the national miners' strike of 1912, Sharp wrote to the Webbs:

> Within a week, the miners' Federation has converted Parliament and the nation to accept a legislative measure for which they would have had to fight ten years if they had relied solely on political action through the Labour Party. It may be that the political socialist movement will be swallowed up in a movement much more vague philosophically, but much more concrete practically, on the lines of industrial unionism. On the other hand it may be that in ten years time socialism will be the dominant force in the country.[52]

The war had done nothing to remove this uncertainty. Arguably, it had compounded it. Yet Winter concludes with a confidence singularly lacking amongst contemporary reformist leaders:

> Not only did the attitude of the Labour leadership to socialism change during the war; so too did the attitude of socialists to the Labour Party. By 1918, the tactical dispute in Britain had been decided for all but a handful of revolutionaries who eventually formed the Communist Party. It became then [in 1918], as it is today [1974], the major institutional focus for socialist political activity.[53]

But where does this leave the trade unions, and syndicalism? For many socialists, particularly in the wake of the 1918 general election, their collective industrial organisations were the best hope for both economic and political change. At the war's end, the Labour Party leadership still felt intensely vulnerable to

---

52 Winter 1974, p. 29.

53 Winter 1974, p. 6. For another example, see Morgan 1975, p. 30. For Morgan, the debate was effectively over by 1918: 'Like Mainwaring, the Welsh miners generally opted for the orthodox view after 1918 – a crusade for a socialist society through the Labour Party'. See also my comments on Morgan and Stead in Chapter 5 below.

the continuing challenge of a radical tradition it would be anachronistic to label communist. 1919 was the key date, and syndicalism was the threat.

The contrast between the parliamentary weakness and industrial strength of labour's forces was widely discussed in labour circles in the year following the war. Robert Williams, for example, scathingly observed that the PLP's only legislative success had been to reduce tax on tea by two pence in the pound:

> Industrial action, on the other hand, has made it possible for the organised workers to obtain the necessary purchasing power with which to buy tea, and now and then to add sugar and milk.[54]

In this period, before Labour had proved itself capable of gaining office, some of the ideas gathered together under the umbrella of direct action could hold more immediate relevance to workers. As Ramsey MacDonald admitted in his book *Parliament and Revolution* (1919): 'To the man who responds day by day to the call of the factory whistle, Parliament too often appears to be an ineffective thing'.[55] In his book *Direct Action* (1920), William Mellor, industrial editor of 'The Daily Herald' wrote that: 'The road to freedom lies not through the polling booth but through the workshop gates'.[56] George Dewar was of the opinion that 'there is all over the country a great number of working men in mines, factories and shipyards, who distrust the parliamentary method'.[57] He went on to explain that this distrust lay in the remoteness of parliament:

> Anyhow, the average working 'hand' in the chief centres of industrial ferment today cannot visualise a solid body of 350 Labour MPs trooping into the 'Aye' lobby in the third reading of a bill, say, to conscript the wealth of the country, or to nationalise the essential industries ... The parliamentary system is too far from his ken to be implicitly relied on ... What relation to his daily life in the pit shaft, by the blast furnace, or at the cogging mill, can a three-line whip or a full dress debate have? That doubt is understandable.[58]

---

54   Williams 1921, p. 142.
55   MacDonald 1919, p. 56.
56   Mellor 1920, p. 51.
57   Dewar 1919, p. 210.
58   Dewar 1919, p. 211.

Labour Party leaders had a problem. 'Even before the war', Ramsay MacDonald wrote, 'impatience was being shown with parliament … as the means of expressing the popular will'.[59] Since then the situation had deteriorated further. Now the Labour Party was faced with the question of

> how to restore reality to politics, how to make Parliament as real to people as it was to the landlords when it was enabling them to enclose the commons and keep up rents, and as it was to the capitalists when it gave them Free Trade and Peace, Retrenchment and Reform.[60]

This was no simple matter in the heady atmosphere of Europe in 1919, when revolutions were sweeping away tyrannies and mass strikes were shaking governments. 'The belief in the Parliamentary road', Henderson confessed, 'appears old fashioned'.[61] Direct action, by comparison, could appear modern and dynamic, especially when bracketed with revolutionary developments in Europe. Again Henderson, speaking on the causes of worldwide unrest in 1919, warned his listeners:

> Do not make the mistake of thinking that the Soviet theory of government is a Russian invention which has no attraction for the workpeople of other countries. It has captured the imagination of a much greater number of people than is commonly realised. The soviet theory and the theory of direct action were of immense significance. They meant that huge masses of people had begun to lose faith in the Parliamentary system and the ordinary machinery of representative government.[62]

The low esteem in which parliament was held was a serious concern for the Labour Party, especially given the impatient clamour for social change. In his 1918 booklet *The Aims of Labour*, Henderson had warned of the threat from direct action should the Labour Party after the war 'by its own weakness, or the stubborn resistance of other parties and classes, [be] unable to fulfil the expectations of its followers'.[63] Therefore, much rested on the outcome of the December 1918 general election. The reconstructed Labour Party, armed with constituency parties, individual membership and Clause Four, hoped to

---

59   MacDonald 1919, p. 1.
60   MacDonald 1919, p. 67.
61   *South Wales News*, 14 July 1919.
62   *Llais Llafur*, 20 September 1919.
63   A. Henderson 1918, p. 61.

make large gains amongst the 12 million or so new voters eligible under the Representation of the People Act.[64] Whilst in normal conditions Labour might have found more cause for celebration in the drastic decline of the Asquithian Liberals, its own performance was not good enough to deflect the threat from direct action. Although it had done well to field 363 candidates, only 57 of these were returned, a mere 17 more than in December 1910.[65] The *Labour Leader* warned the PLP of the likely consequences of any failure to make the most of its newfound position as the main opposition party:

> If they fail to take full advantage of that opportunity, they may destroy the political labour movement in this country. They will drive the trade unionists from political action, in despair of its futility as illustrated by the Parliamentary Labour Party, and they will give an irresistible impetus to the most extreme forms of industrial action.[66]

Unfortunately for the PLP, its small size was made worse by the calibre and political character of its members. In the atmosphere in which the election took place, anyone suspected of pacifism – rightly or wrongly – was punished at the ballot box. Only three ILP candidates were successful. The party was without its most effective performers; Keir Hardie was dead, Henderson, Snowden and MacDonald defeated. Fifty of the MPs were trade union candidates, almost half of these miners, whom Beatrice Webb judged to be 'for general political purposes dead stuff'. Of William Adamson, the Fife miners' leader elected PLP Chairman, she wrote: 'He fumbles in political life as we should fumble with a pickaxe in the dark recesses of a mine, and gets about the same output as we should do'.[67] Following a lacklustre opening parliamentary session, Adamson rose 'with a considerable amount of fear and trembling' to face the Labour Party's 1919 Annual Conference.[68] Delegates were angry that the burning issues of allied intervention in Russia and the continuation of conscription had not even been raised, and that the PLP 'was not acting as the leader of working class opinion in the House of Commons'.[69] A Glasgow delegate pointed out 'the

---

64   Labour Party National Executive Committee [NEC] Minutes, Box iv, 1918, General Election: Report to Organisation Sub-Committee.
65   Clegg 1985, p. 237. Shortly after the election, four other MPs took the Labour whip, bringing the total to 61.
66   *Labour Leader*, 2 January 1919.
67   Mackenzie 1984, pp. 329, 331.
68   Labour Party Annual Conference Report 1919, p. 130.
69   Labour Party Annual Conference Report 1919, pp. 148–51.

inability of the party to do anything on the question of the Glasgow strike and
the use of the military on that occasion'.[70]

In a situation where labour was weak in parliament yet strong in industry,
the centre of gravity passed decisively from the Labour Party to the trade
unions. People now began to talk of a direct action 'movement'. William Mellor
wrote that:

> Without a philosophy, without a coherent aim, ill-coordinated and
> frowned upon by constitutionalists, the movement for direct action is
> growing ... The workers are instinctively turning to direct action, and this
> instinct will triumph over the old traditions and constitutional ways.[71]

But what did people understand by 'direct action movement'? In 1919, there
was no organisational structure to affiliate to, no identifiable leadership, no
equivalent even of the pre-war ISEL. Instead, sources reveal disagreement not
only on the merits of 'direct action', but on the meaning of the term itself.

William Mellor's book allows us to enter the debate at ground level with a
lowest common denominator definition: 'Direct action can, in a general way,
be defined as the use of some form of economic power for the securing of ends
desired by those who possess that power'.[72] Conversely, Gerald Gould, an associate of Mellor's at the *Daily Herald*, defined it in his book *The Coming Revolution of Britain* as action 'for political purposes', and cited as examples strikes to
enforce nationalisation or the removal of troops from Russia, a definition with
which later authors like Coates and Miliband have agreed.[73] Henderson understood direct action to be 'anti-Parliamentary', but as Hinton has rightly stressed:
'it was by no means confined to those who rejected parliamentary action altogether. Even within the syndicalist movement few ... were dogmatic in their
hostility to parliamentary politics'.[74] Many ILPers, for example, could balance
a continuing faith in electoral politics with approval for direct action over certain issues. The editor of the *Merthyr Pioneer* also saw a connection between
direct action and parliament, describing it as 'the entry of the organised trade
unionists into political life, armed with the industrial weapon of the strike, and
determined on its use for the purposes of class-biased legislation'.[75] Herbert
Booth, agent for the Forest of Dean miners told them:

---

70   Labour Party Annual Conference Report 1919, p. 129.
71   Mellor 1920, p. 139.
72   Mellor 1920, p. 15.
73   Gould 1920, p. 7. Coates and Topham 1991, p. 713; Miliband 1964, chapter 3.
74   Hinton 1983, p. 93.
75   *Merthyr Pioneer*, 2 August 1919.

He did not personally look for the emancipation of the workers from sending Labour men to Parliament, but rather from the combined and united efforts of the workers through their trade union organisations, by which means they could compel Parliament to legislate in accordance with their wishes.[76]

Sexton, a bitter opponent of direct action of any kind, told the House of Commons that in his view there were two distinct types of 'direct actionist':

There is the whole-hogger; he who believes in the abolition of Parliament entirely, and the substitution of an Industrial Union of Workers, taking control of industry. Then there is the Red Actionist, who would combine political with direct action, and would use the weapon of direct action in order to compel Parliament to hurry up and concede its demands.[77]

More than any other figure, group or publication, George Lansbury's *Daily Herald* helped to popularise the idea of direct action in 1919. The paper's popularity was shown by a fund raising demonstration in March 1919 when, despite snow and freezing temperatures, over 100,000 marched to Hyde Park. The paper had begun its life as a strike sheet before the war, and in 1919 it was being funded by a combination of readers' workplace collections and, for the first time, a number of union executives. Its daily circulation in 1919 was around 300,000, and its influence was such that at the 1919 TUC conference, Clynes could accuse it of trying to dictate the policy of the entire labour movement.[78] In the early months of 1919, it championed the idea that the industrial power of workers, exercised through a general strike, held the key to the transformation of society. Although it remained vague on details, it felt that direct action, if properly co-ordinated, could achieve 'peaceful revolution'. However, there is no evidence to suggest that Lansbury ever entirely turned his back on the parliamentary road in this period. Rather, he seems merely to have de-emphasised it for a time.[79]

To the left of Lansbury stood the revolutionary syndicalists and industrial unionists who were the prime movers behind the rank and file organisations in industry. Although the particulars of their strategies varied, and there were

---

76   *South Wales News,* 6 May 1919.

77   Vol. 119 HC Deb. 5s, c.1238.

78   Labour Party Annual Conference Report 1920, p. 175; TUC Annual Conference Report 1919, p. 227.

79   Any edition of the *Herald* or *Daily Herald* in the early months of 1919 contains these ideas.

nuances between the positions of pure syndicalists and industrial unionists, their perspectives were inherited from those outlined by the ISEL, in which the workers' industrial organisations would provide not only the means for over-throwing capitalism (whether by the general strike or encroaching control) but also the basis for the organisation of the post-capitalist economy and society.[80]

Direct action thus evades easy definition—it was all things to all people. Miliband has stressed his view that 'the issue was not revolution and social-ism, but direct industrial action for limited and specific purposes'.[81] There is no doubt that this was often true, especially when one focuses, as Miliband largely does, on the Council of Action in 1920 whose stated aim was to force the withdrawal of troops from Russia. However, in 1919 many saw the 'direct action movement' as possessing a more thoroughgoing, insurrectionary quality. Cer-tainly the language and rhetoric of direct action was littered with revolutionary imagery. In part, and Miliband makes this point himself, the association of dir-ect action and revolution was deliberately made, to ward people away from both.[82] Henderson, for example, appealed for a vote for Labour in 1918 on the grounds that it was the only guarantee against direct action and the 'horror' of revolution, 'of barricades in the street and blood in the gutters'.[83] 'Direct action' and 'Bolshevism' were often used interchangeably at this time. Lloyd George stated that: 'Direct Action is ... Bolshevism pure and simple'.[84] Shaw of the weavers' union told the TUC conference in 1919 that Lenin was 'an apostle of direct action for political purposes', and that the aim of the direct action move-ment was to establish soviets in Britain.[85]

Despite the red scare rhetoric, there were plenty who did embrace direct action as the British vehicle for revolution. Many in the rank and file commit-tees described themselves as 'Britain's Bolsheviks', and they were not, as we shall see in the case of mining, without influence in 1919. It was because of them that direct action was sometimes understood to be a specifically unofficial phe-nomenon; both Smillie and Brace of the MFGB EC defined it as such on various occasions.[86] In addition to these there were a much larger number, best rep-resented by the *Daily Herald*, who had a hazy notion of what they meant by revolution, but who felt that the changes in society which direct action could

80   Coates and Topham 1970, pp. 1–138.
81   Miliband 1964, p. 66.
82   Miliband 1964, p. 69.
83   Henderson 1918, pp. 57, 61.
84   *Western Mail*, 11 July 1919.
85   TUC Annual Conference Report 1919, p. 118.
86   Vol. 112, HC Deb 5s, c. 337; Labour Party Annual Conference Report 1919, p. 118.

make were of a sufficient scale to deserve the term 'revolutionary'. 'Trade unionism is defensive in purpose', wrote William Mellor, 'direct action, by implication
is both offensive and revolutionary'.[87]

The debate in the labour movement about direct action at the war's end was
not an abstract one. 1919 witnessed an extraordinarily high number of strike
days, 34,969,000 in total, or an average of about 100,000 workers on strike every
day of the year.[88] A number of features marked out the unrest. In the first
place, the majority of strikes were concentrated in two waves – from January
to March, and from July to August. As Wrigley has written: 'The key feature of
the industrial unrest ... was the conjunction of so many challenges in a very
short period'.[89] Secondly, although the strikes were uncoordinated, more often
than not they shared the same sorts of demands as to hours and wages, thus
raising the possibility of a united front of trade unions against the government.
Thirdly, many of the strikes were unofficial, a fact which contributed to the
atmosphere of insurgency and caused the government serious problems in
devising a strategy to deal with the unrest. Tom Jones, Deputy Secretary to the
cabinet, wrote to Lloyd George in Paris that:

> Much of the present difficulty springs from the mutiny of the rank and file
> against the old established leaders [who] no longer represent the more
> active and agitating minds in the labour movement.[90]

Churchill complained that the 'curse of trade unionism is that there is not
enough of it, and it was not highly enough developed to make its branch secretaries fall into line with the head office'. He added: 'With a powerful Trade Union,
either peace or war could be made'.[91] Fourthly, the restructuring of industrial
relations during the war meant that in many cases, especially with the larger
and more powerful unions, strikes were directed against the government as
much as against employers, thus encouraging a politicisation of the unrest.

Taken together, these features meant that the strikes appeared not as disconnected and sectional, but as part of a more general revolt. *The Worker*, at
the height of the first wave of unrest, wrote:

---

87 Mellor 1920, p. 79.
88 Department of Employment, British Labour Statistics 1971, Table 197, p. 396.
89 Wrigley 1990, p. 112.
90 Middlemas (ed.) 1969, 8 February 1919, p. 73.
91 Cited in Cronin 1984, p. 21; TUC Annual Conference Report, 1919, p. 97. The Parliamentary
    Committee of the TUC issued a circular warning against unofficial strikes in February, to
    little effect.

There have always been discontented workers who think of a different type of society. But the present discontent differs from much past discontent in the numbers, the social importance and the potential power of the discontented. There were never so many men in open revolt against the conditions imposed upon them by the existing system of society as there is today.[92]

The dangerous social implications of the strike wave were made worse by the serious unrest in the armed forces caused by slow demobilisation. The first Cabinet meeting of 1919 was interrupted by 1,500 soldiers, who marched to Downing Street and demanded an audience with Lloyd George.[93] On 3 January, a mutiny broke out amongst soldiers stationed at Folkestone and spread in a flash to another dozen camps:

Ten thousand soldiers marched through the town, held a mass meeting at which they formed a soldiers' union, elected 140 men to act as clerks, took over the Demobilisation Department, and in one day issued all the necessary pass papers, ration books and railway warrants for the whole camp. By the Sunday the camp was clear.[94]

It was, as the *Herald* eagerly pointed out, an object lesson in the 'direct action method'. Two weeks later, General Childs told the Cabinet that whereas previously 'we had a well disciplined and ignorant army ... now we have an army educated and ill-disciplined'.[95] Anxiety about the state's ability to suppress unrest led the War Office to circulate a questionnaire to all army commanders as to the suitability of their troops to such a task, a document spectacularly leaked by the *Daily Herald* in May.[96] The situation was compounded by serious unrest in the police force, where the National Union of Police and Prison Officers was threatening to strike for recognition.[97]

These features of the unrest greatly influenced the debate about direct action inside the labour movement, one which increased in intensity as the graph of industrial action peaked in February/March and again in July/August.

---

92   *The Worker*, 15 March 1919.
93   W.C. (514), 8 January 1919, CAB 23/9.
94   *Herald*, 11 January 1919.
95   W.C. (522), 30 January 1919, CAB 23/9. For the fullest account of the unrest in the armed forces, see Rothstein 1980.
96   *Daily Herald*, 13 May 1919.
97   See Wrigley 1990, pp. 53–79.

Direct action was at the heart of the debates at both Labour Party and TUC conferences in 1919. Although the main arguments revolved around resolutions concerning the use of direct action for the specific purpose of forcing the government to abandon its intervention in the war against the Bolsheviks, the arguments are equally applicable to the question of direct action over domestic questions in general. As Smillie said at the TUC:

> My own feeling in the matter is that many of those who have spoken against direct action this week are as anxious to avoid direct action upon purely trade union questions as upon political matters. It is direct action of which they are afraid, and not the particular class of question upon which direct action is taken.[98]

Smillie was right. Labour officials were deeply and publicly anxious that a general strike in the conditions of 1919 could quickly escape the control of trade union leaders, with dangerous consequences. Clynes told the TUC that, in his opinion, a strike of millions of workers of even limited duration and aims would quickly escalate:

> It is far easier to get your men out than to get them back ... You cannot bring millions of men out to begin a great struggle like this without anticipating a condition of civil war. Your government would not be standing idly by.[99]

Clynes was echoed by figures within the leadership of the Miners' Federation. Hartshorn, for example, warned that in the event of a Triple Alliance strike, whether over mines nationalisation or the Russian question, 'within a week or ten days revolutionary conditions will have developed in this country to an extent that nobody will be able to control the situation'.[100] As we have seen, some labour leaders played up the threat of revolution and civil war to discourage direct action; however, it is worth making the simple point that for this tactic to be at all effective, the threat had to be credible. Clynes was not met with derision by TUC delegates because nothing in his argument seemed beyond the realm of possibility.

---

98   TUC Annual Conference Report 1919, p. 336.
99   TUC Annual Conference Report 1919, p. 338.
100 *Rhondda Leader*, 23 August 1919.

Similarly, when Stuart-Bunning was forced to explain the Parliamentary Committee's refusal to convene a labour conference in the spring for the purposes of discussing direct action over Russia, he outlined what the alternatives had been. If the conference had rejected a proposal to ballot union members, or such a ballot had been lost, 'it would have been a miserable and tragic fiasco'. But this was not the worst possible outcome: 'But it might have succeeded! The national strike might have been declared! What then?' The government, he said, must have fought:

> If the government fought it meant revolution! The project therefore resolved itself into a desperate gamble with the lives of men, women and children ... The Parliamentary Committee might well hesitate to call a Congress.[101]

Genuine fear or convenient alibi? The fact that the question needs asking is telling enough. Aside from, but related to, this worst case scenario of civil war and revolution, two other concerns underpinned the moderate labour leaders' objections to direct action. One was, quite simply, the possibility of defeat in a large scale conflict with the government. Trade union leaders in particular were worried about the possible implications for their carefully developed and nurtured organisations. William Brace MP, the SWMF President, admitted to delegates at the Labour Party conference in 1919 that:

> He was afraid, seriously afraid, that if they used the industrial weapon because they could not have their own way politically, they would smash their Trade Unions which were the foundation not only of their industrial but their political power.[102]

J.H. Thomas too was worried that 'if we lose there may easily be a period of reaction and oppression for many years to come'.[103] On the other hand, there was an equal concern lest direct action should result in victory over the government. Arthur Henderson conceded Robert Williams's claim that the Triple Alliance 'might make and unmake governments and terminate dynasties', but he argued forcefully that 'success under [the present] conditions may be secured at too high a price'. Not only, he explained, would victory encourage unofficial action

---

101   TUC Annual Conference Report 1919, p. 52.
102   Labour Party Annual Conference Report 1919, p. 121.
103   Vol. 112, HC Deb 5s, c.1492.

and a further breakdown of trade union discipline, but it would also have dangerous constitutional implications.[104] Jimmy Thomas put the same point in the House of Commons:

> However strong the trade union movement may be – and it is strong. However powerful the trade union movement is – and it is powerful, it is not stronger, more powerful or more important than the state as a whole. In other words, whilst we must be prepared to fight and defend our rights as trade unionists and workers, we can only defend those rights when they are consistent with ... our position as citizens of the state as a whole.[105]

For men like Henderson, Thomas and Clynes, direct action was anathema to social democracy. Sexton complained of the direct actionists that they fashioned their version of democracy on 'the mob', which was 'a bigger tyrant and a greater despot than the capitalist system'.[106] Clynes agreed, asserting that:

> [D]irect action is a blow at democracy ... The will of the people finds enduring and beneficial expression only when that will seeks social change by reasonable and calculated instalments, and not by violent act of revolution.[107]

It was the particular role of the Labour Party to give expression to 'the will of the people' through the constitutional means of the ballot box and parliamentary legislation. To give ground to the idea that social change might come about via the organised industrial power of workers would be to undermine the very constitutional foundations upon which the Labour Party rested. Jimmy Thomas put it bluntly:

> The question as to whether we are to use the industrial weapon as distinct from the political weapon raises in acute form the issue, that if that is going to be Labour's policy in the future we may as well abolish the Labour Party and the whole political machinery at once. Do not let us humbug or play with the thing; let us be quite frank and say, 'In our judgement, we have found a new and more effective weapon, and a surer weapon.' But

---

104   *South Wales News*, 14 July 1919.
105   Vol. 112, HC Deb 5s, c.339.
106   Vol. 112, HC Deb 5s, c.362.
107   *Merthyr Pioneer*, 17 August 1919.

do not let us delude ourselves by saying we want to run both together. The two things are absolutely irreconcilable, and cannot be run together in that sense.[108]

George Lansbury, writing in the *Herald*, could not conceal the exasperation he felt with this brand of reformist dogma. In early February, he wrote:

Do none of these Parliamentarians, these preachers of constitutional methods, these finger-shaking pundits, realise that the old world is breaking up under their feet, that a new spirit is moving on the face of the waters, that a new spring is in the air? It is not too much to say that the constitutional method is on its last trial and the sands are running out.[109]

There were some leading figures in the labour movement who found themselves agreeing with Lansbury. Both Smillie and Williams of the Triple Alliance at various times advocated the use of direct action precisely because, they argued frankly, unless the leaders went some of the way with the rank and file militants, they would lose control of the movement. In February, Williams was calling for a National Convention of labour's forces to draw up a programme of demands and initiate action from above in order 'to prevent unofficial strikes, to avoid the growth of anti-officialism amongst the best of our men and women … We must lead or simply be brushed aside'.[110] Smillie made the same point when he proposed to the Labour Party conference that it endorse direct action as a legitimate strategy:

It would be safer for the Labour movement of this country to meet with the Trade Union movement, and calmly and constitutionally discuss the question and decide upon action, than wait until a revolution breaks out in some other part of the country which might sweep from one end to the other.[111]

The right within the trade union leadership protested that the atmosphere of 'unofficialism' was precisely why direct action was too risky a gamble to take. The fear was that once underway, a large scale strike would develop a

108   TUC Annual Conference Report 1919, p. 113.
109   *Herald*, 8 February 1919.
110   Ibid.
111   Labour Party Annual Conference Report 1919, p. 119.

momentum of its own, and that this would favour the militants in the rank and file groups. William Brace warned: 'Starting a war is easy; stopping it once it has started is an entirely different matter'.[112] Sexton railed about 'letting mad dogs loose', telling the Labour Party Conference:

> They were letting loose an element that was rife today in the Trade Union Movement, that would take every advantage of the confusion, and make it impossible for them [the union officials] to exercise any controlling power. They were letting loose an element they could not control ...[113]

Of the Labour Party leaders, Ramsey MacDonald was most alive to the possible dangers inherent in the uncompromising attitude of Clynes, Henderson and Thomas. Even though he had been an implacable opponent of syndicalism during the 1910–14 unrest, he now felt that the Labour Party was in danger of being left behind by 'the new ideas [which] have sprung up regarding the relations between political and industrial action, new theories of the state ... [and] new philosophies of mass action'.[114] MacDonald sought to legitimise direct action in constitutional terms; pointing to the manner in which the election had been conducted by Lloyd George, and the subsequent broken promises of the government, MacDonald held that parliament had forfeited all 'moral authority'.[115] Whilst direct action aimed at overthrowing the state was incompatible with British socialism, he argued: '[It] is no mean weapon against a Parliament elected by fraud ... There must be means found for challenging the abuse of its power by such a parliament, and "direct action" is one of them'.[116] MacDonald, like many of his colleagues, believed there was a danger that direct action could escape the control of responsible leaders and lead to a revolutionary situation. However, his greater fear was that if Labour and trade union leaders took up a position of simple hostility to the notion of direct action, then the initiative would pass to the militants in the rank and file committees. 'Above all', he wrote, 'we must keep in touch with the industrial movement', and so retain an influence over what he called 'the active spirits' who were behind the unrest.[117] Labour politics were in a state of flux in 1919; they had not yet settled down into the now long familiar pattern

---

112   *Rhondda Leader*, 22 March 1919.

113   Labour Party Annual Conference Report 1919, p. 119.

114   MacDonald 1919, p. 2.

115   *Forward*, 4 January 1919.

116   *Socialist Review*, vol. 16, no. 88, January to March 1919, p. 18.

117   *Labour Leader*, 2 January 1919.

of Labour hegemony. Far from it, for in the wake of the December 1918 general election, the project of gradual social change via parliament and the state could appear almost irrelevant to a working class conscious of its own industrial power.

Direct action thus posed a threat to Labour and trade union leaders in 1919 on several fronts. There was the fear that a general strike even for limited goals might spin out of control and develop in a revolutionary way. There was the fear that defeat could lead to the overnight loss of painstaking organisational gains made over the course of several decades. Even those in the leadership who advocated direct action often did so in order to be able to better control the movement and prevent it from falling into the hands of less responsible elements. The challenge posed by the industrial unrest was, therefore, a complex one; anyone searching for a simple division inside labour politics around the question of reform or revolution will miss much of the real essence of direct action's political significance in 1919. Whilst the issue was sometimes posed in these terms, the broad spectrum of ideas assembled under the banner of direct action meant that very often this dichotomy was blurred, distorted, or simply overlooked.

The debate in 1919 was not, in the main, around the question of 'reform or revolution?', but rather of 'direct action or parliament?', which is not necessarily the same thing. But whatever interpretation one puts on direct action, whether 'do-it-yourself' reformism or the revolutionary general strike, the notion that the collective self-activity of workers held the key to economic and social change struck at the heart of the Labour Party's political philosophy. 'You can't make a silk purse out of a sow's ear', MacDonald wrote in 1919, meaning that the working class was incapable of accomplishing the socialist transformation of its own volition.[118] Direct action might have a place in rescuing the constitution from the clutches of Lloyd Georgian tyranny, but the long-term future still lay in patient educational and propaganda work so that one day Labour representatives could secure *for* the workers that which they could not secure for themselves.

The turn towards direct action thus caused a serious and protracted debate within the labour movement, one which both influenced the campaign for the Miners' Charter and was influenced by it. In the prevailing industrial unrest, a national miners' strike would be no ordinary affair; rather, it would pose a threat to British capitalism, its social order, and the political system which underpinned them. As such, it was perceived as a serious danger not just by

---

118   MacDonald 1919, p. 65.

members of the government, but by labour leaders concerned to protect their own socialist project to be pursued via parliament and the state.

Conversely, it was looked forward to by direct actionists, eager to capitalise on their strong industrial position. In March 1919, the *Herald* wrote that '[t]he whole Labour movement looks to the miners to hew out of the capitalist system a platform from which it can make a great leap forward'.[119] Revolutionaries like John MacLean went further. For him, and for an influential group within the MFGB, a miners' strike over hours and wages might, in the context of mass demobilisation, pull in millions of workers from other industries with potentially revolutionary results: 'All revolutions have started on seemingly trivial economic and political issues. Ours is to direct the workers to the goal by pushing forward the miners' programme and backing up our "black brigade"'.[120] Of course, what the actual outcome of such a strike would have been is a matter of conjecture. The point is, rather, that for contemporaries the stakes were extremely high and the situation precarious and unpredictable. Unless one grasps this, and the political controversies that flowed from it, one can understand little of the significance of 1919, nor the miners' place within it.

---

119   *Herald*, 22 March 1919.
120   *The Call*, 23 January 1919.

CHAPTER 2

The Miners' Federation of Great Britain: Bureaucratic Reformists, Militant Miners and the Development of the Miners' Charter

> The coal miners, above all other classes of workers, have the power to bring about a general strike. Today in 1919, we say the destiny of the nation is in the hands of the workers, particularly the coal miners. How will they shape that destiny?[1]

∴

In common with the rest of the labour movement the MFGB was profoundly affected by the war, emerging from it with a new leadership structure, new perspectives and new priorities. However, as with the labour movement in general, it is not possible to talk of a universal experience of war within the MFGB. In the first place there was considerable regional variation; the need for increased output, and for ever more volunteers and conscripts, affected the rhythms of work and life in the coalfield communities to different degrees. Even within coalfields there were divergent local experiences. In South Wales, for example, the commotion of war was greater in the densely populated steam coal areas like the Rhondda, which were vital social and economic zones for the war effort, than in, say, the more outlying and scattered anthracite producing areas.

Secondly, and more importantly for our purposes, there was a sharp differentiation between the wartime experiences of leaders and led within the MFGB. The outlook of those who spent the war in union headquarters, on conscription platforms, or closeted in conferences and committees at Whitehall, was changed in a different way to those who spent it at the front or at the coalface. Both groups were in the thick of things, but in highly contrasting ways, and these different experiences produced different sets of goals, and different methods and strategies for achieving them.

Alan Campbell, in tracing the development of consciousness amongst Scottish miners between the 1870s and 1920s, identified from 1911/12 onwards two

---

1   *Solidarity*, February 1919 (The organ of the National Shop Stewards and Workers' Committee Movement).

'ideal typical orientations to union activity', namely the 'bureaucratic reformist' and the 'militant miner'.[2] Although, as Campbell himself readily admits, these can at best be only 'a skeletal summary of the components of the miners' consciousness', they nonetheless help to capture in broad outline an important aspect of the reality of the politics of mining unionism in our period.[3] In the following section we will look at how the experience of the war had by 1919 shaped the outlook of both bureaucratic reformist and militant miner, and how the Southport programme drawn up in January of that year, known as the Miners' Charter, was in fact an uneasy amalgam of their different perspectives for post-war change.

1   Post-War Perspectives of the Bureaucratic Reformist Leadership

The immediately striking thing about the MFGB in this period was its size. By the end of the war it was organising about one million men, making it by far the largest trade union in Britain, and reflecting the industrial predominance of coal. Indeed in 1919 roughly one in eight of the population either lived in, or came from, a mining community.[4]

Although it had not been particularly successful in terms of material gains, the 1912 strike over a minimum wage saw the MFGB come of age as a national union capable of forging a single fighting unit out of its nineteen affiliated district organisations. However, the process of integration and centralisation remained ongoing after the war, and much of the power structure of the MFGB continued to lie with the district unions, which controlled their own funds (contributing relatively small sums to the parent body), and had their own leadership structures and constitutional arrangements. Regional autonomy was complicated by a high degree of regional diversity; the weight which the giant South Wales Miners' Federation with its 200,000 members carried inside the MFGB dwarfed the comparatively miniscule Forest of Dean or Kent associations, with their 5,000 and 1,500 members respectively.[5] Mining conditions varied widely from region to region; the high, wide seams of south Yorkshire and the North-East Midlands bore little resemblance to the heavily faulted seams of the South Wales coalfield, or the 'rat-holes' around Accrington in North-East Lancashire. Working conditions, hours and wage levels differed

---

2   Campbell 1989, pp. 8–23.

3   Campbell 1989, p. 9.

4   Coal Industry Commission 1919, vol. ii, Reports and Minutes of Evidence on the Second Stage of the Inquiry, p. 479. (Hereafter cited as CIC 1919, vol. ii).

5   MFGB 1919.

from coalfield to coalfield. Collective bargaining was carried on not at national level, but at five regional conciliation boards. Furthermore, local conditions and traditions fostered considerable variation in the political character of union politics from district to district. Nottinghamshire's reputation for conservatism and moderation contrasted sharply with South Wales's militant image for example.

Nor does this complete the picture of diversity and differentiation. Within individual coalfields, size of colliery, type of seam, character of ownership and management, and residential patterns varied considerably, and produced a range of political and industrial characteristics; areas with reputations for militancy and moderation could lie side by side within the same district. And still further desegregation is possible. Even within the same colliery, miners might face widely differing conditions. At one part of a seam the coal might give itself up fairly easily, allowing a decent wage to be earned, whilst at another the collier might be working in water under a three foot high roof. Each district, each colliery had its own maze of price lists, bonuses and working practices for each of the numerous grades of men and boys involved in getting out the coal.

The governing structures of the MFGB were designed to incorporate or accommodate this heterogeneous membership. The Executive Committee (EC) was comprised of full-time officials of the affiliated unions, elected by the MFGB Annual Conference. Some areas like South Wales or Durham might have two or three representatives on the EC at any one time, whilst the smaller unions might be lucky to have one. Whenever a major issue arose, the EC would summon a delegate conference, so that the opinions of the various districts could be expressed and a national policy acceptable to all (or most) could be hammered out. Again, before taking action or making settlements, it was usual for national ballots to be taken. This regular recourse to conferences and ballots has led some to over estimate the degree of democracy inside the MFGB.[6] Diversity and parochialism within the Federation did help to militate against an overly rigid top down approach, but it is important to stress that both the EC and the delegate conferences were dominated by officials from the district unions. There were rank and file delegates present at conferences, but their voices were seldom heard from the floor, and their opinions were often drowned out when delegations cast their votes en bloc. Furthermore, the lower grades of mine workers were under represented within the union, both at a

---

6   See, for example, Clegg 1985, pp. 177, 279; Neville 1974, pp. 532–3. R.P. Arnot also tends to play up the democratic qualities of the MFGB (Arnot 1953).

local and national level; the colliers were top dog in the pit and in the lodge, while the surface workers, for example, had a constant struggle to make themselves heard.[7]

Union activists celebrated the fact that by 1919 the Lib-Lab era had passed, and the MFGB was well into what might be called its second phase. The men who had led the Federation since its foundation in 1889 – Pickard, Cowey, Edwards, Mabon and Ashton – had all departed. Since 1909 the MFGB had been affiliated to the Labour Party, and in the course of the subsequent decade it was the solid Labour figures of Smillie, Herbert Smith and the younger Frank Hodges who rose to prominence. However, Lib-Labism was not completely consigned to the past. Hodges, after all, only replaced Thomas Ashton as Secretary in January 1919, and other Lib-Labs would survive into the 1920's. Cann, Buckley, Bunfield, Straker and Brace, who had been present at or near the MFGB's inception, were all still serving on the EC in 1919. Now into their sixties, they had made the conversion to Labour from Liberalism, but many of their views about the conduct of industrial and political affairs had remained unchanged. Furthermore, these men did not exhaust the list of old Lib-Labs who continued to hold considerable power within the Federation, as we shall see when we look at the leaderships in Nottinghamshire and South Wales, and to a lesser extent Scotland and Yorkshire.

In an interview conducted by Hywell Francis in 1973, R.P. Arnot gave an excellent insight into the nature of these men who still formed an important part of the national leadership. Arnot recalled how he had approached Smillie at the end of a MFGB EC meeting in 1917, seeking the Federation's help in a campaign to stop the imprisonment of Bertrand Russell for his anti-war views. Smillie told him to open the door into the room where the EC members were still chatting to each other. 'Cake through and look at them', he said. Arnot was instantly struck by the 'collection of fossils' that confronted him: 'A more venerable set of portent, grave and reverend seniors you never saw'. Smillie pointed out Sam Finney, the ageing leader of the Midlands Federation, as an example. When Bertrand Russell's case had been raised he had simply said 'Folks are no' put in prison unless they have done something wrong'.[8]

Most EC members were Labour in politics, and moderate in trade unionism. Cape, Sutton, Hall, Finney, Roebuck and Hoskin were typical; all in their fifties or sixties, they owed much to the pre-1914 style of leadership. Vernon Hartshorn, aged 47, was slightly younger, more dynamic, but hardly less moderate. His col-

7   MFGB Records 1918–20.
8   Interview with R.P. Arnot, 6 March 1973, South Wales Miners' Library, SWML, RPA/53/8–9.

leagues from South Wales, George Barker and James Winstone, were marginally to the left of the main group, but there was no figure like A.J. Cook or Arthur Horner in the leadership; the syndicalists and future communists were very much present in the MFGB at the war's end, but had yet to break through into its leadership either at district (South Wales aside) or national level.[9]

The dominant figures in the national leadership were Robert Smillie (President), Frank Hodges (Secretary) and Herbert Smith (vice-President).[10] Hodges, still only 33 years old in 1919, was the rising star of the younger generation of South Wales militants who had been associated with the Unofficial Reform Committee (URC) before the war. He had long since parted company with the syndicalists, and had only secured the SWMF nomination for Secretary as the moderate candidate against Noah Ablett. However, his youthfulness, his syndicalist past, and his espousal of guild socialism meant that Hodges was still seen very much as a dynamic man of the left. R.P. Arnot remembered how 'it was an immense change' when someone as 'modern' as Hodges burst onto the scene. Not only was he a socialist, he was 'exceptionally clever, able and accomplished'. He was cosmopolitan and dashing: 'He could very rapidly pick up every subject; he was right in the forefront of whatever was being discussed politically in the country and whatever was the latest fashion'. He was, according to Arnot, unquestionably the most talented man of his generation of trade union leaders: 'He was the ablest, he was a brilliant speaker. He was a born Queen's Counsel'. It was obvious that Hodges's career still had far to go. Feted by other labour leaders who were 'stunned' by his performance on the Sankey Commission, 'he was told there was no stage he might not reach', and Arnot personally believed that Hodges 'was quite convinced ... that he would be the next Prime Minister'. Although nobody in 1919 could predict Hodges's future fall from grace, his obvious ambition and careerism caused those who knew him in South Wales to raise question marks over his left-wing credentials.[11]

However, even the brilliant Hodges stood in the shadow of Robert Smillie. 62 years old in 1919, his socialist pedigree was unmatched among trade union leaders. Working alongside Keir Hardie, he had been a pioneer of both mining trade unionism and socialist politics in Scotland from as early as 1880, and was a founder member of both the Lanarkshire Miners' Union (LMU) and the ILP. Following the death of Enoch Edwards in 1912, he had taken over

---

9   Bellamy and Saville 1972–7, vol. 1. Biographical details of leading figures in the areas under
     study in this book have also been gleaned from other sources, notably local newspapers,
     and are included in the relevant chapters.
10   See Chapter Nine below for a discussion of Herbert Smith.
11   Interview with R.P. Arnot, 6 March 1973, South Wales Miners' Library, SWML, RPA/53/9–10.

as President of the MFGB, but he had already effectively been in-situ during the six week long strike earlier that year. During the war he enhanced his reputation by refusing either to sign the Treasury Agreement or to become Food Controller in the Coalition Government. His anti-militarist ILP position on the war was tinged with a pacifism that enraged patriots but endeared him to the left. He was, along with Sidney Webb, the key figure in the War Emergency Workers' National Committee (WEWNC) which enabled Labour to survive the war politically, and he chaired the famous Leeds Convention in 1917. When Robert Williams told that gathering that the government had 'taken your own leaders from your ranks and used them against you', a voice rang out 'Not Smillie!'[12]

By 1919 Smillie was a legendary figure in the labour movement. Probably the key element in his popularity was the fact that he seemed to have maintained his integrity despite his accession to the upper circles of the labour bureaucracy; the fact that he still lived in a miners' row in Lanarkshire was an important part of the Smillie mythology. His commitment to the cause of labour was unquestioned, even in the non labour press.

There is of course no living man who desires more ardently the radical improvement of the people; no man who, with much ability, strives more greatly for it; no leader who can possibly match him in influence and power.[13]

The *Nottinghamshire Guardian* similarly observed that:

Other men have the shadow of power – the ability to pass resolutions and to issue orders through trade unions. Mr. Smillie has the substance. He can command the devotion of his followers. He is trusted. His word is law ... he has achieved in the labour world the sign mark of popularity and respect – he has ceased to be 'Mr. Robert Smillie' and has become 'Bob Smillie' to a million men.[14]

As well as being heralded as a leader of integrity and principle, Smillie was often portrayed as an uncompromising militant. John MacLean, who had time for

12   Bellamy and Saville 1972–7, vol. 3, pp. 165–73. Winter 1974, chapter 7; Williams 1921, p. 83.
13   *Western Mail*, 1 April 1919. It should be noted that this appraisal of Smillie came after he had exerted his influence to have the Sankey Commission accepted by the MFGB, an act which endeared him to a newspaper which was seen as the voice of the South Wales coalowners.
14   *Nottinghamshire Guardian*, 16 August 1919.

neither timidity nor caution, spoke of his 'greatness and grandeur', calling him 'the mightiest fighter the workers of Scotland had ever had'.[15] Robert Williams asked: 'What working class leader could be more indicative of the spirit of proletarian revolt than Smillie?'[16] Smillie was more circumspect about his militant image, describing himself as 'an evolutionary revolutionist', defined as one who wanted to see dramatic changes in the way industry was run, but who did 'not desire to see an armed or bloody revolution in this country'.[17] In reality, Smillie was a pragmatist, prone to restraint in both industrial and political affairs. Alan Campbell accurately describes him as 'cautious' and 'strategic'.[18] His success as a labour leader was due in no small measure to the fact that he had never allowed his socialist beliefs to prevent him from constructing working relationships with those to his political right. Both in his early days in the LMU and National Union of Scottish Mineworkers (NUSMW), and as President of the MFGB, he had worked closely with Lib-Labs and Labour right-wingers, and had, where necessary, made compromises. The war was a case in point. Both as President of the MFGB and as Chairman of the WEWNC, he had held together uneasy alliances of jingoists and anti-militarists, concentrating not on doctrinal disputes, but on practical measures to defend the material conditions of the miners and the working class.[19] It was his skill as a broker between left and right, militants and moderates, which enabled him to retain the confidence of people from opposite sides of the political spectrum within the labour movement. As R.P. Arnot has perceptively written:

> And if in mining matters of acute controversy he often seemed to adopt a cautious attitude and on occasion to hold a balance between left and right, this brought him no ... reproach, but an increased influence as one to whom each side looked for aid against its opponents.[20]

Although he was on the left of the MFGB leadership in 1919 and had clear political differences with EC members like Finney, Bunfield and Cape, Smillie nonetheless had much in common with his colleagues. He functioned as part of the national leadership, never separately from it, staying within operational parameters acceptable to himself and those well to his right. He was, like all

15   *The Call*, 9 August 1917.
16   *Herald*, 1 February 1919.
17   TUC Annual Conference 1919, p. 336.
18   J. McIlroy et al. 2009, p. 174.
19   Winter 1974, chapter 7; Arnot 1955, chapters 4–7.
20   Arnot 1953, pp. 126–7.

the EC members, a bureaucratic reformist, in the sense that they 'all shared an ideological commitment to political reformism ... coupled with a bureaucratic trade union practice'.[21] The pursuit of political reform through parliamentary legislation had been a central plank of mining trade unionism as far back as the middle of the nineteenth century. The Miners' National Union, for example, was founded in 1863 expressly 'for the purpose of gaining legislative change'.[22]

Parliament featured more prominently in mining politics than in that of any other occupational group. Partly this was due to the sheer size of the industry and its importance to the national economy. Parliament had always been prepared to devote considerable time and attention to mining affairs, even in the classical age of laissez-faire. Royal Commissions and Select Committees on mining resulted in landmark pieces of legislation in 1872, 1887 and 1911, which laid down regulations and guidelines for the operation of the industry. Apart from helping to improve the appalling safety record, the granting of miners the right to elect and pay for their own checkweighers was instrumental in the creation of modern mining trade unionism.[23]

Prior to the third Reform Act of 1884–5, the miners' unions had sought to influence business at Westminster with lobbying and petitioning. With the enfranchisement of large numbers of miners, they were now in a position to elect a considerable number of MPs in those constituencies where miners dominated the electorate.[24] The great engineering lockout of 1897–8, and the backlash against trade unions in the courts around the turn of the century, further convinced the fledgling MFGB that major change in industry would come about by parliamentary rather than industrial methods. The Act of 1908, which reduced the collier's hours to eight per day, was claimed as a major vindication of this strategy. Even though it had taken twenty years of agitation to achieve, it had nonetheless held the inestimable advantage, from the leadership's vantage point, of avoiding the risks to union organisation which industrial action entailed. In this respect, Smillie's experience of union building in Lanarkshire had made him as fervent an advocate of the benefits of parliamentary legislation as any miners' leader. He and his socialist colleagues had faced enormous difficulties in establishing permanent union organisation in the last quarter of

---

21   Campbell 1989, pp. 9–10.

22   Arnot 1953, p. 43.

23   Arnot 1953, pp. 43–50.

24   Gregory 1968, p. 16. Miners were in a better position than most workers to elect their own MPs because of their concentration in certain constituencies.

the nineteenth century. Scottish mining unionism, and Lanarkshire's in particular, had 'been characterised ... by a succession of short-lived federations
of local unions established on the cyclical upswing of the coal market' which
'were then destroyed during strikes to resist wage reductions when the market
declined'.[25] As Gregory has pointed out:

> To the early socialists this state of affairs was both a challenge and a
> stimulus: finding themselves at the head of unstable and ineffective local
> unions they turned all the more readily to political action to offset their
> industrial weakness.[26]

From 1912, Smillie had been one of the key architects of a union strategy which
sought progress through a judicious combination of industrial and parliamentary means. When recourse was made to a national strike on the sole occasion
of the 1912 minimum wage dispute, meticulous care was taken to make sure
that it remained under the tight control of the leadership, and it was called off
when parliament conceded the principle (though not the detail) of a minimum
wage.[27] In the wake of the strike it was business as usual, with the Federation pressing for a range of legislative enactments, from a scheme to provide
comprehensive workmen's compensation, to the nationalisation of the coal
industry.[28] In addition, the aim of the MFGB leadership was to continue to nurture and develop union organisation, in particular to increase the Federation's
cohesiveness and degree of centralisation, and win the right to represent all
miners in collective bargaining at the national level.

National wage bargaining and nationalisation both seemed to be a long way
off in 1914. Indeed, state ownership had been discussed at MFGB conferences
as far back as 1894, and by now the annual resolutions in its favour were ritu-
alised gestures.[29] The war intervened to dramatically change the political and
economic landscape, highlighting as it did 'the immense strategic importance
of coalmining in the economy'.[30] Coal was vital not just to domestic industrial
production (including munitions), but also for shipping and foreign exchange.
In 1915, Lloyd George, then Minister of Munitions in Asquith's government,
said:

---

25   Campbell 1989, p. 11.
26   Gregory 1968, p. 90.
27   Arnot 1955, chapter 6.
28   The first nationalisation Bill was drafted by the MFGB in 1913.
29   Kirby 1977, pp. 31–2.
30   Kirby 1977, p. 24.

> In peace and in war King Coal is the paramount lord of industry ... It is our real international coinage. When we buy goods, food and raw materials abroad, we pay not in gold but in coal ... we cannot do without coal.[31]

Pronouncements of this sort left the miners in little doubt as to their importance. This could be a source of pride, but set against rapidly falling real wages and very high profits it could also become an irritant. It only took a slight ebbing of the initial high tide of patriotism for industrial unrest to creep back into the coalfields, threatening the continuous production upon which the war economy rested. In the interests of stabilising industrial relations and (to a lesser extent) rationalising production, one of the first acts of the Lloyd George government in December 1916 was to introduce state control of the mining industry. Responsibility for the industry rested with a newly created Coal Controller's Department within the Board of Trade, headed by Sir Guy Calthrop.

Although initially anxious that state control had been introduced to suppress rather than appease industrial discontent, it soon became clear to the MFGB leadership that its benefits far outweighed its disadvantages. Lloyd George promised Smillie that the Munitions Act would not be applied to the mining industry, and furthermore suggested that state control might well be the first step towards full nationalisation. In the meantime, the central plank of state control – the creation of a profits pool funded by a percentage of the owners' 'excess profits' – represented an enormous advance for the union. Although the primary function of the profits pool was to subsidise those collieries which might otherwise have gone out of production, its corollary was to destroy the owners' claim that national wage bargaining was impossible due to regional variations in productivity and profitability. For the first time in its history, the MFGB was able to make claims for national wage rises. And for the first time in its history it received them, first in 1917 and again in 1918. Simultaneously, the MFGB also received national recognition from the government, its wage claims having been submitted directly to, and resolved by, cabinet. The centralising effects of national bargaining led the MFGB to restructure its leadership in the last year of the war, and Smillie and Hodges were elected as the union's first full-time national officers.[32]

As the war drew to a close, the retention of state control became the first priority of the MFGB EC. In its view this was essential both for the continuation of national bargaining and for the greater goal of nationalisation. The issue of

---

31   Armitage 1969, p. 103.
32   MFGB Special Conference 14–16 January 1919, pp. 6–8.

the union and the state dominated the MFGB Annual Conference in 1918. As James Winstone said, in moving the resolution on nationalisation from South Wales: 'The time has arrived ... now is the time and this is the place'.[33] Hodges added: 'The war has so changed the position that what might have taken place in a period of ten or fifteen years is now possible within a comparatively short period'.[34]

The resolution on nationalisation (and several others besides) was, in keeping with conference tradition, aimed at parliamentary legislation and 'cooperating with the national Labour Party to ensure the passage of a new Bill into law'.[35] Shortly before the general election of 1918, Hodges spoke of the prospects for the realisation of the MFGB's demands:

> To accomplish this tremendous task without social disorder, the Miners' Federation hopes to have its strength accurately reflected in Parliament. It is possible that the Miners' Federation could accomplish this by a great National Strike. But it is in the interests of the nation that it should be accomplished without a strike. National stability is essential in the years following the declaration of peace. The trade unions, acting through Parliament, can alone accomplish this in anything like order.[36]

The war had shifted the goalposts in favour of the MFGB, but it had done nothing to disabuse the bureaucratic reformist leadership of the risks inherent in national industrial action. Direct action was not discounted by the EC in the run up to the campaign for the Miners' Charter, but parliamentary and electoral politics continued to be its strategic orientation, even after the disappointment of the December elections. Below them, however, amongst the rank and file there were influential groups for whom direct action represented the only road to success.

2 Post-War Perspectives of the Militant Miner

In his use of the term, Alan Campbell defines the ideal-typical 'militant miner' as one who believed in the socialist revolution, involving the socialisation of

---

33   MFGB Special Conference 9–12 July 1918, p. 44.
34   MFGB Special Conference 9–12 July 1918, p. 117.
35   MFGB Special Conference 9–12 July 1918, p. 53.
36   *Merthyr Pioneer*, 14 December 1918.

the mines, and who was a member of the Communist Party.[37] This is appropriate for the 1920s, but not for our brief post-war period before the party's formation. For our purposes the militant miner as an ideal type in opposition to the bureaucratic reformist first appeared in South Wales in 1910–11, and was a syndicalist rather than a communist. Whilst the 'communist' label calls forth a fairly clear set of principles and beliefs by which we can recognise Campbell's militant miner, the term 'syndicalist' describes a much broader and more heterogeneous group.

What we will describe as the 'syndicalist' current in the coalfields in fact contained considerable ideological diversity. Militant miners in our period might describe themselves as syndicalists, revolutionary-syndicalists, industrial-unionists, industrial-socialists, or other variations thereof, a fact which points to the cross-pollination which took place between the different strands of the left at this time.[38] Whereas some pure revolutionary-syndicalists rejected political action entirely, many who saw themselves as being within the syndicalist fold did not. Industrial-unionists, for example, were prepared to accept some measure of auxiliary political (parliamentary) activity. Others might vote Labour at elections but play no organised role in party politics, while still others might find no contradiction between holding syndicalist ideas and ILP membership. As one student of mining syndicalism has written:

> Individuals could selectively subscribe to more than one set of principles, agreeing or disagreeing with whatever aspect suited or failed to fit in with their particular views and creating, in effect, their own personal composite ideology ... Given this openness, or more pejoratively, this inconsistency, the difficulty of applying hard and fast labels becomes apparent.[39]

Campbell offers the Scottish miners' leader and communist Willie Allan as typifying his militant miner, but for our period it was unquestionably Noah Ablett. One of the founders of the Unofficial Reform Committee (URC) in South Wales in 1911, he was the dominant figure amongst the militants throughout the rest of the decade, and although he was a full-time Agent in 1919, he remained a leading exponent of the ideas encapsulated in *The Miners' Next Step*.[40] His own political make-up reflected the ideological inconsistencies of the militant

---

37 Campbell 1989, Figure 1, p. 23.
38 Davies 1991, pp. 4–28.
39 Davies 1991, p. 28.
40 See Chapter Five below.

miner which he represented. At the 1912 TUC conference, he described himself as a syndicalist, but one who did not rule out working class political action. However, only the week before, he had written to the *Rhondda Socialist* saying that people should view him not as a syndicalist but as an industrial unionist.[41] In line with *The Miners' Next Step*, he placed a low value on electoral activity, but he was not averse to supporting Labour candidates at election time.[42] On the other hand, he did not see political action as meaning only electoral activity, and was a fervent advocate of working class education, a pioneer of the Plebs League and Central Labour College (CLC) classes which so influenced his generation of militants.[43]

Before the Communist Party hardened the ideological diversity within the syndicalist camp into a doctrinal division over reform or revolution, Ablett's eclecticism captures the essence of the militant miner. Whilst he might hold elements of reformism in his political consciousness, the militant miner's main orientation was, like Ablett's, firmly towards his trade union and direct action as the means to social and economic transformation. This belief that the main impulse towards socialism would come from below, via the collective power of workers exercised through their industrial organisations rather than through the ballot box, was the hallmark of the militant miner in 1919 and of the rank and file movement to which he belonged.

The rank and file movement first emerged in South Wales at the beginning of the second decade of the twentieth century.[44] The South Wales Miners' Federation (SWMF) had been a slow starter in mining trade unionism. It was not formed until 1898, and during its first decade of existence it acquired a reputation for moderation within the MFGB. However, from around 1907, industrial relations in the South Wales coalfield began to deteriorate as the owners sought to combat declining productivity by holding down labour costs.[45] Lib-Lab leaders who had successfully established the SWMF in extremely difficult conditions and with recourse, on occasions, to quite militant action, had by now become conservative and set in their ways, clinging tenaciously to collective bargaining methods and mechanisms which were rapidly becoming anachronistic in the changing conditions in the coalfield. Between 1903 and 1910 for example, the Conciliation Board failed to settle 231 out of the 391 disputes

41   Davies 1991, pp. 30, 64–5.
42   Interview with D.J. Davies, 3 November 1972, SWML, DJD/16/6.
43   Egan 1986, pp. 19–28.
44   See Chapter Five below for more on the URC in South Wales.
45   Williams 1973, pp. 46–50.

referred to it, and as the end of the decade drew nearer, increasing numbers of these unresolved disputes ended in lockouts or strikes.[46]

Matters came to a head over a lockout at the Naval Colliery of the Cambrian Combine group of collieries in the Rhondda in September 1910. The long and bitter struggle over payment for abnormal places marked a watershed in mining trade unionism in South Wales and beyond.[47] As Chris Williams writes, the dispute 'simultaneously exposed the stark opposition of labour and capital in a manner that drastically undermined the viability of Lib-Lab notions of industrial cooperation and wider social harmony'.[48] When the miners were eventually forced back to work by the leaders of the SWMF and MFGB in June 1911 on terms which had been rejected in October 1910, the hiatus between rank and file militants and moderate Lib-Lab officials hardened and became permanent.

In the last months of the Cambrian Combine strike, rank and file militants were pushing for escalation to a national strike for a minimum wage. However, their experience of the official leadership had convinced them that a successful campaign would have to come from outside the official machinery. In the second half of 1911, they held an ongoing discussion about how to sharpen their activity and increase their effectiveness in the light of the previous year's experience. The upshot was the creation of the URC at the end of 1911, and the publication of its ideas and programme in *The Miners' Next Step*.[49]

Woodhouse describes the make-up of the URC at its formation:

> As it came into being at the end of 1911 the URC therefore represented a hard core of unofficial leaders in the Rhondda and their immediate supporters in the Plebs League and on the Cambrian Combine Committee, together with the extended network of contacts throughout the coalfield who had been associated in one way or another with the drafting of *The Miners' Next Step* or who sympathised with the general aims of the URC. The organisation of the URC was consequently of the loosest form. W.H. Mainwaring kept a book of about two hundred addresses of contacts in South Wales and in the MFGB generally, and it was through these that *The Miners' Next Step* was distributed and the particular policies of the URC on specific issues taken into the lodges.[50]

---

46   Woodhouse 1969, pp. 4–9.
47   Woodhouse 1969, pp. 44–63.
48   Chris Williams, in Cambell et al. 1996, p. 125.
49   Woodhouse 1969, pp. 72–6.
50   Woodhouse 1969, p. 76.

The starting point of *The Miners' Next Step* was its attack on what it called 'the Conciliation Policy' for its failure to secure acceptable wage levels, and for preventing effective trade unionism through its ' "tying up" and "Delay" character'.[51] Furthermore, because of its need for full-time professional union negotiators, the Conciliation policy created a caste of leaders who stood aloof from and above the miners, and who 'become "gentlemen", they become MPs, and have considerable social prestige because of this power'.[52] The result was that: 'The leader has an interest – a vested interest – in stopping progress' by maintaining the status quo, and limiting the degree of control exercised by the rank and file, both within the union and at the coalface.[53] The opening section of *The Miners' Next Step* shared some common ground with Robert Michels's attempted theorisation of the problems inherent in stable, permanent working class organisations; the inevitability of the development of oligarchical leadership with distinctive interests from those of the rank and file, its infusion with bourgeois modes of thought and behaviour, the necessarily corrupting nature of power; all these were major themes.[54]

In language which owed much to the influence of Tom Mann's *Industrial Syndicalist*, the basic remedy offered by *The Miners' Next Step* was the transformation of the union into an industrial organisation which 'recognising the war of interest between workers and employers, is constructed on fighting lines, allowing for a rapid and simultaneous stoppage of wheels throughout the mining industry'.[55] In the first place this would be achieved by a change in policy; the adoption of the demand for a minimum wage was vital, as it made conciliation unnecessary and clarified the battle lines. 'A man either receives the minimum or he does not. There is nothing to conciliate or negotiate upon'.[56] Secondly, the pamphlet argued for sweeping constitutional changes to the SWMF. A process of democratic centralisation would overcome institutionalised sectionalism and bureaucratic domination by abolishing the districts and placing control of the organisation in the hands of the rank and file. The principles of the proposed constitution were that '[t]he Lodges have supreme control', '[t]he Executive becomes unofficial', and the 'Agents or organisers become the servants of the men'.[57]

---

51   'The Miners' Next Step', in Coates 1974, p. 18. (Hereafter cited as TMNS).
52   TMNS, p. 18.
53   TMNS, p. 19.
54   Gilbert 1992, pp. 78–9.
55   TMNS, p. 23.
56   TMNS, p. 24.
57   TMNS, pp. 22–6.

Armed with 'a militant aggressive policy' the union would be characterised by enhanced solidarity, and by the far greater involvement of the rank and file in running their own affairs.[58] Great emphasis was placed on the educational and confidence building aspects of struggle, even for immediate and limited goals:

> Every fight for, and victory won by the men will inevitably assist them in arriving at a clearer conception of the responsibilities and duties before them. It will also assist them to see that so long as shareholders are permitted to continue their ownership, or the State administers on behalf of the shareholders, slavery and oppression are bound to be the rule in industry. And with this realisation, the age-long oppression of Labour will draw to its end.[59]

As this suggests, *The Miners' Next Step* rejected the nationalisation of state socialists as bureaucratic and autocratic. Its objective was industrial democracy:

> Every industry thoroughly organised, in the first place to fight, to gain control of, and then to administer that industry ... This would mean real democracy in real life, making for real manhood and womanhood. Any other form of democracy is a delusion and a snare.[60]

Once again, the road to control lay in the miners' struggle with the owners. The pamphlet suggests that this would not, initially, take the form of a full frontal assault on the direct question of ownership. Rather, it envisaged a series of lesser and partial struggle over wages and conditions which would, by simultaneously eating into the profits of the owners and increasing the fighting ability, spirit and confidence of the miners, lead incrementally to workers' control.

The role of strikes in preparing the rank and file morally and politically for the socialisation of the mines is an important and much overlooked theme of *The Miners' Next Step*. The influence of De Leon has often been commented on, but this emphasis on the importance of industrial action (collective restriction of output as well as strikes) in developing the consciousness of workers goes back to Marx. Commenting in 1853 on wage strikes in Lancashire, Marx wrote that he was:

---

58   TMNS, pp. 28–9.
59   TMNS, p. 30.
60   Ibid.

convinced that the alternative rise and fall of wages, and the continual conflicts between masters and men resulting therefrom, are, in the present organisation of industry, the indispensable means of holding up the spirit of the labouring classes, of combining them into one great association against the encroachment of the ruling class ...[61]

Without these economistic strikes, he wrote:

the working classes ... would be a heart-broken, a weak-minded, a worn-out unresisting mass, whose self-emancipation would prove as impossible as that of the slaves of Ancient Greece and Rome.[62]

The Miners' Next Step shared this view of the value of partial struggles for developing class consciousness. Where it parted company from Marx was that it saw no role for a political party of the working class to assist this process and direct the development of consciousness to a revolutionary level. For the South Wales syndicalists, this was superfluous to requirements; their attitude was that militant industrial action would carry all before it and overcome all political obstacles in its path.

This then was the method of the militant miner who established the rank and file movement in the coal industry; by transforming the union organisation, and investing it with a fighting spirit, one could transform both the miners themselves and the industry and society in which they worked and lived.

URC members carried their message beyond South Wales, sending 'missionaries' to other coalfields to explain the ideas and strategy of *The Miners' Next Step*, and to urge branches to take up the campaign for the minimum wage.[63] Whilst it would not quite be accurate to say that the URC's efforts led directly to the national strike of 1912, one can nonetheless trace the dispute's origins back to the unofficial agitation which began in the Cambrian Combine strike. This was the limit of the URC's pre-war success, however. Efforts at establishing Reform Committees were made in Scotland and the North-East at this time, but they came to nothing, and in South Wales the advent of war led to the collapse of the URC.[64]

The crisis of the URC was due in part to its organisational flaccidity, but it was mainly a function of its political heterogeneity. In 'normal' conditions,

61   Marx and Engels, *Collected Works*, vol. 12, p. 169.
62   Ibid.
63   Woodhouse 1969, p. 73.
64   Pribicevic 1957, pp. 302–3.

political differences amongst URC supporters could be glossed over or avoided by a syndicalist concentration on union reform. The war demanded a political response, and the organisation was unable to accommodate the polarisation which ensued within its ranks. Whilst some like W.F. Hay took up an unwavering opposition to the war, other militants like C.B. Stanton became ultra-patriots, making joint activity on industrial questions impossible. Whilst the URC's collapse was temporary in nature, it nonetheless signposted the fragility of the unofficial movement at moments of political crisis, and exposed the weak flank of syndicalism. In theory, 'politics' could be ignored or at least subordinated to straightforwardly industrial issues. In practice, at certain junctures politics became paramount and divisive, undercutting the industrial strategy of the miners' reform movement. These political fault lines would again come into play at crucial moments in the campaign for the Miners' Charter in 1919.

The demise of coalfield-wide rank and file organisation after 1914 did not prevent the continued activity of groups of militants at lodge level, and these played a crucial role in the South Wales strike of July 1915. This was the key episode in the mining industry during the war, shaping the contours of national industrial relations between miners, owners and government to the end of hostilities and beyond. In fact, the strike can be seen as the opening act in the power struggle between the MFGB and the government, which culminated in the crises of decontrol in 1920 and the lockout of the following year. For this reason, the strike ranks equal in importance to the wartime strikes in engineering, and yet curiously it has been almost completely overlooked by historians.[65] It is tempting to include a full account of the strike here, but this would mean straying too far from the present narrative, and so we must confine our discussion of it to those areas which shed light on the campaign for the Miners' Charter in 1919.

As we have seen, the main outcome of the 1915 strike was government control of the industry, a development which, for the MFGB EC, transformed national wage bargaining and nationalisation from lofty ideals into realisable goals. However, for others the strike's victory represented a defining moment for different reasons. 1915 was critical in shaping the post-war perspectives of both bureaucratic reformist *and* militant miner.

In the first place, whilst the MFGB leaders derived an enormous strategic advantage from the strike, had it been left up to them it would not have

---

65   Apart from a brief article by an economist shortly after the strike, there is only one (highly flawed) article in print which deals with the events of July 1915 (Carter 1915). O'Brien 1984–5, pp. 76–104.

happened. Much has been made of the MFGB's refusal, in contrast to other unions, to comply with the Treasury Agreement or the Munitions Act, described by Smillie as the 'Workers' Slavery Act'.[66] However, the refusal of the MFGB to submit to government compulsion did not mean that it stood outside of the patriotic consensus which had been established between government and unions, for whilst there were important nuances in its wartime strategy, its attitude was never oppositional.

The position of the MFGB was demonstrated in the course of a series of meetings which took place between Smillie and Lloyd George, then Minister of Munitions, in the early summer of 1915, at which they discussed an arrangement whereby the miners could be excluded from the Munitions of War Act. Whilst Smillie and the MFGB Executive were implacably opposed to an Act which would give the government powers of compulsion over the miners, they were nonetheless well disposed to delivering voluntarily what Lloyd George wanted; sustained and, if possible, increased output of coal to feed Britain's war economy, and avoidance of strikes and absenteeism. Smillie stated publicly that the union was willing 'to do everything within reason to secure a continuous, steady, and if possible an enlarged output of coal'.[67]

Under Smillie's guidance, the MFGB leadership's strategy attempted to balance a dual commitment to class and nation, and to assert their reciprocity. In terms of resisting government compulsion, this formula achieved a measure of success. Beyond this point the reconcilability of these two commitments proved evasive. In practice, the perceived need to serve the nation placed severe constraints upon the leadership's ability to sanction industrial action over the pressing issue of wages. Consequently, in April 1915 the MFGB conference decided to shelve its demand for a national wage increase, and shortly afterwards the EC instructed affiliated unions to pursue a claim for a 20 percent rise via the district Conciliation Boards.[68] Although acceptable rises were negotiated in most cases, this could not conceal what had been a major climbdown by the MFGB leadership, which had aborted its planned drive for national wage bargaining. Furthermore, the decision to pursue district increases had involved abandoning the miners in South Wales, where the owners were clearly not going to make any reasonable settlement.[69] At this stage of the war, the MFGB was far from having things its own way in industrial relations. The *Herald*

---

66   *Herald*, 26 June 1915.

67   *Times*, 26 June 1915.

68   MFGB Conference 21–23 April 1915.

69   *Western Mail*, 20 April 1915.

expressed the disappointment of the left: 'There is little doubt that the miners have been defeated by the coal owners and the government'.[70]

The SWMF leaders' approach to wartime trade unionism chimed with Smillie's. Vernon Hartshorn maintained that strikes were to be avoided as long as the war lasted, but 'the industrial truce meant the status quo in industry, not the prostration of labour before capital'.[71] However, as the national leadership had found, it was easier to balance a commitment to their members and to the national interest in theory than in practice. By 1915, the South Wales coal trade was booming; steam coal from the Central Valleys was the best in the world and was the sole source of fuel for not only the British, but for the French and Mediterranean naval fleets as well. The precious nature of South Wales's steam coal drove its price through the roof, resulting in huge profits for the owners.[72]

However, with an eye to renewed competition from the growing American steam coal industry come the end of the war, the Monmouthshire and South Wales Coal Owners' Association (M&SWCOA) would only offer the SWMF a 10 percent increase on the 1879 standard rate, which had been rendered obsolete by the inflated price of coal. In a nutshell, the owners' offer excluded the miners from any share in the wartime bonanza in the South Wales coal trade.[73] The SWMF leadership appealed to Walter Runciman at the Board of Trade to intervene and force the owners to make an acceptable offer, but to no avail.[74] By July 1915, it was becoming abundantly clear that the alternative to strike action was, indeed, prostration before capital.

As a confrontation loomed, the government and the press sought to present the dispute as a test of the miners' patriotism. The *Western Mail*, for example, carried a cartoon of a miner in German military uniform holding a gun marked 'strike' to the head of the First Lord of the Admiralty.[75] On 14 July, following the announcement by a SWMF conference of its intention to strike the next day, the government upped the stakes by proclaiming the coalfield under the penal section of the Munitions Act. This rendered any strike illegal, and a

---

70   *Herald*, 1 May 1915.

71   Woodhouse 1969, p. 116.

72   Report of the Departmental Committee appointed to inquire into the conditions prevailing in the coal industry due to the war. Cd.8009 xxxviii, Part Two. Minutes of Evidence, p. 12. Prior to the war, the average price of good steam coal was anywhere between 9s11d and 18s 1d per ton. By 1915, even a bulk buyer like the Admiralty was paying 25s per ton and the steam coal left over for sale on the market was fetching massively inflated prices.

73   Carter 1915, p. 457; *Herald*, 22 May 1915.

74   *Western Mail*, 28 June 1915; *Western Mail*, 13 July 1915.

75   *Western Mail*, 14 July 1915.

Munitions Court was set up to prosecute miners who defied the ban. Copies of the Proclamation were posted at every pithead.[76] The government's decision was greeted favourably by the press: 'To capitulate to such brazen intimidation would not only be a calamity, but a crime', announced the *Morning Post*, for whom '[t]he only possible course is to confront the unconscionable conspiracy against the common weal and to break it ... An end to feeble trucking'.[77] The *Times* agreed:

> The Government could have done no less than they have, if they are to govern at all. Any other answer to the decision of the delegates at the Cardiff Conference on Monday would have amounted to abdication. The decision was a flat defiance, not of the coal owners, but of the Government.[78]

What had been a local dispute over a district wage agreement had now become a 'test of government', a confrontation over the right of the state to compel workers in wartime.[79] When Lloyd George was dispatched to Cardiff, his instructions from the Cabinet, should the miners fail to respond to his appeals, were clear:

> The Government would at once take all the necessary measures to paralyse the strike, by preventing meetings, speeches, and the receipt of strike pay ... inciters to a continuance of the strike or to the obstruction of measures taken by the Government, would be prosecuted and punished, and the question would be seriously considered of strengthening the law by making such offences treasonable as acts of aiding and abetting the enemies of the King.[80]

By making strike action treason, and using state force in the name of the 'national interest', the government hoped it would be able to induce a change of heart in the SWMF. As far as the union leadership was concerned, it was entirely successful. Sensitive to accusations of disloyalty, and concerned to respect the rule of law, the EC majority tried to persuade the delegate conference on

76　*Times*, 14 and 16 July 1915.
77　*Morning Post*, 14 July 1915.
78　*Times*, 14 July 1915.
79　*Times*, 14 July 1915.
80　CAB 41/36/34, 19 July 1915.

12 July to abandon plans for a strike.[81] Although it was defeated by a large majority, and instructed by the conference to call a strike on 15 July, only three out of over twenty SWMF EC members were prepared to stand by the membership.[82] The majority travelled to London in defiance of conference instructions, to tell Runciman that 'the EC as a body were prepared to render the Government any assistance they could in the direction of bringing about a termination of the strike'.[83] Despite bombardment from the government, press, and its own leaders, the rank and file, 'rejecting all reiterated appeals to patriotism', stuck to the decision of the delegate conference. On 15 July, all 200,000 South Wales miners defied the Proclamation and struck work for a new wage agreement.

This was not an expression of any particular disloyalty or lack of patriotism; South Wales miners volunteered in huge numbers for the armed forces and those who remained behind cut down their levels of drinking and absenteeism.[84] In the rush to the colours, the pre-war militancy of the coalfield appeared to have been completely emasculated. The patriotic feeling which was so essential in lubricating the wheels of the wartime industrial truce in mining had been consciously encouraged by government in all the coalfields, but nowhere more so than in South Wales, which was of such high strategic importance to the war effort. A. Clement Edwards, Liberal MP for East Glamorgan, carried a typical message. Shortly before the strike he told his constituents:

> The coal you get is the lifeblood of our Fleet. It has enabled our Navy to keep the enemy cruisers to distant seas. It has enabled troops to be brought from the uttermost ends of the earth. It has enabled us to send our fighting forces with munitions and equipment by hundreds of thousands to the shores of France and to the Dardanelles with swiftness and safety ... In a word, the coal you get has been the salvation of our national strength.[85]

As Runciman's Departmental Committee had shown, the miners had everywhere tended to respond positively to patriotic appeals of this kind, allowing

---

81   SWMF Conference 12 July 1915.
82   *South Wales News*, 15 July 1915. The three in question were Barker, Hodges and Ablett.
83   *Western Mail*, 17 July 1915.
84   Report of the Departmental Committee, Part One, First General Report, May 1915, Cd.7939 xxvii, pp. 6–18. Second General Report, Cd.8147 xxxvii, pp. 1–2. Cd.8009 xxxviii, Part Two, Minutes of Evidence, p. 121.
85   *Times*, 15 July 1915.

the industrial truce to function relatively smoothly. However, from the miners' point of view, this unwritten truce contained a clause which insisted upon a genuine 'mutuality of sacrifice' between owners and men.[86] George Barker said that when convinced of the national need, the Welsh miner 'would go down the pit in just [the same spirit] as if he were putting on his uniform'.[87] However, with the rejection of their wage claim, and in the face of flagrant 'profiteering' by the owners, appeals ceased to operate as an effective constraint upon the miners. In fact, they could become a catalyst to action.

In the first place, the South Wales miners' role in the war effort had left them in no doubt as to their industrial strength should they choose to exercise it. Secondly, patriotism as an ideology was open to class-based interpretations. The miners were able to reject accusations of disloyalty because they had their own particular interpretation of what constituted patriotic behaviour. Noah Ablett said: 'We are emphatically not pro-German, but we are working class'.[88] An 'experienced miner' explained their attitude to the owners and the government in an interview with the *South Wales News*:

> They allowed matters to drift until it was too late. They counted upon our patriotism and thought we would not dare come out in time of war. Now we are out they call us traitors and such-like names. They are the traitors, for they have driven loyal subjects to revolt. We have a duty to perform to the 50,000 miners who have gone out to the front – my two dear boys among them. We must see to it that the conditions upon which they will return to work after the war is over is at least as good as they were before they went away. If we colliers don't perform that duty we will be traitors to our sons and brothers.[89]

Defence of pay, living standards and working conditions won in past struggles was in fact seen as a patriotic duty of equal importance as the maintenance of coal production. As far as the miners were concerned, by their profiteering the owners had broken the conditions upon which the industrial truce had been founded, and this released them from their obligation not to strike. Furthermore, in this situation it was possible for the miners not simply to fend off accusations of treason, but to level the same charge at owners and government.

---

86   The phrase is Asquith's; *Times*, 21 April 1915.
87   Cd.8009 xxxviii, p. 211.
88   *Herald*, 24 July 1915.
89   *South Wales News*, 19 July 1915 (the newspaper is usually known as the *South Wales Daily News*, but the 'Daily' was dropped in this period).

A *Times* correspondent who spent time amongst the Rhondda miners during the strike noted that there was a widespread belief that 'the Government are only the tools of the hated capitalist'.[90]

Just as government efforts to brand the South Wales miners as traitors backfired, so too did its attempt to coerce them under the Munitions Act. In the mining communities of South Wales, there was a strong tradition of coveting liberty, and pride in the status of being 'free men'.[91] The experience of police and army intervention in the Cambrian Combine strike of 1910–11 had stiffened their hatred of state repression. Ironically, this experience helps to explain the enthusiasm of the South Wales miners for enlistment in 1914; the war was seen as being fought for liberty and against 'Prussianism' and tyranny.

The government's action in proclaiming the coalfield was an affront to the miners' traditions and values; far from weakening their resolve, it strengthened it. The *Merthyr Pioneer* gave voice to the miners' concerns, describing the Munitions Act as 'the most sinister piece of legislation enacted during the last fifty years of this country', as it 'makes bondsmen of freemen'.[92] This was a constant theme in speeches at mass meetings during the strike. James Winstone said: 'The Government does not understand the psychology of the South Wales miner. They may destroy him, but they will never coerce him'.[93]

There was a conviction that to submit to the government now would be to lose far more than a wage agreement; it would mean a return to a form of serfdom. Frank Hodges, at this time Agent for the Garw district, said: 'Above all the Welshman considers that he is a free man, and that he would be enslaved if he worked under the Act'.[94] Ablett felt that to give in 'would be to set back the position of the workers for a century'.[95] In a speech to the SWMF conference on 15 July, Idris Davies encapsulated all the issues when he told delegates: 'If you are going to work, you are going to work in a worse state of slavery than our boys are fighting to relieve'.[96]

Once the miners had resolved to defy the government, there was nothing it could do to coerce them. As the *New Statesman* pointed out: 'The government

---

90   *Times*, 16 July 1915.
91   *Merthyr Pioneer*, 31 July 1915.
92   *Merthyr Pioneer*, 31 July 1915.
93   *Western Mail*, 14 July 1915.
94   *Western Mail*, 17 July 1915.
95   *Western Mail*, 17 July 1915.
96   *Merthyr Pioneer*, 17 July 1915.

may set up a Munitions Court and inflict fines, but it will not be possible to recover the money, and you cannot imprison whole towns'.[97] As for rumours that the government was considering sending in troops, a mid-Rhondda miner voiced the general attitude. 'What can the Government do? They may take us to the tub and put a shovel in our hands, but they cannot make us load'. The *Times* correspondent noted that: 'This assumption of the impotence of the Government is almost universal among the men'.[98]

The South Wales miners were invulnerable to government tactics. Within hours of Lloyd George's arrival in Cardiff on the evening of Monday 19 July, he recognised the hopelessness of the situation, and promptly capitulated. By 20 July, a proposed deal had been drawn up, which was accepted by a SWMF Conference the following day. *The New Statesman* commented sardonically on Lloyd George's volte-face:

> This week, the miners having destroyed his Act by treating it with the contempt it deserved, he has secured peace by going down to Cardiff, and with the greatest possible éclat, giving way to the strikers on practically every point.[99]

The miners won a 50 percent increase on the 1879 standard, as compared with the 10 percent they had been offered, with the award applicable only to SWMF members. The *Merthyr Pioneer* called it 'the best working agreement that has ever been obtained in the coalfield'.[100] W.H. Mainwaring wrote: 'The miners have not received all they demanded, but taken as a whole, and considering the opposition, it must rank as one of the greatest victories in the history of trade unionism'.[101]

The strike of July 1915 represented a major turning point in mining politics during the war, both in terms of the relations between the government and the MFGB, and with regards to the internal politics of the MFGB itself. As such, the strike formed a crucial element in the contextual framework within which the Miners' Charter was first developed, and later fought for. Above all, and ironically enough, the strike's victory not only gave rise to government control and the perspectives of the MFGB leadership for post-war reconstruction in

---

97   *New Statesman*, 17 July 1915.
98   *Times*, 16 July 1915.
99   *New Statesman*, 24 July 1915.
100   *Merthyr Pioneer*, 31 July 1915.
101   *Plebs Magazine*, vol. 7, no. 7, August 1915.

industry, but was also simultaneously an important reference point for the very different perspectives of the militant miner.

In the first place, the strike sheds new light on the commonplace view, best summed up by Susan Armitage:

> Nothing during the war seriously disturbed the miners' conviction that they held a right to make a special claim on their own behalf and to have it accepted. It was just a question of applying enough pressure and the government would give in.[102]

Her point is well made, but only because of the government's humiliation in July 1915. Only after the strike did the power relations between government and MFGB decisively shift in the latter's favour. That government control was established on terms favourable to the MFGB, and that it was followed by a policy of almost ritual concessions until the war's end, was attributable above all to the fact that rank and file miners had defied their leaders and taken direct action.

This was not lost on the militants who had led the strike. At a rank and file meeting in October 1915 chaired by Ablett, speakers proclaimed the strike as a victory for direct action and the 'anti-political' policy of the pre-war URC. Hodges said:

> The miners in South Wales had learnt to discard the politicians and Parliament and to place their whole reliance upon their own Federation. Recent events had shown that they were wise in that decision for, despite the politicians, Parliament and the press they had secured benefits that were undreamt of years ago. The SWMF was the pioneer trade union in initiating several movements that would eventually work the downfall of the capitalist system.[103]

Significantly, the strike represented not just an economic but also a political victory against the government and state. Government intervention in industry during the war may have encouraged the state socialist project, but the government's attempt to coerce the miners in July 1915 had alerted many to the negative aspects of state intervention. Page Arnot's assessment of the impact of the Proclamation of the coalfield was that the miners:

---

102 Armitage 1969, p. 114.
103 *Llais Llafur*, 9 October 1915.

were furious, and in their fury more and more began to think that those in the valleys who said the government and the state was just an executive organ of the capitalist class had the truth of it.[104]

The introduction of conscription for miners in 1917 compounded fears of a 'servile state' which, far from possessing emancipatory potential for the working class, could be an oppressive Leviathan. This aspect of the wartime state generated alarm amongst widespread numbers of miners, giving the syndicalists a constituency in which ideas could take root.[105]

The point is not that government intervention during the war led to a wholesale rejection of state socialist ideas amongst the rank and file of the MFGB. Far from it, for beneficial aspects of control provided powerful arguments for this project. What is important to grasp, rather, is firstly that the state only began to respond positively to the MFGB *after* direct action had been taken by a large and important section of its membership. Even miners who retained faith in progress via state intervention in 1919 were aware of this. Secondly, the fact that the wartime state had presented itself as a Janus-faced creature capable of both benign and malign behaviour gave rise to *both* the reforming perspectives of the MFGB leadership *and* the militant miners' conviction that the only way forward lay with direct action and workers' control.

The 1915 strike was a signal reminder that beneath the veil of wartime class co-operation the antagonistic relationship between labour and capital was still alive. Because of their self-assumed role as guardians of industrial peace in the national interest, union officials had been shown to be markedly less willing than their members to resort to strike action. Furthermore, while the well of patriotism had not run dry, the whole episode had caused the water level to drop significantly. This created a space in which the unofficial leadership could regroup inside the SWMF, and laid the seeds for the second coming of the URC when 'war weariness' became acute in 1917–18.

The Miners' Charter was the programme around which the unofficial movement coalesced in the closing stages of the war. In its demands were reflected both a continuation of the method laid down in *The Miners' Next Step*, and the confidence which the 1915 victory had given militants. The demand for a minimum wage was replaced by a programme designed to seize the initiative

---

104   Arnot 1967, p. 71.

105   SWMF Records 1915. Although ultimately unsuccessful, the URC mounted a campaign against conscription in South Wales, forcing the issue to a ballot in which they received a respectable 28,903 votes for strike action.

in the face of imminent demobilisation, and prevent an employers' counter-offensive by ensuring ex-soldiers were re-absorbed into the industry on the miners' terms.

The centrepiece was for a massive reduction in hours for all grades of mine-workers, to a five-day, thirty-hour week. The Charter envisaged no loss of earnings to compensate; rather it staked a claim for the abolition of piece rates, to be replaced by a £1 day wage, with payment on the basis of six shifts for every five worked. The Charter also included subsidiary demands for full rates of wages to be paid to miners injured at work. In line with *The Miners' Next Step*, these demands, bold as they were, were only the means to an end. For URC members the hope was that the fight for these reforms would culminate in the miners taking over the collieries and running them for themselves.[106]

The Charter, which first appeared in print in the *Merthyr Pioneer* in 1918, was added to in the summer of 1919 with the publication of *Industrial Democracy for the Miners*. Written by W.F. Hay, a co-author of *The Miners' Next Step*, its tone reflected the mood and aspirations of the militants in the immediate post-war period.

'Nationalisation', it predicted:

> is obviously bound to come almost immediately. As scientific socialists we know that nationalisation or state capitalism, the last permutation of which capitalism is capable, is bound to be short-lived. It is therefore all the more important that we should force our demand for democratic control upon the capitalist state when nationalising any industry.[107]

Industrial Democracy for the Miners represented not so much a theoretical development of *The Miners' Next Step* as a practical supplement to it, giving a detailed blueprint of exactly how workers' control could operate.

The years 1917 and 1918 saw a rejuvenation of militants in South Wales and beyond, and by the war's end there were rank and file movements in several important coalfields. Within days of the armistice these were putting pressure on their district and national leaders to adopt the demands encapsulated in the Miners' Charter, and to start the fight for a new order in the mining industry.

---

106   *Merthyr Pioneer*, 3 August 1918, 'Manifesto of the Unofficial Reform Committee of the South Wales Miners'.

107   Hay 1974, p. 105.

3   From Southport to Sankey

During three days of debate at the Special Conference in Southport from 14–16 January, the MFGB finalised the details of its post-war programme. The official Miners' Charter consisted of a package of demands for a six-hour day, a 30 percent wage increase, maintenance on full rates of pay for unemployed soldier-miners and for those injured at work, and nationalisation with joint control. Never had a national union put forward such a dramatic, aggressive and ambitious set of demands. All the more remarkable then that it fell well short of what the rank and file movement was demanding. With regards to hours, the MFGB's proposals were for a six-hour day from bank to bank. In practice, once travel time underground was factored in, for many this meant a seven-hour day. A six-day, 42-hour week was twelve long hours more than the unofficial movement wanted. The 30 percent wage rise also left militants dissatisfied, as it would favour the better-off grades of mineworkers and leave the piece rate system intact.[108]

Despite complaints that the unofficial Charter had been watered down, the official version still represented an audacious bid to prevent peacetime demobilisation from undermining the enhanced power of the wartime MFGB. If successful, it would amount to the most far reaching changes the industry, or indeed any British industry, had ever seen (something that remains true to this day). The Miners' Charter as agreed at Southport, whilst falling short of the URC's version of the previous year, bore the imprint of the rank and file militants who had given it its name. It represented from the outset a compromise between militants and moderates within the MFGB.

It was universally accepted in the union that the miners were to be a major force in post-war politics; what was at issue was how its power should be exercised, and to what end. The battle over this question between official and unofficial leaderships was the central feature of the campaign for the Charter. The campaign was not one in which national conferences handed down decisions to the passive membership. Rather it was characterised by internal conflicts which on several occasions threatened to overwhelm the leadership and push the union into a national strike. The large vote for acceptance of the Sankey Commission's Interim Report and the eventual collapse of the Charter into the 'Mines for the Nation' fiasco cannot mask the militancy of the rank and file during 1919. On the contrary, from January until the end of the Yorkshire strike in mid-August, there were a series of rank and file revolts against the drift towards

---

108   MFGB Special Conference 14–16 January 1919, *passim*.

conciliation and compromise, as militants sought again and again to stop the rot. These revolts provide the focus for much of this book. A brief look at the timetable of events at a national level will provide the political and chronological framework with which to understand them.

Minutes of MFGB Conferences, and negotiations with the government throughout 1919, demonstrate that the EC, Smillie and Hodges included, were determined that the Miners' Charter should be won without recourse to a national strike. The lack of detailed EC minutes, and the disappointingly thin autobiographies of both Smillie and Hodges, make it impossible to say what private misgivings were brought to bear by individual leaders, but suffice to say the avoidance of a strike was a central concern of the EC throughout.[109] From Southport in January onwards, it was motivated more by the desire to avoid a confrontation with the government than it was to fulfil the wishes of its members.

The EC's first goal was to 'whittle down' the union's demands to a level where negotiations would be at least possible.[110] It successfully managed to restrict the wage claim to 30 percent in face of stiff opposition from Scotland, Lancashire, South Wales and Notts. The adoption of the six-hour day/30-hour week demand in a form which actually meant a seven-hour day/42-hour week, again encouraged EC hopes.[111] However, on 10 February, Robert Horne, the Minister of Labour, informed the miners that the government's offer was limited to one shilling a day and a Commission to inquire into the Charter. Smillie was dismissive: 'The replies of the Government have not come within measurable distance of the claims we have put forward'.[112] Nonetheless, at the MFGB Conference on 12 and 13 February, the EC was careful to keep all options open, and to avoid pinning its colours to the mast of industrial action. While it was forced by the government's dismissive response to recommend a ballot of the membership on a national strike, it argued vehemently against Conference carrying a recommendation in favour, out of concern for the effect this would have on public opinion. Smillie also hinted that the EC might have accepted a Commis-

---

109   Smillie 1924; Hodges n.d.

110   MFGB Special Conference 12–13 February 1919, p. 10. The phrase is Smillie's.

111   MFGB Special Conference 14–16 January 1919, pp. 73–99 for debate on wage claims. Scotland and Lancashire wanted 50 percent, whilst South Wales and Notts were for a large flat rate increase (MFGB 31 January 1919, Conference between the Government and the MFGB Executive, p. 3). Smillie was not exaggerating when he told Robert Home that: 'It required all the influence of the Executive to reduce our claim to the point at which it was put as a claim'.

112   MFGB, Conference between the Government and the MFGB Executive, 10 February 1919, p. 9.

sion if the government had guaranteed that concessions to the miners would have resulted. He also made it clear that he did not want the miners to ballot 'on the understanding that the Triple Alliance is coming out. I want them to ballot with the full knowledge that they may have to fight alone'.[113]

In the event, delegates overturned the EC, and conference issued a strong recommendation for a strike vote. The ballot result, announced on 25 February, showed an overwhelming majority in favour, and notices were lodged to expire across the country on 15 March.[114] Meanwhile, Lloyd George returned to London from the Paris Peace Conference. On Friday 21 February, he invited the MFGB leaders to Downing Street, where he renewed his offer of a Commission of Inquiry. His proposal was that three miners' representatives should participate in a Commission to be established by special Act of Parliament, with powers to compel witnesses to attend. At this stage he envisaged that the Commission would present an interim report on hours and wages by 31 March, and that this would be followed by an extended period in which it would investigate the issue of nationalisation.[115]

Lloyd George took the opportunity to graphically illustrate the alternative. The 1912 strike had been serious enough, but it had taken place during 'a prolonged period of peace and prosperity'. Should there be a strike now, it would take place:

> when the nation is burdened and crippled with the gigantic cost of this great war, when its industries are practically at a standstill, when not merely a strike, but the prospect of a strike is stopping business, and arresting the free start of industry, when unemployment is on the increase.[116]

Furthermore, he was explicit in his estimation that such a strike would quickly assume the character of a struggle over state power:

> But there is another reason which makes it even more serious. Then it was a conflict between mine owners and miners. Now the Government are directly responsible for this industry ... The responsibility will be the responsibility of the Government, and if there is a conflict, it will

---

113   MFGB Special Conference 12–13 February 1919, pp. 19–22.
114   MFGB Special Conference 26–27 February 1919, p. 5. The figures were 615,164 to 105,082.
115   MFGB, Conference between the Prime Minister and the Executive, 21 February 1919, pp. 17–18.
116   MFGB, Conference between the Prime Minister and the Executive, 21 February 1919, p. 4.

not be a conflict between mine owners and miners, it will be a conflict between one industry and the whole of the State. I cannot conceive of anything graver than that. The State could not surrender, if it began, without abdicating its functions. It is not a question of this Government or that Government but of Government – every Government. All industry would come to a standstill and come to a standstill soon. But does anyone imagine that would be the end of the conflict? It would be the beginning and not the end of it. The Government must face that, and the country must face it, and the miners must face it – we must all face it.[117]

Lloyd George continued with a scarcely veiled threat, repeated in the House of Commons the following week, that in the event of a strike the government would use its power to control rations and starve the miners into submission, a threat which reflected behind-the-scenes preparations.[118]

According to Nye Bevan, Smillie later told him: 'From that moment on we were beaten and we knew we were'.[119] Opinions differ on whether or not this was the decisive moment at which the EC abandoned any plans for the strike. We will return to this question later. For the moment it suffices to note that Lloyd George had been given ample encouragement from Smillie and the EC that the MFGB was looking for a way out of the looming confrontation, and that should he re-pitch the offer of a Commission, it would receive a careful hearing from the miners' leaders.

Smillie had been careful not to rule out the possibility of a national strike, and at times he had candidly expressed the opinion that it was the likeliest outcome.[120] However, he had also made it clear that he did not relish the prospect. In his opening statement of the campaign for the Charter, he confessed that 'the days that are to come ... I fear are to be dark days'.[121] He told delegates at the Special MFGB Conference on 12–13 February:

> Nobody knows better than we do, we, who have been in this thing year after year, who have come through strikes locally and nationally, nobody knows better than we do the grave responsibility that rests on all the men

117 MFGB, Conference between the Prime Minister and the Executive, 21 February 1919, pp. 4–5.

118 MFGB, Conference between the Prime Minister and the Executive, 21 February 1919, p. 5. 112 HC Deb 5s, 24 February 1919, c.1449. Desmerais 1975, pp. 1–15.

119 Bevan 1952, p. 21.

120 MFGB Special Conference 14–16 January 1919, p. 60.

121 MFGB Special Conference 14–16 January 1919, p. 13.

in this room today. We know how necessary it is to think out not merely the beginning of what is to take place, if it takes place, but the end of it.[122]

Along with the usual concern that strikes, once underway, might spin out of EC control, Smillie spoke often of his desire to avoid disruption of industry. He spoke of the miners' moral duty as 'honest citizens of the State' to exercise their industrial power with responsibility and restraint: 'I would not be a party', he told Lloyd George, 'and no member of this Executive would be a party, to taking the nation by the throat and saying, "The time has now come when we are efficiently strong to enforce our claims"'.[123] Again, prior to the meeting with Lloyd George, he told MFGB delegates:

> Personally, I sincerely trust that at this late hour we may get a settlement of our claims without the necessity of a strike. I have no desire ... to hold up the industries of this country merely for the fun of it, or to prove our strength. I would rather cut off my right hand than take any step to injure our fellow workers or the general industries of the country.[124]

The EC was apparently divided over how to respond to Lloyd George's offer and in private EC members confided that they thought the MFGB Conference which met on 26–27 February would simply reaffirm the original demands and finalise details of the strike.[125] However, both Smillie and Hodges threw their weight behind acceptance of the offer, providing that the MFGB were allowed to nominate half the Commission. Over the previous days Lloyd George had agreed to shorten the timescale of the proposed Commission's preliminary hearings, which would now have to be completed by 20 March. Smillie stressed to delegates that participation would not mean 'to give up your powers to fight if you fail to gain your ends', merely to delay a strike by a week or so.[126] Hodges went further, suggesting that the delay would 'bring us in harmony with the Triple Alliance' and make a general strike more likely should the Commission fail the miners.[127] Before the vote Smillie made an impassioned appeal to the delegates to divert the union from its collision course:

---

122   MFGB Special Conference 12–13 February 1919, p. 11.

123   MFGB, Conference between the Prime Minister and the MFGB Executive 21 February 1919, p. 7.

124   MFGB Special Conference 12–13 February 1919, p. 47.

125   *Western Mail*, 27 February 1919.

126   MFGB Special Conference 26–27 February 1919, pp. 16, 48.

127   MFGB Special Conference 26–27 February 1919, p. 48.

Is it a fact that any man in this room wishes to have a fight with the Government or the country solely for the purpose of proving our strength? Not one of us wish that. If we get into a fight, it must be to secure improvements for our people, and if we can get these improvements without a fight, then in God's name, do we want a fight if we can secure them without? Do we want to impose misery on the nation?[128]

Against expectations, Smillie and Hodges managed to swing conference behind the Commission. By insisting on a card vote, they were able to present an image of unanimity, when large numbers of delegates were opposed to participation on the grounds that it was tantamount to accepting binding arbitration.[129]

Subsequent developments at the national level are familiar. The Sankey Commission sat in its first stage from 3–20 March. The majority report, signed by the miners' nominees, was passed over by the government, but it agreed to implement the report signed by Justice Sankey and the three non-mining industrialists which recommended a seven-hour day for underground workers (with a possible reduction to six after July 1921), a 46½-hour week for surface workers, and a flat rate 2s increase per shift for all grades. Furthermore, it held out the hope that in its second stage nationalisation might be conceded as 'the present system of ownership ... stands condemned'.[130] Following MFGB Conferences on 21 and 26 March, between which the EC conducted fruitless negotiations with the government in an attempt to secure further improvements, the membership was balloted with a strong recommendation for acceptance, to which it complied with a huge majority of 693,084 to 76,992.[131]

Only a few weeks earlier a national strike, perhaps of the whole Triple Alliance, had seemed likely, for the drift towards compromise at the top of the MFGB was not matched at the bottom. All along the line, pressure had been exerted by large sections of the rank and file to 'get to business' with the government and launch the strike.[132] Local leaders like Tom Greenall, James Brown, George Spencer and Vernon Hartshorn all came into conflict with the EC at various points. Such was the strength of feeling in some areas that at the conference of 12–13 February delegates from South Wales and Nottinghamshire

---

128   MFGB Special Conference 26–27 February 1919, p. 49.

129   MFGB Special Conference 26–27 February 1919, see for example the contributions of S.O. Davies, J. Potts and Herbert Booth, pp. 30, 33 and 41.

130   CIC 1919, vol. i, p. viii.

131   MFGB EC Meeting 15–16 April 1919, p. 2.

132   MFGB Special Conference 26 February 1919, p. 15; see the contribution of the rank and file delegate, A. Jones from South Wales.

had attempted to put a resolution, ruled out by Smillie, to dispense with a ballot and call an immediate strike.[133] At each conference, delegates took the floor to express their concern that unless the MFGB EC stopped its hesitation it would be impossible, in the words of the Scottish leader James Brown, to 'restrain the men from taking personal action'.[134] When the EC had urged participation in the Commission, militants amongst the delegates predicted 'trouble with the unofficial rank and file movement'.[135] Noah Ablett spelled out the situation as he saw it:

> There has been a great clamour from every section, and it may be considered necessary to appease the Government of public opinion, but in my opinion it is infinitely more important to appease our own men. We have come to the parting of the ways in the trade union movement. We shall be up against unofficial strikes if we are not going to take a lead, and not alter at all but endeavour to enforce our demands; if not, we shall have trouble in the coalfields, and, in my opinion, there will be justification.[136]

Dominated as they were by agents and officials, the MFGB conferences were at one remove from this 'trouble in the coalfields', which had in fact been underway since New Year's Day. More than one in five of total strike days across all industries in 1919 were accounted for by mining, despite the fact that there was no national strike.[137] Between January and March alone there were 54 serious disputes in mining, involving almost half a million miners for an aggregate of over two million working days.[138] Even these figures do not capture the full extent of the remarkable wave of militancy which swept many districts of the MFGB during 1919. Right up until August this kept alive the possibility of a different outcome to the campaign for the Miners' Charter. In fact, without a focus on the (mainly) unofficial strikes that took place in most districts during these months, little of the significance of the events of 1919 can be appreciated. It is to a selection of these that we now turn our attention.

---

133 MFGB Special Conference 12–13 February 1919, pp. 23–4.
134 MFGB Special Conference 14–16 January 1919, p. 90.
135 MFGB Special Conference 26–27 February 1919, p. 21.
136 MFGB Special Conference 26–27 February 1919, p. 19.
137 Department of Employment, British Labour Statistics 1971, Table 197, p. 396.
138 Board of Trade, *The Labour Gazette*, January to April 1919.

CHAPTER 3

Fife and Lanarkshire

Ironically enough, in a month or so thousands of men who had voted in Lanarkshire constituencies for Lloyd George were either actively engaged in, or looking with more than benevolent neutrality upon, a tumultuous, unofficial strike having ill-concealed revolutionary objectives.[1]

∴

The counties of Lanarkshire and Fife, which stretched from west to east across the 'waist' of central Scotland, were the most important areas of the Scottish coalfield in 1919, home to 55,000 and 23,000 miners respectively.[2] These were organised in their own distinct county unions, which, although affiliated to the NUSMW, retained a high degree of autonomy, with their own leadership structures, rules, and funds. They also had their own distinctive political and industrial traditions, reflecting the different economic, geographical and social configurations of each area.[3] In the opening section of this chapter, we explore the characteristics of mining trade unionism in each county, in particular those which had, by the war's end, encouraged the creation of revolutionary-led rank and file committees of considerable size and influence. Before January 1919 was out, these committees would make a spectacular bid for the leadership of the Fife and Lanarkshire miners, by advocating joint action with the rejuvenated Clyde Workers' Committee in a general strike of the industrial workers of central Scotland for shorter hours.

The Forty Hours strike, as it became known, has been the subject of considerable historical research and controversy. However, the debate has focussed upon the Clyde engineers' part in the struggle, whilst that of the miners has been largely overlooked.[4] In fact, the miners' contribution to what was one

---

1 MacDougall 1927, p. 771.
2 Campbell 1989, p. 10; Supple 1987, p. 23.
3 A. Campbell, in McIlroy et al. 2009, p. 175.
4 Campbell 1989 is the most important exception to this rule. The strike features as an important episode in an overarching account of the shifts in consciousness of Scottish miners in this period.

of the most notorious episodes in the twentieth-century history of the British working class should not be underestimated. No account of the 40 hours movement that leaves them out can be complete.

Firstly, it proves that the movement for shorter hours in central Scotland was much wider than has previously been acknowledged; the strike earned the prefix 'general'. Secondly, it supports John Foster, against revisionist historians like Alistair Reid, in his argument that the movement for 40 hours was 'openly political' in its intent to halt a shift in the balance of class forces from labour to capital in the aftermath of the war and keep alive the prospects of socialist reconstruction.[5] Finally, the episode was a dramatic announcement, just two weeks after the MFGB's Southport conference, of the impatient mood in the coalfield, and of the potential for class conflict inherent in the demobilisation crisis.

1   The Formation and Growth of the Reform Committees

The road to permanent, stable, coalfield-wide union organisation in Scotland had been long and arduous. For much of the last quarter of the nineteenth century, such organisation as there was remained localised and ephemeral, with amalgamation or federation into county and national unions extremely difficult to accomplish. An attempt to form a Scottish Miners' National Federation in 1886 floundered within a year, and although the Scottish Miners' Federation established in 1894 proved more durable, it remained fragile; by the end of the decade, its membership still represented less than one-fifth of all underground workers.[6]

The key weakness of Scottish mining trade unionism in this formative period lay in the elusiveness of effective union organisation in the largest county, Lanarkshire.[7] The barriers to unionisation here in the western part of the west-central coalfield were manifold; the small scale of the majority of mining concerns helped to foster the collaborative and self-help tendencies which existed within the tradition of the independent collier right into the twentieth century; aggressive employers habitually sought to take advantage of economic downturns to attack the nascent unions which had been formed on the back of the previous upturn, and were often successful in turning strike defeats

---

5   Foster 1990–1, p. 53.

6   Arnot 1955, chapters 4 and 5.

7   Arnot 1955, p. 91.

into routs.[8] Probably the most serious obstacle was the ethnic and religious division within the workforce. The period in which the west-central coalfield underwent its greatest expansion had coincided with the peak years of Irish immigration, and sectarian conflict had been a persistent feature ever since. On occasions, this was expressed in outbreaks of violence, but even in normal times, at pithead meetings in Lanarkshire it was not unusual to see miners assembled into two groups, with the 'Billies' standing remote from the 'Dans'.[9]

Weak union organisation and a mining workforce largely hostile to Home Rule meant little or no incentive for Liberal Associations to accommodate miners' candidates at elections. The ethnic and religious divisions in the work-force thus impeded the development of a strong Lib-Lab tradition in Scottish mining unionism.[10] When, in the wake of what has been called 'the crisis of the independent collier', it became clear that labour aristocratic policies of enfor-cing apprenticeships and restriction of output were no longer practical, it fell largely to ILP socialists to shoulder the responsibility of devising a workable strategy for trade union development.[11] By largely skipping the stage of Lib-Labism, Lanarkshire's trajectory of union formation was very different from other mining areas in Britain. Bob Smillie and Keir Hardie were to Scottish min-ing unionism what Mabon and Pickard had been to South Wales and Yorkshire. The key figures in the eventual establishment of a genuine county union, the Lanarkshire Miners' Union (LMU) in 1896, were, unlike elsewhere, socialists, and they remained in control during our period. When Smillie resigned his pos-ition at the head of the LMU at the beginning of 1919 to take up his full-time post as MFGB President, the gap he left behind was filled by a troika of veteran ILPers, James Tonner (President), James Welsh (vice-President), and Duncan Graham (General Secretary).[12]

In one sense therefore, from a position of relative backwardness, Scottish mining unionism, once established, leapt into the political forefront of the MFGB. Indeed, it was from Scotland that the first explicitly socialist resolutions, on social ownership of the means of production, were put forward at an MFGB conference, at Leicester in 1897.[13] However, socialist politics did not equate

---

8   Campbell 1989, pp. 10, 16; Arnot 1955, chapters 4 and 5.

9   Campbell 1994, p. 3; Campbell, in McIlroy et al. 2009, pp. 177–8.

10   Gregory 1968, pp. 90–5.

11   Reid 1978.

12   National Library of Scotland (NLS), PDL 31/2, Lanarkshire Miners' County Union, Minutes of Council Meetings, 1919; see Knox 1984, pp. 177–8 and 271–3 for entries on Duncan Graham and James Welsh.

13   Arnot 1955, p. 93.

to militant policies in the LMU. Experience of union fragility in the industrial conflicts of the late nineteenth century had engendered a cautious approach to industrial action, and nurtured a conservative and bureaucratic outlook. By the war's end, the LMU Executive and Council were dominated by men whom J.D. MacDougall described as 'the baldheads – respected church elders, JPs and County Councillors'.[14] The ILP membership of Tonner, Graham and Welsh went hand in hand with industrial moderation. In the January 1918 edition of *Socialist Review*, Welsh denounced the 'wild cries' of the militants in the coalfield, and called for a 'soberer outlook'. His rightward trajectory continued in the 1920s; a part-time novelist, his work *The Moorlocks* (1924) was a thinly veiled attack on the Minority Movement, and he sided with Adamson in the internal strife of 1927–8.[15] Although he still retained a relatively good reputation with the militants, Duncan Graham too was in the process of moving to the right. Entrenched in the bureaucracy since its inception, he became increasingly remote from the rank and file after his election as MP for Hamilton in December 1918 and he too ended up supporting Adamson in the 1920s.[16]

Antagonism between officials and rank and file was already becoming a feature of LMU affairs before the war, and was underpinned by the structural decline of the coalfield. In 1913, Lanarkshire's output accounted for just over half of Scotland's total, but the underlying trend was down.[17] With some of its most accessible coal reserves already exhausted, since the beginning of the century it had been increasingly dependent on thin seams.[18] In an attempt to boost productivity, owners had been introducing cutting machinery, and by 1920 one-third of Scottish output was being mechanically extracted, compared with only 6.8 percent in northern England.[19] Significantly for events in 1919, there was a particular concentration of large, mechanised collieries around Blantyre, Cambuslang and Hamilton in the Central Clyde valley, in relatively close proximity to the engineering militants of the CWC.[20] For the workforce, the introduction of machines meant deskilling, intensification of labour and closer supervision. Furthermore, the high level of investment created pressure to keep unit costs low, leading to long hours and depressed wages. As Campbell points out:

---

14   MacDougall 1927, p. 763; Campbell, in McIlroy et al. 2009, p. 174.
15   Knox 1984, p. 272.
16   Bellamy and Saville 1972–84, vol. 1, pp. 133–4.
17   Supple 1987, p. 23.
18   Campbell 1994, p. 2.
19   MacDougall 1927, p. 763.
20   Campbell, in McIlroy et al. 2009, p. 175.

a further corollary of intensively mechanised cutting and conveying was the physical concentration of different grades of underground workers in closely supervised work units which generated common grievances.[21]

In these conditions, the more militant and anti-collaborative currents within the tradition of the independent collier could find expression, and there developed a pattern of regular lightning strikes, a form of resistance not easily controllable by the union centre.[22] These laid the basis for Lanarkshire becoming one of the most strike-prone coalfields in Britain, matching and sometimes outstripping South Wales for militancy.[23] This caused a strained relationship between officials and members. The first organised expression of rank and file discontent with the bureaucracy came in 1912, in the wake of the unsatisfactory conclusion to the national minimum wage strike. There were widespread grumblings about the democratic limitations of a union constitution 'which seemed to be designed to afford ordinary members a merely nominal share in the control of the Association'. Particular criticism was levelled at the method by which officials were elected; rather than there being a ballot vote of the membership, they were appointed by Pit Committees, which themselves were 'appointed in a far from satisfactory way'.[24] In August 1912, anti-bureaucratic sentiment finally gave rise to a 'Miners' Indignation and Reform Committee' which held a demonstration in Hamilton addressed by W.F. Hay. A central demand of the South Wales Unofficial Reform Committee – that all agents and officials should be subject to periodic re-election and be barred from holding Executive positions – was adopted by the meeting.[25] An attempt was also made by George Harvey, a founder member of the Plebs League and Socialist Labour Party activist, to establish an alternative 'Northern Counties Mining Industrial Union' in the North-East and Scotland.[26] Neither of these organisations was able to sink roots at this stage, but intra-union tension remained, and would be sharpened further in the latter stages of the war.

Turning now to the Fife coalfield, which stretched along the coast and its hinterland to the north of the Firth of Forth, from Clackmannan in the west

---

21   Campbell 1994, p. 2.

22   MacDougall 1927, pp. 764–5. MacDougall describes a 'rebellious, independent and stubborn spirit ... intolerant of the well-meaning efforts of the leaders to attain whatever justice could be got by methods of diplomacy and conciliation'.

23   Campbell 1994, pp. 2–3.

24   Plebs Magazine, vol. ix, October 1917, no. 9.

25   Campbell 1989, pp. 12–13.

26   Pribicevic 1957, pp. 302–3.

to Leven in the east. Conditions for union growth were more favourable here than in Lanarkshire in the nineteenth century. In part this was due to the later development of the coalfield; it was still expanding beyond the turn of the century. Another reason was that Irish Catholics formed a much smaller minority here than in Lanarkshire, and the workforce was relatively free of the ethno-religious divisions which hampered organisers in west-central Scotland. Taken together, these factors allowed the establishment of a different tradition of industrial relations. The Fife and Kinross Miners' Association, established in 1870, was to all intents and purposes the only really established union in Scotland for much of the 1880s and 1890s. Whilst in other areas fledgling unions were being torn apart by hostile employers, in Fife stable conciliation machinery was being established. As a consequence, Fife was far less prone to militancy than was Lanarkshire before the First World War.[27]

However, in common with Lanarkshire, mechanisation had a detrimental effect on labour relations, and additional pressure on unit costs was generated by the coalfield's reliance on the increasingly competitive export market.[28] In the second decade of the twentieth century, these economic factors began to undermine the conciliatory policies of the Fife union leaders, who were, in contrast to the rest of Scotland, Lib-Labs. When the 'ultra-respectable'[29] John Weir died in 1908, William Adamson took over the reconstituted Fife, Kinross and Clackmannan Miners' Association (FK&CMA) as General Secretary, a post he retained in 1919 despite his accession to the Chairmanship of the PLP. Adamson, who had been elected as Labour MP for West Fife in December 1910 without any perceptible alteration to his Liberal politics, was the dominant figure in the FK&CMA, which he viewed as a personal fiefdom. Approaching his sixties at the end of the war, he was in many ways an anachronistic figure; his commitments to Baptism, temperance, and self-improvement went undiminished in the changed social and political conditions of the 1920s. So too was his commitment to conciliation and arbitration in industrial matters and his outright hostility to strikes which often pushed him closer to the owners than to his members. A letter from Charles Callow, a senior official in the Fife Coal Company, to Adamson in December 1918, which assumes he can be relied upon to help prevent a strike, is testimony to his collaborationism.[30]

---

27   Arnot 1955, pp. 70–4. Campbell 1994, pp. 2–4.
28   Macintyre 1980, p. 53.
29   Arnot 1955, p. 74.
30   National Library of Scotland (NLS), Acc.4311, item 2.50. For biographical details, see Knox 1984, pp. 58–61; Bellamy and Saville vol. 7, pp. 4–6.

Future miners' leader and communist Abe Moffat, a young revolutionary in Fife in our period, remembers how unpopular Adamson had become amongst the rank and file:

> He was very, very close with the coalowners, socially and otherwise. That was the main thing which the miners didn't like about Adamson; his very close association with the coalowners and particularly with the Reids of the Fife Coal Company. He was always against strike action, always against progress.[31]

As Stuart Macintyre has written, Adamson's 'supine approach was overtaken by the rapid tempo of change in the mining industry in Fife', and the growing militancy of sections of the rank and file increasingly led to conflict within the FK&CMA.[32] However, despite the growing chorus of criticism, Adamson continued to block any moves towards militancy. Though his base was rapidly shrinking, his position was shored up by the more conservative miners in the older and smaller pits which were scattered through the rural settlements to the west of Dunfermline, and around the coastal villages in the most eastern part of the coalfield. The more militant miners were to be found in the central area of the coalfield, in the large, modern pits around Dunfermline, Cowdenbeath, Lochgelly and Bowhill. These miners made up the majority in Fife, but the union constitution prevented them from making their weight felt inside the FK&CMA.

Their first difficulty was that branches were based upon place of residence rather than colliery, undermining strong union organisation at workplace level. The second was that each of the 54 branches, regardless of size, had an equal vote at the Executive Board of the union. This favoured smaller, usually conservative branches some of which had only a few dozen members, against larger more militant ones where there could be up to two thousand. When larger branches sought official backing for industrial action, small branches often combined to deny them on the Executive Board. With the help of the union constitution, and agents and henchmen like Joe Westwood, Adamson was able to maintain his grip on the union machine. As such, the fight for militant policies and changes to the union constitution went hand in hand. The

31   Abe Moffat, 'The Fife Miners and the United Mineworkers of Scotland', Transcript of an interview by Paul Long, 18 January 1974, in Scottish Labour History Society Journal, no. 17, 1982, p. 13.

32   Macintyre 1980, pp. 52–3.

key demand in this regard was for the branch financial vote on the Executive board, a method which would give the greatest say to branches with the
most dues-paying members. Adamson resisted this with increasingly authoritarian methods, and this resulted in the later clashes with communist miners.
The well-documented splits and breakaways, which dogged mining unionism
in Fife, and the rest of Scotland in the 1920s, had deep roots and were prefaced
by the revolt of January 1919.[33]

Rank and file and leadership alike had been keen supporters of the war in its
early stages, with something like a quarter of all Scottish miners joining up by
August 1915. As hostilities dragged on this was to change, amongst the rank and
file at least. By August 1917, the cost of living had soared 80 percent above its
pre-war level, whilst Scottish miners' wages had risen by only 43 percent. Serious shortages of basic foodstuffs like butter, cheese and sugar compounded the
discontent. Meanwhile, the pacifist *Forward* was busy exposing the huge scale
of 'profiteering' by merchants and shipping companies.[34] Whilst enthusiasm
for the war waned amongst the ranks of the miners, there was no such drop-
ping off amongst the leaders, and when exemption was removed from younger
miners in the summer of 1917, Fife officials helped to enforce the new conscription regulations by participating in the Scottish colliery recruiting courts.

As officials of the NUSMW and the county unions embraced a patriotic truce
with owners and government, wartime grievances found little or no redress
through official structures. In Lanarkshire in the summer of 1917, the first moves
were made to reconstitute the rank and file committee that had made a brief
appearance before the war. The main centres of such activity were in the
large pits around Blantyre, situated on the south-eastern edge of Glasgow's
urban sprawl in the central Clyde valley, and in Coalburn further south. In
Blantyre, these militants were successful in persuading the District Committee
to call a one-day strike against profiteering and inadequate food rations, which
involved about 2,000 miners in twelve pits.[35] In June, a prominent member
of the Blantyre Reform Committee who was a delegate to the LMU Council,
successfully proposed a resolution for a county-wide one-day strike over the
same demands, and on 2 August, all 50,000 Lanarkshire miners struck, with
thirteen separate demonstrations calling on the government to put an end to
profiteering. John Maclean was buoyant:

---

33   Campbell in McIlroy et al. 2009, pp. 173–87; Macintyre 1980, pp. 52–3; MacDougall (ed.)
      1981, pp. 19–20.
34   Campbell 1994, p. 13; MacDougall 1927, pp. 766, 771. *The Call*, September 1917.
35   MacDougall 1927, p. 768.

[It is] certainly the most important [strike] ... in the whole history of the working class in Scotland. It easily transcends the spontaneous strike on the Clyde that forced the Government to give us the House Letting Act.[36]

For Maclean, the strike's significance lay in the fact that, whilst not against the war itself, it contained the seeds of such a movement. In the wake of the strike in the Coalburn and Blantyre districts, the Reform committees were successful in organising large mass meetings which declared opposition to the extension of conscription, threatened further strikes, and demanded immediate peace on the lines laid down by the February regime in Russia.[37]

In the last week of August, members of the various local rank and file committees came together in Hamilton and established the Lanarkshire Miners' Reform Committee (LMRC). About one hundred miners were in attendance, mostly in an individual capacity, although there were delegates from branches in Coalburn and Blantyre. The LMRC's policy was borrowed from *The Miners' Next Step*; the ultimate aim was the direct control of the mines by the miners, and this was to be achieved via the transformation of the union, whose strategy 'should be based on the principle of the class struggle'. To this end the Conciliation Board would be abolished, agents and officials would be elected annually by ballot vote of the membership, and a lay executive would be established. 50,000 copies of the LMRC's manifesto were printed and distributed throughout the coalfield, and plans were made for an ambitious education programme for the coming winter.[38]

The LMRC consolidated and grew over the coming months. Although it was unable to prevent an extension of conscription, it was in the forefront of a successful campaign to oust the ultra-patriotic Secretary of the LMU David Gilmour from office for taking up a government post on the Labour Advisory Board. The growing hostility to the war was also shown when Lanarkshire miners voted in favour of peace by negotiation by 18,767 to 8,249 in a ballot.[39] Campaigns such as these attracted further support to the LMRC, which by April 1918 could claim the direct affiliation of 24 union branches, and, according to MacDougall: 'Its connexions extended to active minorities in the remaining hundred branches, and it acquired enormous influence amongst the mass of ordinary union members'.[40]

---

36   *The Call*, 9 August 1917.
37   *The Call*, 23 August 1917.
38   *The Call*, 6 September 1917.
39   MacDougall 1927, p. 770.
40   MacDougall 1927, pp. 768–9.

The war-weariness and anti-war sentiment which swept Lanarkshire from 1917, spawning the LMRC, did not leave Fife untouched. By October 1917, a Fife Miners' Reform Committee had also been established, centred on the new mining town of Bowhill. In the spring of 1918, Fife too experienced an upswing of militancy, with a number of unofficial strikes at Low Valleyfield, Kelty and Bowhill, in which FMRC members played leading roles. In June 1918, through selling literature and collecting money, the FMRC was able to organise a county-wide tour by James MacDougall, John McLean's close associate, which took in a score of towns and villages. This led to the affiliation of more branches, including Bowhill Branch with its 1,500 members.[41]

By the end of the war therefore, there were vibrant Reform Committees active in large numbers of pits in both Fife and Lanarkshire with county structures that allowed them to mount propaganda campaigns on a coalfield-wide basis, organise a significant protest movement against the conditions of war and its prosecution, and occasionally to lead industrial action. The armistice presented these robust rank and file movements with new possibilities. The gap between the economic and social aspirations of the membership and the bitter conditions which confronted all those who had endured the war on the home or foreign fronts, helped to swell the ranks of the activists and bring simmering anger to the boil. The personal experience of the future miners' leader and communist Abe Moffat, was not untypical:

> Well, it was really after the First World War when I became active in the trade union movement. Like many other men I was fighting for my King and country, and we were told then that we were fighting to make this country fit for heroes to live in, but we discovered after the war that we had to be heroes to live in it! And that obviously changed my mind.[42]

The housing conditions of the miners in Fife and Lanarkshire were appalling, characterised by poor quality miners' rows, in which overcrowding and unsanitary conditions generated illness and disease. In the middle ward of Lanarkshire, for example, 35,000 miners and their families lived in 17,000 houses. In Lanarkshire, out of the 188,531 children born between 1891 and 1910, a staggering 22,279 died before reaching twelve months of age. This squalor would soon be given notoriety by the NUSMW vice-President John Robertson at the

41   MacDougall 1927, p. 769; *The Call*, 15 August 1918; MacDougall n.d., pp. 19–22.
42   Abe Moffat, 'The Fife Miners and the United Mineworkers of Scotland', Transcript of an interview by Paul Long, 18 January 1974, in Scottish Labour History Society Journal, no. 17, 1982, p. 7.

Sankey Commission.[43] Conditions in many of Scotland's mines were equally abysmal, and in 1919 alone, 180 miners were killed and 11,075 injured.[44] Work and housing conditions provided plenty of ammunition for the reform committees. The chasm between the lofty promises of Lloyd George and the grim realities of life lent discussions about social change an urgency that militated against the smooth functioning of collective bargaining, exacerbating tensions between an impatient rank and file and a plodding bureaucracy. Encouraged by the reform committees, there was a dramatic leap in union activity at the end of the war. James MacDougall recalled: 'It was a veritable revival. The members commenced to attend the branch meetings as they had never done before'.[45]

Miners displayed a high degree of solidarity and militancy. On Saturday 11 January for example, 2,000 miners in Holytown, Lanarkshire, went out on unofficial strike against the eviction from a colliery house of Willie Hughes, a mine manager who had been victimised for his association with the LMRC. By Monday morning, the Bellshill, Viewpark and Rosehall collieries were out, and soon 10,000 miners were involved. John Robertson, in the course of trying to persuade the men back to work at a large mass meeting, twice referred to the 'talk of revolution' spreading amongst the men, pleading for constitutional action. He only raised a cheer when he threatened to call out every miner in Britain over the Hughes eviction.[46]

Robertson himself drew attention to the shift in mood that had taken place amongst the rank and file in a speech to the NUSMW annual conference in 1919:

> In past years there were often complaints about the apathy and indifference of the rank and file. He would be a bold man who would accuse the rank and file of apathy and indifference at the present time. Never before in their history was such a keen interest taken in the work of the Union by their members. Social and industrial questions occupied their attention to a greater extent than ever before. Every progressive leader welcomes this change, welcomes the push they are getting from the rank and file. The command to the leaders was either 'get on or get out'. There was a tendency, with the better organisation and increased activity, for the members to take precipitated action.[47]

---

43   CIC 1919, vol. i, pp. 345–54; for housing conditions in Fife, see Macintyre 1980, p. 50.

44   South Wales Coalfield Archive, SWMF Statistical Department, E242.

45   MacDougall 1927, p. 768.

46   *Bellshill Speaker*, 17 January 1919.

47   NLS, PDL 45, NUSMW, Proceedings of Annual Conference, 13 August 1919.

The industrial militancy of the immediate post-war period was accompanied by a radical swing to the political left; indeed political radicalism and industrial militancy fed off each other. John MacArthur, later to become one of the leaders of the breakaway Reform Union in Fife, was a prominent activist and revolutionary in the FMRC in the eastern part of the coalfield in 1919. His memoirs leave us a vivid picture of the politically turbulent atmosphere in which the young reform committees were operating:

> The Great War and the first year or two after it was a period of trade union and political ferment for me, just as it was for so many other workers in Britain. There was our local activity in Fife. There was the Clydeside shop steward's agitation and the rent strike, the Irish revolt, the impact of the Russian Revolution, the crushing of the Spartacist's revolt in Germany and the killing of Karl Liebknecht and Rosa Luxemburg, the returning soldiers, the serious discussions as to whether workers could achieve their emancipation by parliamentary or anti-parliamentary means. These things led to tremendous fervour among the militants with whom I was associated, but also among the workers as a whole.[48]

In this situation political discussion took on a new vitality, shaking off much of the sterile abstraction that had plagued the left in the past. Although there was no strong political organisation in east Fife at the time, Levenshorehead saw meetings every Saturday and Sunday night at which speakers from all sorts of different traditions spoke to large audiences. The radicalism of the Fife mining areas in these months is captured by MacArthur in his recollection of a debate between Jack Leckie, an enigmatic revolutionary, and Tom Kennedy, MP for Kirkcaldy from 1921 and on the right of the Labour Party. The meeting, titled 'Reform or Revolution?', took place in Buckhaven's biggest hall, and was packed to overflowing.

> Leckie rolled up his sleeves, and pranced up and down the whole length of the platform, shouting what had Kennedy's pals in Germany done to his pals Rosa Luxemburg and Karl Liebknecht? He said he knew what he would like to do – and put his fist under Kennedy's nose.[49]

Leckie won the debate overwhelmingly on a show of hands.

---

48   MacDougall n.d., p. 17.
49   MacDougall n.d., p. 23.

Support for the Russian and German revolutions helped to encourage interest in the idea of direct action, and there was no shortage of syndicalist literature in the coalfields. In particular, pamphlets by the American Industrial Unionist Daniel DeLeon, and George Harvey's shilling book *Industrial Unionism and the Mining Industry* were widely circulated.[50] However, the most important figures in spreading and politicising the reform committees were the revolutionaries John Maclean and James MacDougall.[51] Maclean had been campaigning amongst Lanarkshire miners for opposition to conscription and war since early 1916, and MacDougall, who worked at the pithead at Blantyre after his release from prison in 1917, was instrumental in building the reform committee there.[52] Maclean was in favour of industrial unionism as well as Marxist political action, and in the reform committees he saw the possibility of combining theory and practice. The Scottish Labour College classes which he and MacDougall ran in the mining communities in Lanarkshire and Fife gave a thorough grounding in Marxism to a whole generation of young activists, several of whom became communists and leading union figures in the 1920s and 1930s.[53] Furthermore, by bringing activists together, the classes acted as nuclei for the emergent reform committees from 1917 onwards.[54] MacLean's influence among militant miners was registered when he was named official spokesperson of the reform committee movement in January 1919, a position which he declared himself as proud of as his appointment as Glasgow's Revolutionary Consul to the Bolshevik government.[55]

The importance of the relationship between Maclean and the militant miners of Fife and Lanarkshire has been noted before.[56] However, the way in which Maclean's influence amongst the miners paved the way for their involvement in the January 1919 movement for shorter hours in Scotland has not. Both the FMRC and the LMRC had established links with the Clyde Workers' Commit-

---

50   *The Call*, 15 August 1918. 150 copies of Harvey's book were sold on James MacDougall's tour in Fife.

51   Campbell 1989, p. 14 emphasises this point.

52   MacDougall 1927, p. 767; *The Call*, 9 August 1917.

53   MacDougall 1927, p. 775; Smith 1978, p. 113. Among Maclean's students were Willie Allan, who became the Secretary of the breakaway United Mineworkers of Scotland, John Bird and Jock MacArthur.

54   *The Plebs Magazine*, vol. ix, December 1911, no. 11. There were already seventeen classes underway in Fife and Lanarkshire by the end of 1917, all of them in towns and villages in which there was a reform committee presence.

55   *The Call*, 23 January 1919.

56   Campbell makes this point (1989, p. 14); see also Howell 1986 and Ripley and McHugh 1989.

tee in 1918, when they were active in the campaign to free John Maclean from prison, and subsequently in his campaign in the general election in December. It had taken delegations to demonstrations in Glasgow to demand MacLean's release, and to celebrate the victory when it came. MacArthur describes the experience:

> It gave us a great feeling of taking part in a big movement; the whole agitation was on a much bigger scale than anything we had experienced in the mining villages and small towns of Fife.[57]

These campaigns exposed militant miners to the wider working class movement. The connections made during 1918 meant that when the agitation developed for the Forty Hours strike in Glasgow, MRC members in pit villages as remote as Wemyss and Leven in east Fife could be involved from the start. In Lanarkshire, links were even closer, as most of the collieries were located on the fringes of Glasgow's sprawling urban conurbation, where miners rubbed shoulders with workers from other industries.

A second important factor was Maclean's perspectives for the post-war class struggle. As an internationalist his priority was to prevent the crushing of the Russian Revolution, the gatekeeper of the developing world revolution. Those who were advocating a general strike to defend Russia were 'too idealistic' according to Maclean, as 'the workers are not generally of our way of thinking, and so are unable to see that their material interests are bound up with Bolshevist stability in Russia'. The burning question was 'how to start the fight [and] get the mass on the move?' The answer was provided by the demobilisation crisis:

> The demobilisation has already created a menacing unemployment problem. We can get the support of the unemployed if we can suggest a means whereby they can get a living. The only possible solution is a drastic reduction of hours per week. This reduction will appeal to the employed if they are assured of at least a pre-war standard of living. Here we have the economic issue that can unify the workers in the war against capitalism.[58]

For Britain's leading revolutionary, the Miners' Charter was the catalyst that could initiate that war. In the context of mass demobilisation, a national miners' strike for shorter hours would have the potential to pull in millions of

---

57   MacDougall n.d., p. 21.
58   *The Call*, 23 January 1919.

workers from other industries, and could provide the basis for control of production through workshop committees. Maclean was not alone amongst the Clyde revolutionaries in looking to the miners to play a vanguard role. Harry McShane's memoirs recall the delight with which the revolutionaries in the CWC greeted the formation of the reform committees, and it was clear that the CWC itself adopted the demand for the 30-hour week because this was the demand of the unofficial Miners' Charter.[59] Right up to the commencement of the strike on the Clyde, the CWC's paper *The Worker* was pushing for the demand to be reduced from 40 to 30 hours in order to stay in line with the miners:

> If we are to have that liberty which our brothers fought for and which we have sweated for during the past four years we must line up with the miners. Our demand and theirs must be the same, and we must stand shoulder to shoulder if we are to win.[60]

Maclean himself was in favour of delaying the strike on the Clyde until March in order to wait for what he felt was the inevitable national miners' strike. However, once it became clear that there was no stopping it, he gave it his full encouragement, observing that 'historical events never start and shape themselves as we plan them'.[61]

The Clyde Workers Committee and the miners' reform committees could not be constrained by Maclean's timetable. The pressure of demobilisation on the Clyde was more intense than anywhere else in Britain. Housing was almost unbearably overcrowded *before* the war, but as a major munitions producing centre, thousands of additional workers had crammed into the area since 1914. With countless thousands now returning from France:

> [M]ore and still more men and women find themselves on the street, and as a result we shall see a decline in wages with ever increasing unemployment, unless we are prepared to regulate the hours of work to meet the situation.[62]

As impatience mounted, news came through of the general strike across the Irish Sea in Belfast. When the CWC met in conference on Saturday 25 January

59   Smith 1978, p. 109.
60   *The Worker*, 25 January 1919.
61   *The Call*, 30 January 1919.
62   *The Worker*, 25 January 1919.

to finalise plans for its strike, to commence on the following Monday, the LMRC was well represented. Meanwhile, the FMRC was already leading an unofficial county-wide strike in Fife.

2   The Reform Committees Lead a Mass Movement: The January
        Strikes

As is often the case with mass movements that develop wider political content and significance, the strike wave of January 1919 had its immediate roots in sectional demands. In April 1918, a committee of the surface workers had drawn up a list of demands for an eight-hour day, a wage increase, and extra payment for overtime. All branches of the union endorsed the claim, as did the Executive Board, but negotiations stalled and no steps were taken by the leadership to break the logjam. By September, the surface workers, impatient at official inaction, began to lodge notices unofficially.[63]

At this stage the MFGB intervened and persuaded the Scottish union to accept a deal they had agreed with the Coal Controller, which involved a 49-hour week, exclusive of meal times. The surface workers gave their qualified consent. Their condition was that the owners allow them to work the shift from 6am to 3pm so as to finish at the same time as the rest of the miners. The owners, however, insisted that tradesmen and mechanics must remain behind to sharpen tools, repair machinery, and so on, which meant a 7.30am to 4.30pm shift.[64]

Adamson predictably persisted with negotiations, though it was clear the owners were not going to budge. In the first half of January 1919, he tried to reopen a channel to the Coal Controller, who pushed him back to local negotiating machinery. On 20 January, the matter was discussed at the Fife Conciliation Board, but it came to nothing.[65] Adamson of course was set against any kind of action and according to *The Worker* he had 'put the case for the colliery companies brilliantly' at mass meetings.[66] By this time the surface workers had been in dispute for almost ten months and their patience was exhausted. A member of the FMRC wrote:

---

63   NUSMW Minutes of Executive Meeting 12 September 1918.
64   *West Fife Echo*, 29 January 1919.
65   *Dunfermline Journal*, 25 January 1919.
66   *The Worker*, 8 February 1919.

Since April 1918 the surface workers' dispute has been 'put forward', 'negotiated', 'conciliated', 'remitted to executives', 'remitted to districts', 'postponed for a fortnight', 'postponed for another fortnight', and then 'just for
two days'.[67]

On 22 January, the surface workers in the Cowdenbeath district held a mass
meeting and voted for an indefinite strike to begin the following morning.
Significantly, the decision to strike had now gone well beyond the desire to
knock off at 3.00 pm; it was, as Bob Selkirk explained, equally 'to force the
Companies to employ demobilised men' to do the extra work which they were
insisting upon.[68] This was the opportunity for the FMRC to fill the vacuum left
by the officials and advance itself as a genuine alternative leadership. On 23
Thursday, it organised a mass meeting of all grades in Cowdenbeath Public
Park, which voted unanimously to strike in support of the surfacemen.[69] That
evening, the FMRC organised similar mass meetings at Lochgelly, Glencraig and
Bowhill, all of which voted to come out. From Blairhill colliery in west Fife
to Bowhill in the east, a stretch of nearly twenty miles, every pit struck work.
The next day the strike was spread to Wemyss and Buckhaven in the eastern-
most parts of the coalfield by a combination of FMRC led mass meetings and
picketing. Robertson and Adamson scrambled to call their own meetings at
Bowhill, Cowdenbeath and Lochgelly in an attempt to head off the strike, but
to no avail. By Friday night, 20,000 were out on unofficial strike, practically the
entire Fifeshire coalfield.[70]

The FMRC found itself at the head of the mass movement for three main
reasons. Firstly, it had proved flexible enough to respond quickly to the mood
of the miners. Over the previous weeks, the committee had been pushing for
action over its unofficial version of the miners' charter; a six-hour day, a five-
day week, and a £1 per day minimum wage. One such meeting had taken
place at the Gothenburg Hall in Kelty on Sunday 12 January, addressed by
Robert Lamb of Gallatown and Charles Tuke of Blairhill.[71] Whilst there was
considerable interest in the programme, no action was forthcoming. FMRC
members were quick to sense, however, that the mass of the miners were

---

67   Ibid.
68   *The Worker*, 8 February 1919.
69   *Glasgow Daily Record*, 24 January 1919.
70   *Glasgow Daily Record*, 25 January 1919; *West Fife Echo*, 29 January 1919; *Leven Advertiser*,
      30 January 1919.
71   *West Fife Echo*, 15 January 1919.

more likely to strike in support of the lesser demands of the surface workers, so they moved to agitate around this issue.

Secondly, and quite simply, they were prepared to act decisively. While the officials temporised and called for talks and more talks, the FMRC called for immediate all-out strikes in support of the surface workers. And thirdly, they proved to have a network of militants sufficient for the task. The work which the committee had done in leading protests against price rises, food shortages and conscription during the war, and the meetings and literature sales they had organised, meant that when faced with an opportunity, they had enough bodies on the ground to make an effective transition from propaganda to agitation.

Its ability to lead significant action against the officials quickly identified the Miners Reform Committee as an alternative leadership in the coalfield. Over the weekend of 25–6 January, the FMRC achieved a dramatic escalation; at packed meetings in Cowdenbeath, Lochgelly, Glencraig and Bowhill, its activists persuaded the miners to extend their demands, winning overwhelming majorities for its unofficial programme. Before the general strike had even commenced on the Clyde, the Fife miners were out for a 30-hour week and a £5 weekly wage. On Tuesday, 10,000 marched behind the FMRC's banner and unanimously declared for the new demands to be added to those of the surface workers. The *West Fife Echo* reported that the escalation was enough to 'take one's breath away'.[72]

In Lanarkshire, John Maclean took up the cudgel. On Friday 24 January, in his first big public appearance since his release from Peterhead, Maclean spoke to a large meeting of miners organised by the LMRC in the Motherwell town hall. In 'an impassioned address' Maclean explained the demands of the reform committee, and called on the Lanarkshire miners to strike. 'He looked to the miners of Lanarkshire to come out with the men on the Clyde, and show the absolute solidarity of the working class'.[73] On Monday morning, in areas where the LMRC's support was strongest, miners struck in support of the Forty Hours strike which began that day. The following day, a one-day strike called by the LMRC, again in solidarity with the Glasgow workers, received more widespread support. This was then extended into an indefinite strike for the reform committees' demands by means of mass meetings and pickets over the following days.[74]

72   *West Fife Echo*, 5 February 1919; *The Worker*, 8, 15 February 1919; *Fife Free Press*, 1 February 1919.

73   *Motherwell Times*, 31 January 1919.

74   Ibid.; *The Worker*, 8 February 1919.

In less than a week, the MRC had transformed what had seemed a sectional strike of surface workers into one of all miners in Fife and Lanarkshire over hours and wages, and linked it up with the Forty Hours strike in Glasgow. Although the formal demands of the Fife surface workers, the Glasgow engineers, and those of the MRC all differed from each other, they were underpinned by the same motive, namely, resistance to the adverse effects of demobilisation in order to resist a shift in class forces in favour of capital, and keep alive the possibility of post-war social transformation in the interests of the working class.

The success of the reform committees in commanding a mass following in January was contingent upon their ability to convince the miners that to begin the fight over hours and wages without the rest of the MFGB was not a hollow gesture. They achieved this by linking their campaign with the wider movement of Scottish workers. To miners in Lumphinnans or Blantyre, winning solidarity from miners in Bolton or Mountain Ash in the face of MFGB officialdom might have seemed beyond their means. But in Scotland in January 1919 there was a mass movement on their doorstep. Combined, the Forty Hours strike and the miners' strike presented a powerful base from which a general strike against the effects of demobilisation could plausibly be spread to the rest of Britain's miners and beyond. Maclean's strategy appeared to be working.

These links allowed the MRC to overcome the credibility gap that is problematic for any rank and file movement; they gave their militant policies a pragmatic flavour, evoking a response from miners way beyond the ranks of their immediate supporters. In Fife, geographically remote from the main action in Scotland, contact with striking engineers was particularly important, so the FMRC arranged speaking tours of strikers from the Forth and Clyde. On 29 January, CWC speakers received 'tumultuous cheering' from a crowd reported as 20,000 when one inquired: 'Then we in Edinburgh and Glasgow may take it that you will stand with us until all our demands are conceded?'[75] In Lanarkshire, especially in the north of the coalfield, there was a more immediate and tangible connection with the wider strike movement. In Cambuslang for example, which lay on the south-eastern edge of Glasgow, 10,000 workers from several industries were out on strike by Wednesday, 29 January, generating the feeling of an 'industrial uprising' in the town.[76] In these multi-industry towns of the central Clyde valley, strike meetings tended to be non-sectional in nature and make-up, fostering unity between workers from different occupations.[77]

---

75   *Fifeshire Advertiser*, 1 February 1919.

76   *Lanarkshire and Hamilton Herald*, 1 February 1919.

77   Ibid. Meetings were often explicitly billed as 'A Mass Meeting For All Workers In Lanarkshire'.

Involvement in a wider working class movement helped to generate an exceptional militancy amongst the miners, particularly in Lanarkshire, and the strike took on the character of a rebellion, described by one local paper as 'an orgy of unofficialism'.[78] With good reason, Alan Campbell has likened its scenes to those in Zola's *Germinal*.[79] James MacDougall later recalled:

> Notwithstanding some intimidation, the movement was a genuine ebulli-
> tion of the masses. Men might be reluctant to come out against the official
> mandate of the union, but once out they were swept entirely off their
> feet by the emotional current around them. It was not really an ordinary
> strike. It was a religious ecstasy – the joyful rush of the pent-up discon-
> tents of the war years to find a vent. Staid men did unheard of things, took
> extraordinary risks, because they were in the grip of the idea. Fanatical
> orators tore themselves to shreds addressing tense audiences, assembled
> in packed halls or massed in public parks, from early morning till late at
> night. Unpaid pickets marched by night many a rough stage in order to
> stop distant collieries early in the morning. The [Reform] Committee was
> in permanent session at Blantyre ...[80]

One of the most notorious episodes of the strike took place on the night of Thursday 30 January in response to the decision of the Bellshill branches to return to work on the Friday. Between five and six hundred pickets marched from Tollcross and Blantyre to prevent a return to work. At the Hamilton Palace colliery, which lay en route, several hundred pickets broke into the lamp cabin and destroyed all the lamps to prevent anyone who might evade the picket line from descending the pit. They then broke into the colliery offices, smashed windows and doors, ransacked filing cabinets, and tried to break open the safe. A policeman who intervened was assaulted, receiving three head wounds. A group of firemen waiting at the pithead for the cage, were threatened with attack by the crowd, which also threatened to set fire to the winding gear, and the firemen retreated and went home. The crowd then moved on to Bellshill Cross, where there was a clash with 'a strong contingent of police who met them with the baton'.[81]

This incident points to some of the strike's main features; the use of flying pickets for one. Impossible to spread the strike through official channels, the

---

78   *The Mail for Kirkcaldy, Central and West Fife*, 11 February 1919.
79   Campbell 1989, p. 15.
80   MacDougall 1927, p. 772.
81   *Bellshill Speaker and North-East Lanarkshire Gazette*, 31 January 1919, 14 February 1919.

reform committees of Fife and Lanarkshire looked to mobilise the largest possible numbers of miners to do the job, and 'the mob' made regular appearances in the local press. Secondly, it seems that a section of miners were prepared to use violence, or the threat of violence, to achieve their ends. Reported incidents were high, and Lanarkshire newspapers in particular contained numerous stories of 'armed pickets' carrying 'sticks, bottles and knuckle dusters'. In Cambuslang, following two days of only partially successful picketing, some miners allegedly took to carrying firearms, closing down every colliery in the district.[82]

Of course, press reporting of violence is often exaggerated or manufactured during large industrial disputes. However, there is reason to believe that the January strikes in Scotland were genuinely aggressive in nature. Press accounts of particular incidents were independent enough to suggest that at least some of the reported incidents were authentic.[83] Alan Campbell has drawn attention to the existence of a violent subculture in this period, particularly in Lanarkshire, pointing to the proliferation of youth street gangs, and to the militant republicanism of many Lanarkshire miners of Irish descent. Explosives and detonators from the mines often found their way to the Irish Republican Army around this time.[84] Harry McShane points out that 'a surprising number of men had brought back weapons from the war' and that some carried them on the unemployed demonstrations in the early twenties, again lending credibility to reports of firearms in January 1919.[85] Rebellious youth was in the thick of it. Of the nine miners arrested and charged with forming part of a riotous mob at Hamilton Palace and Blantyre Cross on the night of 30 January, all but one 'were very youthful in appearance, two of them seeming to be mere boys'. Indeed, upon their conviction, the jury recommended six of them for leniency because of their young age.[86]

The strikes were not just a challenge to the coalowners and the authorities. They were a direct assault on the miners' officials, who set about trying to undermine the action. The NUSMW Executive issued the first of several statements on 27 January, which expressed their decision to 'entirely disassociate themselves from the present erratic strike movement, and recommend the

---

82   See for example *Lanarkshire and Hamilton Herald*, 1 February 1919; *Glasgow Daily Record*, 1 February 1919.

83   The reports were obviously not simply gleaned from a press agency, or any other single source.

84   Campbell 1989, p. 17; Campbell 1994, p. 9.

85   Smith 1978, p. 111.

86   See the report of the trial in the *Bellshill Speaker*, 3 May 1919.

miners of Scotland to continue at work, pending the reply of the Government'. They urged the miners to cross the picket lines and announced that there would be no strike pay.[87]

The rank and file uprising widened the hiatus between leaders and led in the mining unions almost to breaking point. In Lanarkshire, Manus Duddy of Blantyre told the Executive members on the eve of the strike that 'he did not think [they] realised that the members were crying out for a long time back against the methods and inattention of the agents and Executive to the members'. In his opinion, it was the feeling of alienation from the union that had led the majority of the members to vote against a proposed increase in subscriptions shortly before the strike. Typically, the Executive had simply ignored the ballot result and increased the rates regardless.[88] Similarly, in Fife, Adamson was increasingly willing to disregard the wishes of the members in order to maintain his hold on the FK&CMA. In December 1918, he had attempted to block the election of John O'Neill as delegate of the Buckhaven branch to the monthly conference of the union. O'Neill, a member of the FMRC, had defeated James Neilson, an old Adamson supporter.[89]

The reform committees had responded to these autocratic abuses of power by campaigning for the removal from office of the incumbents. In Lanarkshire, the LMRC had 'for months on end ... demanded the scalp of every miner's agent attached to the LMU'.[90] At the Motherwell town hall meeting on 24 January, Maclean warned:

> Whilst they did not wish to dismiss a single official of the unions ... if any of them did stand in the way of the policy they were fighting for, then they would have to go. A mere handful of men could not be permitted to hold back the aspirations of a million.[91]

The strike was not just unofficial; it was anti-official. Skirmishes between police and pickets made headlines, but the real antagonists were not workers and coalowners or state, but miners and union leaders. As one FK&CMA Executive member admitted, 'the present strike was headed by men who thought the leaders did not die quick enough. They wanted to fill the leader's shoes'.[92] The

---

87   NUS MW Executive Meeting 27 January 1919.
88   LMU Council Meeting Minutes 10 February 1919.
89   MacDougall n.d., p. 26.
90   *Lanarkshire and Hamilton Herald*, 5 February 1919.
91   *Motherwell Times*, 31 January 1919.
92   *Leven Advertiser and Wemyss Gazette*, 30 January 1919.

efforts of Joe Westwood and Wullie Adamson to engineer a return to work led to direct confrontation with the FMRC. When at one meeting a reform committee member leapt to the platform to answer a speech by Westwood, the latter became so enraged that he threw off his coat for a fight.[93] At another Fife meeting, Charles Muir, an Executive member from Bowhill, threatened to bring in the troops against his own members in his capacity as magistrate if the picketing did not stop.[94]

In Fife, the officials enforced a ballot over the sole issue of the surface workers' hours, which resulted in a slim majority of 724 for a return to work. We will return to the details of this controversial ballot below, but it is interesting to note that constitutional union procedures were by now held in such low esteem by the advanced section of the membership that a large crowd from all over central Fife, estimated at between 10,000 and 15,000, gathered in Dunfermline public park and voted to ignore the ballot and continue the strike over the reform committees' demands. They then elected twelve delegates and marched to the union headquarters to confront the officials. Sam Hynds, the Treasurer of the FK&CMA (who was to die two weeks later), addressed the crowd and 'was received with derisive cheering and the singing of the Red Flag'.[95]

Anger with the officials ran even higher in Lanarkshire, where bitterness and hostility erupted into open rebellion. Patrick Powers, a prominent official from Bellshill, was assaulted as an argument with a striker turned violent. Powers sustained a fractured skull and was taken to Glasgow Royal Infirmary in a critical condition.[96] On Wednesday 29 January, a demonstration of 20,000 miners from Cambuslang, Blantyre, Burnbank, and Shettleston marched to the union's headquarters in Hamilton. The demonstration elected delegates to seek an audience with the agents inside, in order to press them to make the strike official. Before long the miners became impatient that their delegates were being kept waiting, broke into the building and occupied it. In the process, the building was partially ransacked. Telephone wires were cut, glass broken, pot plants destroyed, and union documents and correspondence hurled into the crowds in the street below. William Small, the Assistant General Secretary, and Hugh Gallacher, miners' agent, came under intense pressure. According to reports, they were manhandled and even threatened with a revolver when they refused to hand over the keys to the office safe.

---

93   *The Worker*, 15 February 1919.
94   Ibid.
95   *Dunfermline Press*, 1 February 1919.
96   *Bellshill Speaker*, 31 January 1919.

Whatever precisely occurred in the miners' offices is unclear, but by the time the insurgents departed, Small had agreed to send telegrams to all LMU branches declaring the next day's strike official, and to convene a conference between the LMU Executive and the LMRC. Emboldened by the dispersal of the mob, Small had a change of mind. The MRC responded by arriving for its meeting with the officials the following day with another huge demonstration in tow. Once again the offices were occupied, and the Executive was forced to face the reform committee leaders. The meeting lasted for two and a half hours. The upshot was that the strike would continue, with a delegate meeting of the LMU convened for Friday to vote on making it both official and indefinite. It was also agreed by officials that picketing expenses would be reimbursed by the union. To the crowd outside this sounded like official approval was a formality, and cries went up of 'It's a victory'.[97]

The delegate meeting never took place, however. Faced with a mutiny of the membership, the officials locked up the Hamilton offices, cancelled the meeting, and fled the coalfield. Considering Glasgow to be within the LMRC's reach, they circled their wagons in distant Edinburgh, where they convened an emergency meeting:

> [T]o consider the situation in the county of Lanarkshire brought about by the attempts of outsiders to force an unofficial strike, and which made it impossible to hold a conference of the delegates at the union offices without the risk of intimidation and violence.[98]

The elemental anger and militancy of the strikers, harnessed by the MRC, had culminated in an open insurrection against the officials. During the occupations at Hamilton, a red flag was hoisted above the union building as a symbol of the MRC and its policies, and rumours spread around the coalfield that the revolutionaries had 'captured the offices and dismissed the officials'.[99] For a few heady days, the reform committees became the *de facto* leadership in Fife and Lanarkshire, declared war on officialdom, and threatened to take control of their unions.

---

97   *Glasgow Daily Record,* 31 January and 3 February 1919; *Glasgow Weekly Mail and Record,* 8 February 1919; *The Worker,* 8 February 1919; *Bellshill Speaker,* 31 January 1919.
98   *Bellshill Speaker,* 7 February 1919.
99   LMU Council Meeting 10 February 1919; *Bellshill Speaker,* 31 January 1919.

3   The Strikes Collapse

On Wednesday 5 February, two weeks after the surface workers in the Cowden-beath and Lumphinnans area of central Fife had originally struck, the last of the 'young hot-heads' from the same region returned to work. Over the pre-vious three days, the bulk of miners in both Fife and Lanarkshire had drifted back in dribs and drabs. Clearly, the reform committees had experienced a ser-ious and sudden haemorrhaging of support. How, and to what extent, did the officials regain control of such a desperate situation?

The battle for influence over the membership ebbed and flowed before and during the strike, and whilst the reform committees held the upper hand for a time, there were crucial weaknesses in their network which officials sought to exploit. In Fife, the main counter-attack came in the form of the ballot of Monday 27 January, mentioned above. Although it was clear that the main demands of the strike had changed over the weekend to encompass the six-hour day and minimum wage, the FK&CMA Executive decided to ballot over the surfacemen's shift times. On this issue the Executive won a slim major-ity for a return to work, by 6,969 to 6,245.[100] The FMRC was furious, point-ing out that only 50 percent of the membership had voted; the ballot had omitted the strike's real demands and therefore had been the subject of a spontaneous mass boycott.[101] The claim is credible. In the areas where the FMRC had most presence, the storm centres of Cowdenbeath, Lumphinnans and Raith, they were able to win an argument for participation in the bal-lot and a vote to continue the strike.[102] When Adamson and Robertson had addressed mass meetings in this area on the previous Friday evening in an attempt to head off the growing strike, they were met with defiance, large majorities voting to continue the action. Again, as we have seen, on the day the ballot result was announced, the FMRC was able to pull a large protest demon-stration in Dunfermline, with feeder marches from Lochore and Cowden-beath.

However, in other areas the roots of the FMRC did not run so deep, and here the officials were able to drum up support. In Clackmannan, and in the extreme east and west of Fife there was a large majority for the officials. As early as Sunday 26 January, Sam Hynds, John Robertson and Joe Westwood addressed miners from Leven, Methil and Kirkcaldy in Bayview Park, and persuaded

100   *The Worker*, 15 February 1919.
101   *Fifeshire Advertiser*, 1 February 1919.
102   *Cowdenbeath and Lochgelly Times and Advertiser*, 29 January 1919.

them to return to work the next day. A sole member of the MRC leapt to the platform and attempted to argue with the crowd, but the cart was simply pulled away.[103]

A pattern developed whereby officials were able to secure a return to work in certain areas, with the reform committee countering with mass picketing the following morning. Where they were able to mount large pickets, at the Kilsyth collieries for example, they met with success, winning a pithead vote for a continuation of the strike for the full programme.[104] But the emergent FMRC network was stretched thin. While in some areas it was strong enough to maintain the action in face of the ballot result, other areas went back unchallenged. So on 30 January, whilst most miners in west Fife were still on strike, at Lochgelly all pits went back. Nearby Lumphinnans was an FMRC base, but the militants there were busy keeping their own pits out, and were unable to prevent the breach just one mile up the road.[105]

In its post-mortem of the strike in *The Worker*, the FMRC attempted an honest account of its own organisational deficiencies when measured against the union machine. The officials, it conceded, had:

> made ample use of the facilities afforded them by having the finance of the union at their back, and they broke the back of the strike in east Fife, by organising meetings in those districts which the strikers had been unable to reach.

In these districts they had managed to persuade the majority that the main issue was the surface workers' demands, and that they could be settled without a strike.[106]

In Lanarkshire, support for the reform committee had been more evenly spread. Nonetheless, here too there were weak spots. In both Fife and Lanarkshire, reform committee strongholds were in the biggest pits, with a markedly narrower base of support in smaller ones.[107]

The precise circumstances in which the strike ended in Lanarkshire is unclear, but it is difficult to avoid the conclusion that the riotous behaviour at union headquarters in Hamilton, whilst spectacular and successful in the short

---

103   *Leven Advertiser and Wemyss Gazette*, 30 January 1919; *Fife Free Press*, 1 February 1919.
104   *Glasgow Daily Record*, 3 February 1919.
105   *Fife Free Press*, 1 February 1919.
106   *The Worker*, 8 and 15 February 1919.
107   Campbell 1994, pp. 1–12. The strongholds of the Reform Committees in 1919 would correspond to the spread of the Communist Party in the 1920s.

term, ultimately undermined support for the reform committees. The marches on 30 and 31 January were designed to pressurise the officials into backing the action. Assuming that physical violence, both threatened and actual, was used against them, and reports of it are numerous enough to suggest that it was, then this was a serious mistake, for it allowed the officials to drive a wedge between the core of the LMRC and its softer support.

The violence may well have been spontaneous, unendorsed by the reform committee. It seems likely that during the first occupation of the miner's offices at least, events escaped LMRC control. The impatient and unruly younger element seen at the Hamilton Palace Colliery disturbances was present here too. When James Tonner spoke of 'disgraceful incidents ... that had sullied the good name of our movement' at the first LMU Council meeting after the strike, he had the agreement of many delegates who were not altogether opposed to the strike.[108] When John Robertson told a mass meeting at Hamilton Palace Colliery on 2 February how officials had been held at gunpoint, the meeting 'vented its indignation in cries of shame', unanimously voted to arm themselves the next morning in order to get through any pickets which might be in place, and passed a resolution of thanks to Robertson for his conduct in the strike.[109]

Hamilton Palace Colliery was not a LMRC stronghold. To see how the violence at union HQ caused a reaction against the strike even amongst its supporters, one must turn to the meeting at the Motherwell Town Hall on the evening of Friday 31 January. Charles Robertson, an LMU Executive member and Secretary of Motherwell Trades Council gave a speech at the invitation of the Glasgow Forty Hours strike committee. The meeting was in angry mood, as the infamous George Square riot had taken place earlier that day, and many in the audience and on the platform had been involved. Robertson launched an attack on the reform committee.

> The speakers had been telling them of the bludgeoning of the police. But what had some of these unofficial bodies done in Hamilton the previous night? Why, one of the executive of the miners had been threatened with a revolver, and the acting Secretary had been seized by the throat and made to write out letters at their dictation. They were condemning the police for batoning, but here were men bludgeoning their own brothers![110]

108   LMU Council Meeting Minutes 10 February 1919.
109   *Glasgow Daily Record*, 3 February 1919.
110   *Lanarkshire and Hamilton Herald*, 1 February 1919.

Violence by the state against striking workers was to be condemned, but so too was violence of workers against union officials, or 'brother against brother'. On this the LMRC appeared split. Some held the opinion that the officials had become strike-breakers, and deserved any treatment that might be dealt out to any scab or blackleg. Others considered violence against officials to be beyond the pale, transgressing basic rules of union loyalty. Still others took the more pragmatic view that because it went against common trade union ethics, it was a poor tactic; the leadership of the miners was to be wrestled from the officials by example, not force. Griffin's article in *The Worker*, on 15 February, diplomatically states that the 'outburst of the rank and file' was 'perhaps too strongly expressed'.[111] Either way, the episode allowed the officials to begin to staunch the flow by uniting militant and moderate members around basic trade union ethics. It appears that some within the MRC even joined in the call for a return to work. Whatever the difference of opinion within the reform committee, in the hands of those opposed to direct action, the violence against the officials became a stick with which to beat it.

The greater resources of the official machine, the organisational weakness of the reform committees, and the disturbances in Hamilton were all contributory factors to the collapse of the movement. The weakening of the strike in engineering in the aftermath of the George Square riot on Friday 31 January must also have played a part. Amongst miners and engineers, enthusiasm for the strike began to wane almost simultaneously, and both sections of workers returned to work within a few days of each other. Above all, the strike collapsed because it did not spread. Unable to go forward, it could only go back. MacDougall wrote in *The Call*: 'The general strike cannot last very long. That is its very nature. Either the strike will rapidly extend over the whole of Britain or it must be terminated'.[112]

Although the strike had failed to trigger a response from workers south of the border, Scottish revolutionaries were not despondent, preferring to see the episode as part of the unfolding international class struggle.[113] Reform committee leaders shared this optimism. From Fife, J.P. Payne wrote:

> Work has now been resumed, but, for all that, to class conscious rebels the strike has not been in vain, for the reform movement, which is really a revolutionary movement, has received an impetus which, with the

---

111   *Lanarkshire and Hamilton Herald*, 5 February 1919. Manus Duddy and a Mr. Ferguson, both members of the reform committee in the Blantyre district, recommended a return to work.

112   *The Call*, 6 February 1919.

113   *The Worker*, 8 February 1919.

assistance of the gravediggers, will yet sweep officialdom and reactionaries, who are its mainstay, out of existence. The Reform movement is
quickly gathering in the rising generation, which demonstrated a real
rebel spirit during the strike – although beaten by the fogies of the coffin
club school.[114]

For Bob Selkirk, the swiftness of the miners to strike was confirmation that they
had learned from the national strike of 1912, when long drawn-out negotiations
had allowed the owners to build up substantial stocks, and he felt that the strike
in Fife and Lanarkshire would soon be repeated nationally. The militancy of the
strike encouraged him to believe that as the struggle for the Miners' Charter
went on, their own programme would come to the fore:

> The demonstrations and mass meetings of this strike have shown con-
> clusively that the national strike of the near future, if it lasts over a week,
> will finish by the miners taking over the pits and working them for them-
> selves.[115]

In fact, as time would show, the high point of the class struggle in Scotland had
already passed, and neither engineers nor miners here played any significant
part during the rest of 1919. The reform committees in Fife and Lanarkshire, in
conjunction with the CWC, had launched a valiant attempt to catalyse a general
strike over demobilisation, but in doing so they had shot their bolt. Second only
to South Wales in terms of their industrial and political influence, the revolutionaries and syndicalists of the Scottish mines were incapable of influencing
events at a national level during the period of the Sankey Commission. The
sequel to the events in Fife and Lanarkshire was not a national miners' strike,
but the intra-union strife of the 1920s. Perhaps John Maclean had been right
about the premature timing of the general strike on the Clyde. But all was not
lost. As our other case studies suggest, the miners of Britain kept hope alive of
a battle with the government over demobilisation, and a war against capital.

114   Ibid.
115   Ibid.

CHAPTER 4

Nottinghamshire

There was a large, brilliant evening star in the early twilight, and under-
foot the earth was half frozen. It was Christmas Eve. Also the war was over,
and there was a sense of relief that was almost a new menace. A man
felt the violence released now into the general air. Also there had been
another wrangle among the men on the pit bank that evening.[1]

∴

In his study of *The Dukeries*, Nottinghamshire pit villages styled on modern
company towns, Robert J. Waller pointed out that the Notts coalfield 'is often
described as if it were a single, homogeneous unit, moderate in politics, and
inclined to Spencerism and the butty system'.[2] Nottinghamshire had earned
its reputation for moderation long before 'Spencerism' entered the vocabulary
of mining unionism. Alan Griffin, the official historian of the Nottinghamshire
Miners' Association (NMA), described industrial relations at the outbreak of
war:

> By 1914, the Association had settled down into a humdrum existence.
> Membership was expanding gradually as the labour force grew, funds
> were steadily accumulating and relations with the employers were good.[3]

The NMA leadership was ultra-moderate, operating firmly within long-estab-
lished conciliation boundaries, and prepared to follow the larger districts in
the MFGB only when they did not stray too far to the left.[4]

Notts was not one of the districts in which trouble was anticipated at the
war's end. And yet 1919 witnessed a coalfield that dramatically defied all ste-

---

1   Lawrence 1988 [1922], p. 5.
2   Waller 1983, p. 291. For a study which uncovers the social, industrial and political complexities
    of the Nottinghamshire coalfield, and especially Hucknall, see Gilbert 1992, chapters 5 and 6.
    For a view from 1926, see McIlroy's chapter on Nottinghamshire in McIlroy et al. 2009.
3   Griffin 1962, p. 17.
4   Griffin 1962, especially chapters 7–12.

reotypes and expectations, as an emergent unofficial leadership led three mass strikes against both local and national officials. In January, March and July, the moments at which the conciliationist strategy of the MFGB leadership was at its most vulnerable, Notts miners moved into action in large numbers, responding to the calls of syndicalists and revolutionaries who, like their counterparts in Fife and Lanarkshire, were able to usurp the official union leadership.[5] As a paradigm of moderation prior to 1919, and collaborationism post-1926, the experience of the Notts coalfield in our period provides an important comparison to what are conventionally seen as more 'advanced' areas.

1   Striking against the Butty System

On the surface, the NMA that entered the war emerged from it unchanged. For one thing, the same moderate leadership remained in place. J.G. Hancock, Charles Bunfield, and William Carter (Agent, Secretary, and Assistant Secretary, respectively) were Liberals in their political outlook, although Bunfield and Carter had eventually been forced to reconcile themselves to affiliation with Labour. Hancock had been a founder member of the NMA in 1881. Bunfield and Carter's rise through the ranks began in the following decade. All three proudly bore the brand of conciliation and co-operation with the owners, with whom J.G. Hancock was on particularly good terms. William Carter 'hated and detested strikes'.[6] George Spencer and Frank Varley were younger officials, aged 46 and 34 respectively in 1918. Spencer had been elected President in 1912, and became a full-time agent when Varley took over the Presidency at the end of the war. Although Spencer had joined the ILP in 1900, he had supported the war without question, and both he and Varley were well within the mainstream of Labour's political thought. Spencer had been a lay preacher with the Wesleyan Methodists until his mid-20s, and although he was by now an agnostic, he retained a conservative outlook on social issues. Speaking at the Workers' Educational Association in Notts, he said that in his view the greatest contemporary danger to the state was 'the decline of family life and the secularisation of marriage'. In addition to these five, there was a treasurer, L. Spencer, and a subscriptions collector, which made up the full complement of seven staff.

---

5   The unrest which affected Notts in July will be dealt with in Chapter Eight below.
6   *Mansfield and North Notts Advertiser*, 21 February 1919; Bellamy and Saville 1972–84, vol. 1, pp. 304–7, vol. 2, pp. 159–60, 371–2; Griffin 1955; Griffin 1962; *Nottingham Evening News*, 24 May 1919.

George Spencer, Carter, and Hancock were all MPs, representing Broxtowe, Mansfield, and Mid-Derbyshire. Hancock, however, was a Liberal. 'Gladstonian ... in politics, Methodist by religion, and an active temperance worker', he had been unable to make the transition to Labour.[7] When he stood in the Mid-Derbyshire by-election in 1909, the MFGB had just affiliated to the Labour Party, and although officially a Labour candidate, Hancock employed the Liberal election agent, was vigorously supported by Sir Arthur Markham, the coalowner Liberal MP for Mansfield, and utterly ignored the Labour Party. Upon his election he 'assured the Mid-Derbyshire Liberal Association that his personal views remained Liberal'. He had the Labour whip withdrawn from him in 1914, and continued as a Liberal MP, holding the seat until 1923.[8]

Despite being a political heretic, Hancock still enjoyed the support of the NMA in 1914.[9] Griffin argues that between 1914 and 1918, the war had completed the NMA's transformation from Liberalism to Labour so that by its conclusion Hancock presented an anachronistic and isolated figure. Certainly he was soon to become so, but Griffin conflates history.[10] In 1918, Hancock's politics were in retreat inside the NMA, but he still exercised significant power and maintained an important base of support. Nottinghamshire had enjoyed better industrial relations than most areas in the first two decades of the century. It produced for a buoyant domestic market, free from the pressure of international competition that heightened the antagonism between capital and labour in, for example, the South Wales coalfield. Geological conditions allowed relatively easy access to good seams in most parts of the county, and in the Leen Valley, the Top Hard seam was the most profitable in the country, giving the miners who worked there prosperity unknown in coalfields elsewhere.[11] Paternalism, industrial and social heterogeneity, and the absence of a south Wales type mono-industrialism discouraged militancy and solidarity, encouraging the NMA to seek institutionalised stable collective bargaining and conciliation procedures; in a word, Hancockism.[12]

The other key factor underpinning Hancock's position in Nottinghamshire mining trade unionism at the war's end was the continuation of the butty sys-

---

7   Bellamy and Saville 1972–84, vol. 2, p. 160.

8   Ibid.

9   Ibid.

10   Griffin 1962, pp. 55, 38.

11   Griffin 1962, p. 18.

12   Gilbert 1992, pp. 179–80. For a discussion of the literature about the regional specificities affecting trade unionism in Notts, see also McIlroy's chapter on Nottinghamshire in McIlroy et al. 2009.

tem. Prior to the 1880s, this system had been the main barrier to the establishment of permanent union organisation. Historically, the butty drove his day men to higher output, so as to increase his own wages, making him as much an employer as a workman. When Notts miners finally struck in 1844, they did so against the butties as well as against the coalowners themselves. When, by the 1880s, changed economic and social conditions had finally prepared the ground for more thoroughgoing trade union organisation, the most important transformation had been the partial withering of the butty system:

> The butty was now far more of a workman and less of an employer than formerly. Indeed, most of the leading spirits of the 1881 union were themselves butties, (or checkweighmen who had risen from the ranks of the butties).[13]

A process of differentiation had taken place within the ranks of the butties themselves, brought on largely by the increasing size of colliery concerns, and the employment by the companies of specialist managers who partially displaced many of them. The system operated in an uneven fashion by the war's end; whilst in some places the butties had sunk nearer to the level of the ordinary collier, in others they remained relatively rich and powerful, forming a natural constituency for Hancock's outdated philosophy. Although no longer 'the master' in quite the same way as before, the butty was still the contractor directly responsible to the management for the proper working of his stall, and the good conduct of the daymen who dug the coal. The daymen got a set day wage, paid by the butty who continued to be paid by the ton.[14]

The butty was thus divided from the ordinary collier by position, status and income. At Radford colliery, in a good stall, a butty could reportedly make – in an extreme case – as much as £12 or £14 per week in 1919, more than double the average wage of the ordinary face-worker.[15] One miner was scandalised that his brother, a rich butty, could afford to spend the week at the York races, return home on the Friday, pay his men, and still pick up over £12 in earnings for himself.[16] In many pits the butties 'kept the gaffer sweet' by offering bribes, and men complained about the operation of the 'market system', or the buying of

13   Griffin 1955, pp. 175–6.
14   *Nottingham Evening News*, 2 January 1919. This is taken from a description of the operation of the butty system at Radford, a medium-sized pit.
15   *Nottingham Evening News*, 2 January 1919.
16   NMA delegate to 1920 Triple Alliance Conference, quoted in Griffin 1962, p. 55.

the best stalls by butties prepared to pay the highest prices.[17] These practices allowed the mining companies to pick and choose men, putting them out of the normal places and into the abnormal, and vice versa.[18] At a Mansfield Market Place meeting on 1 January 1919, the butty system was described as 'the monstrous practice of one worker making a profit out of the other'. Here, in the aftermath of World War One, the butties were still being described in the language of the early nineteenth century; men who 'had been bosses so long that they did not like the idea of being supplanted'.[19]

The continuation of the butty system into the post-war period not only undermined unity by establishing hierarchies in the pits; these hierarchies were replicated within the union machine itself, via the checkweighmen. Throughout the British coalfields, checkweighmen had played a crucial role in the establishment of permanent union organisation. Appointed and paid by the men, they were independent of the colliery companies, and could expand their role in the pits beyond the simple one of 'taking account of the weighing of the coals' as specified in the Coal Mines Regulation Act of 1860.[20] In post-war Nottinghamshire, the checkweighmen continued to play a prominent part in the NMA's affairs, both as branch officers and as delegates to the NMA Council. However, in Notts the checkweighmen were employed not by the daymen, but by the butties. Where the powerful butties predominated, checkweighmen were answerable to them first and foremost, thus ensuring the butty's continuing power inside the union, buttressing Hancock and his moderate colleagues.[21]

In 1919, Hancockism was not only fortified by the rich butty's checkweighmen representatives. Even at pits where the gap between butty and daymen was at its narrowest, the checkweighmen were liable to throw their weight behind tradition. The reason was the development, from mid-1918, of a campaign for a universal day-wage to replace the butty system. If payment by pro-

---

17   Griffin 1962, p. 54.

18   *Nottinghamshire Free Press and Derbyshire Chronicle*, 10 January 1919 (hereafter *Nottinghamshire Free Press*). Disputes over 'normal' and 'abnormal' places, i.e. 'easy' and 'tough' seams, indirectly led to the formation of the South Wales URC in 1911.

19   *Nottinghamshire Free Press*, 17 January 1919.

20   Griffin 1955, p. 11.

21   Griffin 1962, p. 54. Ned Cowey, a Yorkshire delegate at the 1898 conference of the MFGB said: 'In Nottinghamshire they had what was called the butty system, which was a sweating system ... The system was bad, as it put the appointment of checkweighers in the hands of a few men, and he was bound to do whatever these butty men told him to do' (cited in Griffin 1955, p. 138).

duction was successfully abolished, checkweighers would no longer be needed, and they would be returned to the pit from the relative comfort, security and safety of the checkweigh machine.

The butty system had therefore exerted a long and considerable influence on the politics of the NMA. The partial withering of the system and the general growth of support for Labour meant that younger men like Spencer and Varley were becoming increasingly important inside the union, but the butty/check-weighman axis meant that Hancock had not yet done his dash. There remained a considerable base of support for Lib-Labism within the union; in fact, when Labour withdrew the whip from Hancock in 1915, he came close to achieving the disaffiliation of the NMA from the MFGB's political fund in retaliation. He was thwarted in a coalfield ballot, but significantly he managed to persuade Spencer to campaign in his support.[22]

The upshot was that while Hancock's heyday had passed, his position in the union was still relatively secure at the war's end. Along with Spencer he was nominated by the NMA as rivals to Smillie and Hodges for the MFGB leadership posts in 1918.[23] He also successfully resisted a move to force his resignation following his Liberal candidature at the 1918 general election as the Labourites on the NMA Council closed ranks to protect him.[24] Although Varley, Spencer and their ilk had different political allegiances to Hancock, they chose not to disrupt the status quo within the NMA, preferring to defend the power structure of which they were a part.

The main opposition to Hancock and the Lib-Lab old guard came from out-side officialdom, chiefly from a group of rank and file socialists led by Herbert Booth.[25] In 1915, Booth campaigned vigorously around the county, addressing meetings, and issuing 30,000 leaflets calling for a rejection of Hancock's scheme to withdraw from the political fund. His rank and file committee 'of keen social-ists' emerged from this campaign, and from political classes at which he was the tutor.[26] Success in the campaign solidified the group, which went on to

---

22   NMA Council Meeting 30 January 1915.

23   MFGB Special Conference 14–16 January 1919, pp. 6–7.

24   *Nottinghamshire Free Press*, 10 January 1919.

25   Griffin 1962, pp. 22, 39–40, 143, 276. Booth was out of the county from 1918–22 during which time he was agent for the Forest of Dean miners. When he returned to Notts he became active in the Minority Movement, opposed Spencer's Industrial Union, and eventually was to become agent for the reunited Nottinghamshire Miners Federated Union in 1937. In 1915, he was a member of the ILP, and had just returned from a year of study at the Central Labour College in London.

26   Griffin 1962, p. 39.

campaign for improvements in working conditions and democratisation of the
NMA. Booth's rank and file committee would provide most of the prominent
figures in the unofficial leadership in 1919.

Irishman Jack Lavin led a second group of militants in Nottinghamshire.
Lavin had emigrated to San Francisco in 1906, where he became an active
member of the Socialist Labor Party and the Industrial Workers of the World.
He returned to England shortly before the war, finding work in the Yorkshire
coalfield. In 1915 or 1916, he moved to Notts, working at Welbeck colliery near
Mansfield. Over the next two years he built a 'small but influential' group of
supporters who formed a branch of the SLP. He died of tuberculosis in 1919,
but 'his views gained wide acceptance, and his influence lived on after his
untimely death'.[27] The leading figure in the group after Lavin was Owen Ford,
'a wonderful orator ... the platform idol of the miners', who also worked at
Welbeck.[28] The group was referred to by its members and by the press as
representatives of the Workers' International Industrial Union, rather than
the SLP.[29] The group was extremely active, issuing literature and leaflets, and
holding regular meetings in Mansfield Market Place. In 1919, the IWW sent a
full-time female organiser from America to assist the group. In 1920, it would
go on to form the first Communist Party branch in Notts.[30]

Strictly speaking, the two groups never actually merged; differences between
them over holding union positions, and the parliamentary road to socialism,
meant they retained their separate identities. However, they did form a united
front to fight for the unofficial Miners' Charter. Newspaper reports refer to only
one unofficial group in 1919, variously described as 'the unofficial camp', 'the
unofficial movement', and 'the unofficial committee'.[31] For the sake of simplicity
I have referred to this ad hoc committee as the 'unofficial group'; although it
was in many respects similar to the Reform Committees in other areas, it never
formally constituted itself under that title.

This was the situation then at the end of the war; J.G. Hancock's Liberal politics were being squeezed out by the growing support for the Labour Party inside
the NMA, but as long as the butty system remained, and the Labourites in the

27   Ibid.
28   McIlroy et al. 2009, p. 222. Ford was dismissed as checkweighman at Welbeck and black-
      listed in the aftermath of the lockout.
29   *Nottingham Guardian*, 23 July 1919; *Mansfield and North Notts Advertiser*, 15 August 1919.
30   *Mansfield and North Notts Advertiser*, 8 August 1919; for discussion of Lavin, Ford and
      Booth, see McIlroy's chapter on Nottinghamshire in McIlroy et al. 2009.
31   *Nottingham Guardian*, 22 and 27 March 1919; *Nottinghamshire Free Press*, 31 January, 28
      March 1919; *Mansfield Reporter*, 31 January 1919.

NMA chose to work with him, he was not yet done. Against this moderate bloc
stood the ad hoc unofficial group. It had struck a blow in 1915 over withdrawal
from the parliamentary Fund, but little had been heard of it since then. From
the middle of 1918 however, it reappears at the head of a campaign to replace the
butty system with the universal day-wage. As the campaign gathered strength,
it would reveal the structural antagonism between the NMA and the ordinary
miners, and ignite a conflict that would be felt far beyond the Nottinghamshire
borders.[32]

Needless to say, Hancock and his supporters on the NMA Council were
implacably opposed to the campaign for a universal day-wage.[33] However, by
the end of 1918, the pressure was mounting. The Council reluctantly agreed
to a compromise ballot on the possible replacement of the butty system, and
offered four options in the vote: the continuation of the present system; the
present system 'with some modification'; the all-throw-in system (which entailed the equal distribution of a stall's total wage); or the day-wage.[34] Branches
complained that the volume of resolutions in favour of the day-wage was
being ignored, and there was widespread suspicion that the multiple-choice
ballot paper was deliberately designed to confuse.[35] The local press also carried
numerous reports of mass meetings at which branch and district officials
were accused of obstructing the campaign for the day-wage and being 'out
of sympathy with the rank and file'.[36] At a large meeting at the Hucknall
Empire:

> There was a debate on whether the representatives on the Council truly
> interpreted the wishes of the men. It was alleged that there were too
> many checkweighers amongst the delegates, and as their positions would

32   This antagonism extended to many of the less well-off butties. Evidently the erosion of
     their traditional rights and privileges had persuaded them that their welfare would best
     be provided by seeking unity with the ordinary collier. In fact, one of the leading figures
     in the agitation against the butty system in January 1919, Walter Owen, had himself been a
     butty for some 22 years at Mansfield Woodhouse. See *Mansfield and North Notts Advertiser*,
     3 January 1919.

33   Bellamy and Savile 1972–84, pp. 159–61. Correspondence between Hancock and the man-
     ager of Babbington Coal Company reveals his total opposition to the day-wage.

34   NMA Minute Book 1918. The ballot result showed that 3,579 wanted to maintain the present
     system either untouched or in modified form, whilst 15,776 wanted to scrap it. Of these
     7,499, or 47.5 percent voted for the day-wage.

35   *Nottingham Evening News*, 1 January 1919.

36   *Nottingham Guardian*, 2 January 1919; *Nottingham Free Press*, 10, 24 January 1919; *Mansfield
     and North Notts Advertiser*, 14 February 1919.

be precarious by the adoption of day-work, they used their influence for the contract system, though the workers desired the plan of day-working.[37]

Unable and unwilling to transmit the men's grievances, the NMA underwent a crisis of credibility which came to a head in the weeks after the war. A union based on the power of the butties was never going to fight to scrap the butty system. Nor were checkweighers going to fight for the day-wage. The agitation paralysed the union, enabling the unofficial group to assume the mantle of leadership in the coalfield over this issue just as the campaign for the Charter began.

In the final days of 1918, pressure for action gathered pace.[38] On Friday 28 December, thousands attended a mass meeting in Mansfield Market Place. The main speakers were from the unofficial group – Walter Owen of Mansfield Woodhouse, and Tom and Andrew Clarke of Rufford.[39] They called for, and won, a vote for strike action – against the butty system and for the day-wage – to begin on New Year's Day. On New Year's Eve, mass meetings at Mansfield, Sutton and Huthwaite followed suit. On 1 January, the strike began, with pithead meetings voting to come out.[40] By mid-morning there were strikes at Welbeck, Summit, Mansfield, New Hucknall, Rufford and Sherwood. On the following day, the men at Radford joined the movement. While some pits in the southern areas of the coalfield threw in their lot with the strikers, the centre was in the north, where the unofficial group's influence was strongest and the butty system most notorious.[41]

The strike was completely unofficial. Thousands flocked to Mansfield Market Place where Walter Owen explained that for the unofficial movement, breaking the butty system was only the opening shot in pursuit of a fighting unity that would in turn allow the attainment of greater goals – in particular the unofficial Miners' Charter. 'They wanted a six hour day, and £1 a day, but

37   *Mansfield Reporter and Sutton-in-Ashfield Times*, 31 January 1919 (hereafter *Mansfield Reporter*).
38   *Nottingham Evening News*, 1 January 1919.
39   Andrew Clarke was one of three miners to, uncharacteristically, be awarded a scholarship by the NMA in 1920 to attend the Central Labour College in London. The NMA rarely sponsored miners to attend the CLC because of its association with syndicalism, preferring the WEA. See Griffin 1962, p. 60.
40   *Nottingham Evening News*, 1 January 1919.
41   See speech by G.H. James at Radford, *Nottingham Guardian*, 2 January 1919.

as yet they were a divided mob'.[42] Owen also laid out the unofficial group's demand for a reorganisation of the system of electing NMA officials, so that they would all be elected by ballot and subject to re-election every three years. This demand was raised again and again in 1919, and would be won, in part, in 1920.[43]

Although the strike did not immediately develop into a fight for the unofficial charter, it did have the effect of galvanising the officials into action. Belatedly they scrambled to put themselves at the head of the movement. The demand for the day-wage had united the ordinary colliers, surface workers, and the less well-off butties, and shifted the balance of power in the union away from the wealthier butties and the checkweighmen overnight. The Mansfield Market Place mass meeting voted to send a telegram to the Miners' Offices at Basford, requesting an official to explain the Council's position at a further meeting to be held that afternoon at Mansfield's Victoria Hall. In response, the NMA Council convened an emergency meeting, which decided that it would support the demand for a day-wage for all and stall for time by referring the matter to the MFGB EC. In the meantime, it decided that the all-throw-in system should be adopted immediately in Nottinghamshire. This protected the position of the checkweighers, and might be more agreeable to the owners than the day-wage which would likely decrease output and increase labour costs.[44]

George Spencer and William Carter went straight from the Council Meeting to the meeting at the Victoria Hall, where they only succeeded in securing acceptance for their proposals once Owen decided to back them. Presumably, the unofficial group saw an immediate fight with official union support over the all-throw-in system as the best way to get rid of the butty system quickly, and lay the basis for a unified fight for the Charter.[45]

The strike against the butty system was very brief, lasting only one day in most places, and it did not involve the whole coalfield. Nonetheless, before New Year's Day was out, the long despised butty system was history – or at least appeared to be. The unofficial group had mounted a staggeringly swift and successful strike. At a mass meeting at the King's Palace, Sutton, C. Dean said that 'with the smashing of the butty system they were more united than in the past' and that this was only the beginning. NMA delegate T. Knapton referred

42 *Nottingham Guardian*, 3 January 1919; *Mansfield and North Notts Advertiser*, 3 January 1919.

43 *Nottinghamshire Free Press*, 10 January 1919; *Mansfield Reporter*, 3 January 1919; Griffin 1962, p. 40.

44 NMA Minutes 1 January 1919.

45 *Mansfield and North Notts Advertiser*, 3 January 1919.

to the scrapping of the butty system as 'one of the greatest revolutions that had ever taken place in the mining world'. He had been fighting for its abolition for thirty years.[46]

In fact the butty system had not been dealt a terminal blow. The all-throw-in system was not implemented everywhere; what system was adopted and where depended on local variables such as the level of organisation at the pit, and the nature of the management. At some places, such as Cinder Hill, the strike went as far as achieving the day-wage, whilst others, like Gedling, did not even win the full adoption of the all-throw-in system.[47] Later, as the post-war milit-ancy waned, the butty system would resurface in Nottinghamshire, although in modified form, not to be finally abolished until the 1950s.[48] Nonetheless, before the 1921 lockout it appears that nearly everywhere the butty system had been displaced, and certainly the miners themselves felt that a historic victory had been achieved. The apparent smashing of the centuries old butty system gave the unofficial group enormous prestige. Moreover, the victory had deep implic-ations for the NMA's structure, organisation and democracy, challenging the complex series of relationships upon which the NMA had rested since the 1880s. All at once, the unofficial group had succeeded in breaking the Lib-Lab/butty axis, undermining Hancock, Bunfield and Carter, and discrediting Spencer and Varley, neither of whom had been prepared to challenge vested union interests. Not bad for a day's work. At the very dawn of 1919, before the campaign for the Miners' Charter had begun in earnest, Nottinghamshire, a by-word for modera-tion in the MFGB, had been swamped by an amorphous, though very real, direct action movement.

2   The Demobilisation Crisis

Encouraged by its dramatic success, the unofficial group now dared to dream that Notts might provide the catalyst for a national strike over the Miners' Charter. It had no faith that the NMA leaders would ever endorse any such action, and it was unsurprised when two weeks after the MFGB's Southport Conference they had still not reported back to the branches.[49] The group held a series of unofficial meetings around the county with the aim of rekindling the militancy of 1 January. At Sutton, its leaders told miners:

---

46   *Nottinghamshire Free Press*, 17 January 1919; *Mansfield Chronicle*, 23 January 1919.
47   Griffin 1962, pp. 54–5, 98.
48   Griffin 1955, vol. 1, p. 1.
49   NMA Adjourned Council Meeting 28 January 1919.

The Federation said the reduction of hours was a great problem, but if it was left to them, they would not get the reduction for ten years. They ought to strike now, until they get the six hour day, six days for five, and the £1 day.[50]

These meetings attracted considerable audiences, and the unofficial Charter's demands for a flat rate increase helped to galvanise support from the lower paid workers in the mines – the pit boys, labourers and surface workers who were traditionally under-represented in both the NMA and MFGB.[51] However, whilst the unofficial group won many meetings over to their Charter on a show of hands, they could not move miners into action. The group switched strategy; whilst it continued its propaganda campaign for the unofficial Charter, it began agitating over long-standing local grievances. There were three main demands – an acceptable rate for abnormal places, a main road workers price list, and the abolition of forks and screens from the pits.

Nottinghamshire coalowners had always insisted upon the use of forks and screens for the loading of coal into the tubs, as it ensured that only the best quality lump coal came out of the pits. The miners were bitterly opposed to this on two grounds. Firstly, it caused a considerable lowering of earnings for the piece worker, as it meant that small coal, or slack, was left in the pit with the waste. Secondly, the slack left behind made 'gob' fires much more likely. Shovel filling would have a positive effect on both safety and earnings, so it had long been a longstanding demand.[52]

A minimum wage for abnormal places would help rectify the pay anomaly between pits in different parts of the county. The sense of unfairness about the discrepancy was as strong as that which had driven the Cambrian miners in 1910. Smith said: 'if work was worth 8s3d in one end of the county where they could not get out of the way of coal, it was worth it at the other end, where they could not shift for muck'. 8s3d was decided upon as the acceptable minimum.[53]

The main road workers were the timberers, chargemen, haulage men, horse keepers and anyone else involved in the making and maintaining of pit roads and the conveying of coal. Again, their pay depended upon local conditions:

---

50   *Nottinghamshire Free Press*, 17 January 1919.

51   *Mansfield Reporter*, 31 January 1919; *Mansfield and North Notts Advertiser*, 7 February 1919.

52   Griffin 1962, pp. 50, 98.

53   *Nottinghamshire Free Press*, 14, 21 March 1919.

There had never been an attempt to systematise the pay of main road workers. There was no such thing as a county scale, and at some collieries there was no scale at all, and the men were at the mercy of the manager.[54]

Along with the butty system, pressure for action over these issues had been building since the beginning of 1918, when war weariness had renewed the possibility of successful intervention by militants around bread and butter issues. By the beginning of May 1918, all three demands had got as far as the NMA Council.[55] There they had stuck, however, and by the end of the year the Council had still come to no decision on action.[56]

The fractious state of affairs around these issues had implications for the national campaign over the Charter. Discord between miners and leaders at county level could, under certain conditions, spill over into the realm of national issues and policy. The strike over the butty system, during which workers had voted en masse to support the unofficial Charter, suggested to the unofficial group that its best hope of influencing the post-war national campaign was to initiate action over local grievances. This interplay between local and national issues provides a continuing focus of this study.

On the evening of 21 January, an unofficial mass meeting in Mansfield Market Place was poised to vote on strike action against fork loading, when miners from Bolsover Coal Company's Crown Farm Colliery arrived with the news that 25 men were being made redundant to make way for demobbed soldier-miners. 1,368 men had enlisted from the Company's collieries during the war, and already 340 had returned at Crown Farm alone. Management had decided that the Company would not bear the cost of reabsorbing the demobbed miners, and insisted they displace miners who had begun working in the pit after August 1914.[57]

There was considerable support for such a policy both within the MFGB and the NMA, as post-1914 men were accused of having used the pits as a 'bolthole' in which to hide from conscription;[58] the Bolsover Company's general manager could truthfully say he was acting in accordance with the miners' wishes at Crown Farm.[59] However, amongst the 25 miners being laid off were

---

54   Report of NMA Council Meeting in *Nottingham Evening News*, 31 March 1919.
55   NMA Council Meeting 29 April, 6 May 1919.
56   NMA Minute Book 1919.
57   *Nottinghamshire Free Press*, 24 January 1919.
58   MFGB Conference 14 January 1919, p. 39.
59   *Mansfield Reporter*, 31 January 1919.

several well-known militants who had led the strike against the butty system, at least one of whom was himself a demobbed soldier, not a post-1914 man.[60] Masquerading as a patriotic measure to provide employment for returning war heroes, the redundancies in fact represented a counter-offensive in which one of the most powerful of the Nottinghamshire coal companies was trying to exploit the miners' patriotism, turn the demobilisation crisis to its advantage, and weed out the militants.

As far as the unofficial group was concerned, this was a bullet aimed at the very heart of the miners' post-war programme; at stake was the question of who was to bear the cost of demobilisation in Notts – owners or miners? If the Notts miners were forced to foot the bill in January 1919, the implications for the rest of the British miners could be enormous. The Fife and Lanarkshire MRCs had already launched a fight over the issue of demobilisation in alliance with the CWC; now it was the turn of the syndicalists and revolutionaries of the Notts coalfield to try and do the same.

On stage in the Market Place, the group changed tack, replacing the motion for a strike against fork loading with one calling for an all-out strike against the lay-offs. The crowd, many of whom only three weeks previously had voted for the dismissal of the post-1914 men, supported the strike call, voting to picket out the rest of the coalfield. The next day, Mansfield, Rufford, Welbeck, Warsop Main, and Silverhill were pulled out. By the end of the week, over 20,000 miners were on strike. Once again, the unofficial group stood at the head of a mass movement.[61]

Frank Varley co-ordinated union efforts to contain the strike. Realising the extent of its miscalculation, the Bolsover Company backtracked, agreeing to use the 14 days' notice given to the 25 men to devise new shift patterns to accommodate them; failing that, they would be found jobs in one of the Company's other pits. Varley took the proposal to mass meetings in Mansfield, Sutton, and Nottingham, but at each one he was voted down.[62] In a matter of days, the situation had been turned on its head; a defensive strike against an attack on the Charter had been transformed by the unofficial group into an offensive one in its support. The hand of the group can be seen at a mass meeting at Forest Town in North East Mansfield for example, where Varley had been 'able to carry the men with him, until subsequent speakers introduced other questions such as

---

60   Ibid.; *Nottinghamshire Free Press*, 24 January 1919.
61   *Mansfield and North Notts Advertiser*, 24 January 1919; Frank Varley's speech at MFGB
       Conference 12–13 February 1919, p. 23.
62   *Nottingham Evening News*, 23, 24 January 1919.

the demand for a six-hour day, payment at £1 a day, and so on'.[63] As Maclean had done in Scotland, the unofficial group in Notts were making explicit links between the demobilisation crisis and the Miners' Charter; if labour was to take its place in a home fit for heroes, then they must engage in a fight with capital on this issue.

NMA officials were overwhelmed by the level of support won by their adversaries. In the Kirkby district, on 22 January the unofficial group led a strike of 3,000 miners over the non-payment of the minimum wage for abnormal places, and then won them to join the general movement. On Thursday 23 January, the pits around Sutton and the least militant district, the Leen Valley, were picketed out. 'Motorised' pickets from Mansfield also spread the strike to the Alfreton district, and the jurisdiction of the Derbyshire Miners Association, where around 7,000 came out.[64] At its height, over thirty pits in Notts and Derbyshire were on strike for the unofficial Charter.

On Thursday 23 January, there was a meeting of several thousand in Mansfield's Titchfield Park. Thomas Clarke spoke in favour of an all-out strike, saying that 'the miners had been chloroformed long enough, and they were now kicking over the traces'.[65] Speeches and resolutions show that the strikers were motivated by more than just the local implications of demobilisation. At its most ambitious it was an attempt to short-circuit the national negotiation processes, which was barely underway, and jumpstart a national strike. Only one week after the formulation of the MFGB programme at Southport, the Notts miners at Titchfield Park passed a resolution calling on Smillie to pull out the MFGB on an immediate national strike 'for the six hour day, five day week, and a proper living flat rate wage'.[66]

The following passage, taken from a report of the meeting, suggests that this was no idle resolution; the Notts miners felt that a national strike was within their grasp.

> After this [the above resolution], there was a cry from the crowd for action forthwith, and it was proposed that the Federation should be given forty-eight hours in which to call a national strike. This was passed with acclamation, and it was further resolved that if the Federation did not achieve victory in the time stated, they should call out the Triple Alliance.

---

63   *Nottingham Evening News*, 24 January 1919.

64   *Nottingham Evening News*, 24 January 1919; *Nottingham Guardian*, 25 January 1919; *Mansfield Reporter*, 31 January 1919.

65   *Mansfield and North Notts Advertiser*, 24 January 1919.

66   *Mansfield Reporter*, 24 January 1919.

One man proposed that the Alliance should be given a time limit of seven days, but on its being objected that this would not give sufficient time for so huge a strike to be organised, the period was extended to fourteen days and carried by a great majority.[67]

Another unofficial meeting in Sutton Market Place that afternoon passed an identical resolution.[68] With a total strike of Notts miners fast approaching, on Saturday 25 January the Bolsover Company performed a U-turn and agreed to reinstate the 25 men. The mass meeting called for Saturday morning to discuss the offer was remarkable for its size and militancy, and the detailed reports of it carried in the local press allow us a fascinating insight into the struggle between official and unofficial leaders in 1919.

The venue was to have been the Mansfield Market Place, but the huge numbers which turned up meant that it had to adjourn to a field outside the town, off the Chesterfield Road. By the time a large contingent from Sutton arrived, something in the region of 8,000 miners had gathered. When one considers that very large meetings were held simultaneously at Newstead, Kirkby, and Hucknall, the turn out is even more striking.[69] The officials offered the miners the owners' capitulation over the redundancies, and sought an unconditional return to work. The unofficial group, on the other hand, looked to maintain and build support for immediate action over the Charter demands.

The staging of the meeting symbolised the battle inside the NMA in 1919. The official and unofficial leaders drew up their drays on either side of the field, facing each other. Between them stood the mass of miners for whose allegiance they were competing. On the official platform stood Charles Bunfield, Jesse Farmilo (checkweigher and delegate at Sherwood colliery), and several other members of the NMA Council. Bunfield attempted to speak first, but was forced to give way when across the field Owen Ford and Bromley of the SLP began to address the assembly. The miners turned their way. 'They delivered strong speeches which suited the temper of the crowd'. Ford and Bromley attacked the officials for 'too much government from the top', and for failing to understand that the miners would no longer tolerate being 'a mere commodity on the labour market'. Ford described the strike for what it was – 'an uprising of the workers in which the spirit of revolt was abroad'.[70]

67   *Nottingham Evening News*, 23 January 1919.
68   *Nottinghamshire Free Press*, 24 January 1919.
69   *Mansfield Chronicle*, 30 January 1919.
70   *Mansfield Reporter*, 31 January 1919; *Mansfield and North Notts Advertiser*, 31 January 1919.

Farmilo and Bunfield weighed in once more, trying to pull the meeting back in their direction. Farmilo appealed for a return to work on the basis that the six-hour day was being dealt with by the Federation, but he was shouted down. Bunfield then made a speech noteworthy only for its tactical ineptitude. On the question of shovel loading, still a widely popular demand, he 'reminded the men that dirt and bind were not coal'. Turning to the Crown Farm dispute, he completed his isolation:

> He was glad to know that there had been an organisation to support the twenty-five men who had been given notice. He could not say unjustly given notice because of the fact that the workmen at the Mansfield colliery passed a resolution that the men coming out of the army should return to their old places and now the colliery had carried out the resolution of the workmen.[71]

Bunfield's words were drowned by the howls of the crowd.[72]

For over three hours the official and unofficial platforms fought for sway over the miners. At one point, in exasperation, the officials tried to provoke a retreat from the crowd, proposing a resolution for an immediate strike for the Charter. The proposal was hurriedly withdrawn when a majority voted in favour. Bunfield and his colleagues were utterly out of touch with the mood of the meeting, which from the outset leaned strongly towards the unofficial group. At one stage, the officials were left standing almost alone in the field as the whole gathering moved over to listen to Owen Ford.[73]

Strategic disagreements within the unofficial group, however, meant that it failed to fully capitalise on the situation. The SLP faction was in favour of proceeding with an immediate strike for the Charter, whilst in the ILP faction some were in favour of a temporary return to work so notices could be lodged for a strike for the Charter commencing on 11 February. Hesitancy and division in the ranks of the unofficial group meant that the officials were able to regain some influence; a compromise resolution was passed that notices be lodged immediately of an all-out strike from 25 February if the Charter had not been granted in full by then. The next day, a further four mass meetings held around the county endorsed this position.[74]

---

71   *Mansfield and North Notts Advertiser*, 31 January 1919.
72   *Mansfield Reporter*, 31 January 1919.
73   Ibid.
74   *Nottinghamshire Free Press*, 31 January 1919; *Mansfield and North Notts Advertiser*, 31 January 1919.

The return to work on Monday morning was cold comfort for the officials. Whilst they had prevented immediate action over the Charter, the price they had paid for doing so was the endorsement of the aims of the unofficial movement, and the advocacy of a strike for its demands. Arthur Thompson must have been speaking for many miners when he said:

> The unofficial movement was adopted by him because they must not measure by deeds but by results, and they had done more in the last four weeks than the official movement had done in years.[75]

As things stood, the traditionally moderate Notts coalfield was now the focus of a militant challenge to the MFGB leadership. Over the course of the next few weeks, Notts officials fought to extricate themselves from preventing the strike they had been forced to propose. In the two months following the Crown Farm strike, the tensions of the post-war period reached a peak. Delicate negotiations between the government and the MFGB were rendered precarious by the militant temper of the Notts miners. On the two weekends following the settlement of the Crown Farm dispute, nine well-attended mass meetings were organised across the county, around the theme of 'The Abolition of Mine Slavery', at which demands for action over the unofficial charter were again repeated.[76]

Frank Varley, who of all the district officials retained the closest contact with the rank and file, had little doubt that the militants were capable of relaunching strike action. No left-winger, Varley nonetheless moved under pressure to adopt militant positions at MFGB Conferences during these weeks. At the Special Conference on 12 and 13 February, called after the rejection of the government's initial offer of a Commission, he intervened in an argument over whether there should be a recommendation for a strike on the ballot papers which were to be sent out. Varley questioned whether they could afford the time to ballot:

> We have had some wonderful happenings in Notts. We had 20,000 men on strike for four days, arising out of a question which was originally bound up with the demobilisation proposals, and having satisfactorily settled that matter they drew up a programme. True, it may be said in districts like South Wales and elsewhere that it may be regarded as a mere incident

---

75   *Nottinghamshire Free Press*, 31 January 1919.

76   *Mansfield and North Notts Advertiser*, 31 January, 14 February 1919.

in the march of progress that 20,000 men struck for four days in Notts. We are usually looked upon as of a more phlegmatic turn of temperament, but it is an ominous sign of the times.

He went on to say that already they were finding it difficult to keep the men at work, and doubted whether they could prevent a strike before 15 March, the proposed date for the expiry of notices.[77]

At the MFGB Conference on 26 and 27 February, at which the offer of the Sankey Commission was accepted, Varley supported the resolution from South Wales which called for a rejection, saying: 'We can't go back and tell the men that we've accepted a Commission'. Again, at the Conference held on 21 March, the day after the Sankey Report was issued and accepted by the government, Varley seconded an amendment opposing the Executive's recommendation that the Conference should stand adjourned to allow negotiations to take place to seek further concessions on the report. He urged the delegates to reject the report altogether, and demand that the government implement the majority report signed by the miners' representatives.[78]

The Notts officials were so concerned by their loss of control that they invited Frank Hodges to speak to their members. Before a packed Grand Theatre in Mansfield on 23 February, Varley acknowledged the rupture inside the NMA, admitting that the officials had been forced to acknowledge the membership 'had lost all faith in them'. They asked Hodges to address the Notts rank and file to 'allay the suspicion that the NMA were out of line with the rest of the Federation'. Hodges made no mention of the offer of the Royal Commission that had been made to the MFGB Executive two days earlier. Instead he stressed the 'revolutionary nature' of the union's demands, and promised that he and Smillie would lead a national strike if the government did not back down.[79]

Two days later, the coalfield was quiet; the strike for the hours and wages demands of the unofficial Charter, scheduled to commence on 25 February, did not take place. However, an unofficial strike did occur two weeks later, whilst the Sankey Commission was still in session. Although over local issues, it raised the danger that if the Commission's interim report did not make sufficient concessions, the unofficial group might be capable of staging a repeat performance and escalate a strike from local to national demands. Its track record since New Year's Day, the numerous votes in mass meetings across

---

77   MFGB Special Conference 12–13 February 1919, p. 23.

78   MFGB Special Conference 26–27 February 1919, p. 21. MFGB Special Conference 21 March 1919, p. 27.

79   *Mansfield and North Notts Advertiser*, 28 February 1919.

the coalfield for nothing less than a 30-hour week and the £1 daywage, the positions taken up by Varley at MFGB Conferences, and Hodges's visit to Notts all suggest that this was within the realms of the possible. Had it transpired, the conciliation process might have been derailed, especially considering the strikes which were simultaneously underway in other areas against acceptance of the interim Report.[80] In fact the Notts miners returned to work on 1 April, having won their local demands. In the next section we will examine the multiplicity of factors, local and national, which prevented the realisation of this potential, and which allowed the MFGB's strategy to carry the day.

3   The Problem of Parochialism

The strikes over the butty system and the Crown Farm redundancies had revealed a compound fracturing of the NMA's lines of authority and command, affecting the organisation's ability to fulfil its function as the articulator of struggle at all three levels; Executive, Council, and lodge or branch. These lines only began to be repaired when the organisation made a concerted effort to address itself to the resolution of long-standing local grievances, an area in which the unofficial group had established clear leadership credentials.

It was only after the butty system strike that the Council took its first positive step. On 11 January, a Special Council Meeting voted to ballot the entire county on removing all forks and screens from the pits, and filling with the shovel without any reduction in the getting price. On 28 January, the Council met again, two days after the conclusion of the strike over demobilisation. This Council meeting was very different. In between times, delegate elections had taken place in the branches, many of which had voted for new personnel.

> This was the first meeting since the periodic delegate elections, and there was present a large number of newer and younger men who have been elected in the places of those who have served their respective districts, in some cases for many years.[81]

Sources suggest the hand of the unofficial group lay behind many of the changes. W.H. Holland complained in a letter to the *Mansfield Reporter* that the 'politico-revolutionary element' was to blame for his removal from the Coun-

---

80   See Chapters Five and Six below.
81   *Nottinghamshire Free Press*, 31 January 1919.

cil.[82] Holland was an ex vice-President of the NMA, and a butty. In this period, he established an organisation to resurrect the butty system, affiliated to the right-wing British Workers League. In 1926, this organisation was to provide George Spencer with a network around which to establish his Industrial Union, as well as a programme: separation from the MFGB, non-political unionism, freedom from strikes, and the reconsolidation of the butty system.[83]

The removal of men like Holland, and their replacement by younger delegates, made the new Council more responsive to the membership. Its first act was to hand in notices for the whole county on Wednesday 22 January to secure the abolition of fork loading, over which the ballot had revealed a large majority for strike. The Council also included the demands for a minimum wage of 8s3d for abnormal places, a uniform list for main road workers, and payment for the setting of 'benk bars', or roof support bars on the coalface.[84]

No ground was made when the two sides met for talks on 5 February, so they reconvened in London at the offices of Richard Redmayne.[85] He recommended that the notices should be suspended for a month, during which time all screens and forks would be withdrawn from the pits for a trial run. The month would be used to assess the cost incurred by shovel loading, and to settle the other matters in dispute. A ballot of the branches resulted in a large majority for acceptance of the scheme, and to suspend notices for a month.[86] Ten days later, Lloyd George made his offer of the Sankey Commission. On 25 February, the day upon which the adjourned strike for the 30-hour week and £1 daywage was due to recommence, work continued normally, without any significant protest. In these weeks the unofficial group was quiet. Whereas in January newspaper reports and union records were full of references to its meetings and activities, at this point there is scarcely a mention of it. It appears that the combination of the Council's decision to submit notices for a strike, and the subsequent removal of forks and screens from the pits, combined with the concessions made by the government to the MFGB in the form of the offer of equal representation on a Royal Commission, had restricted the space in which the militant miners could operate.

Despite the apparent calm which had descended on the coalfield, further trouble was not far away. The NMA Council had agreed to suspend notices until 10 March under Redmayne's scheme. Redmayne, however, was a key witness

---

82   *Mansfield Reporter*, 4 April 1919.

83   Griffin 1962, vol. 2, pp. 116–17; McIroy et al. 2009, p. 214.

84   NMA Council Meeting 28 January 1919.

85   Redmayne was Chief Inspector of Coal Mines and Advisor to the Coal Controller.

86   NMA Special Council Meeting 11 February 1919.

at the Sankey Commission, which had been in session since 3 March, and he had been unable to attend to the matters in dispute in Nottinghamshire. Consequently, he requested another suspension of the notices to allow him more time to fulfil his side of the bargain. The NMA met on 10 March and voted by 554 to 129 to accede to Redmayne's request, and suspended notices until 22 March, the date of expiry of the notices in all the coalfields over the MFGB's demands. The resolution continued that even if the national strike were averted, the Notts miners would strike for their own demands. The feeling of the majority on the Council was that a strike before then would compromise representatives on the Sankey Commission.[87]

They were mistaken in their reckoning that as the men had not struck on 25 February, partly in deference to national developments, neither would they strike now. On 11 March, the NMA officials were confident enough that the members would abide by the decision of the Council that they all travelled to London to see Redmayne.[88] On Wednesday 12 March, it became clear that the militancy of January had not dissipated. The Council's decision to suspend notices once again without consulting the membership shattered the fragile unity that had existed since the Council meeting of 28 January. Once again a gap opened up between the rank and file and the official organisation. Once again the unofficial group proved itself able and willing to fill that gap.

Walkouts took place at pits in the Sutton and Mansfield districts. On the morning of 13 March, lodge officials from Sutton convened a meeting at the King's Palace Theatre. Hundreds were unable to get in. Those who did were furious at the Council's decision. J. Percival, a Council delegate and local official, announced that there would be no strike pay and the meeting nearly broke up in chaos. Order was only restored when Ben Smith, Arthur Thompson and F. Deakin took the stage and endorsed the action, calling for pickets to pull out all the other pits in Notts, and demanded strike pay.[89]

By Friday evening, the strike had spread to include miners from Kirkby, Stanton Hill and Selston, as well as all the pits in the Sutton and Mansfield areas. On Saturday 15 March, a huge meeting took place in Sutton, on the waste ground near the town centre known as the Lammas, where 5,000 miners 'filled the hollow and topped the rubbish heap'.[90] Frank Varley spoke, neither condoning nor condemning the unofficial strike. Spencer, on the other hand, tried to 'justify the delegates having their own opinions'. Heckling grew louder

87   NMA Council Meeting 10 March 1919.
88   *Nottingham Evening News*, 12 March 1919.
89   *Nottinghamshire Free Press*, 14 March 1919.
90   *Nottinghamshire Free Press*, 21 March 1919.

until 'he took off his coat and said he was ready for a battle'. His speech was
drowned out when he demanded a return to work. The meeting instead passed
a resolution to hold branch meetings the next day, and mandate delegates for a
Special Council Meeting to overturn the decision of the previous Monday. On
Sunday 16 March, a similar resolution was passed at an unofficial mass meeting
in Mansfield.[91]

Belatedly, the NMA officials tried to rally support by targeting the more
moderate branches and inviting them to pass resolutions in support of their
handling of the dispute. By now the momentum of the strike was too great for
the officials to fight an effective rearguard action, and when the Council met
on Tuesday 18 March, it voted to rescind its earlier decision by 469 to 282. The
next day the strike was officially on, and 40,000 were out.[92]

The strike was a clear-cut victory for the miners, who returned to work on
1 April, having won all of their demands, except for payment for the setting
of benk bars, which they agreed to refer to the Coal Controller. The 8s3d for
abnormal places was conceded, as was a very favourable price list for the main
road workers. Shovel loading was to continue. During the build up to the strike,
the union had added an additional demand for a basic scale for clerks, and a
three-tier grading system was conceded by the owners. This was the first time
that miners and clerical staff had taken joint action in Notts, and one of the
reasons that an offer made earlier in the dispute was rejected was that it did not
include anything for the clerks, indicating the extent to which the hierarchies
and sectionalism that men like Holland had relied upon had broken down
inside the pits. The strike had lasted almost three weeks, and the NMA paid
out £ 62,246 in strike pay, the largest in its history at that time.[93]

After the Council had made the strike official, nothing of special note
occurred. In the Leen Valley, the miners used their leisure time 'to go for walks,
or exercise their whippets'.[94] There were no further clashes between the rank
and file and the NMA, as any offer was referred to the members for consideration before the Council made a decision. As Spencer said: 'The final arbiters
are to be the men themselves. They have shouted so much about mandates and
that sort of thing'.[95]

---

91   Ibid.; *Mansfield and North Notts Advertiser*, 21 March 1919.

92   NMA Special Council Meeting 18 March 1919.

93   NMA Accounts March and April 1919; *Nottingham Evening News*, 31 March 1919; NMA
      Council Meeting 31 March 1919; *Nottingham Guardian*, 4 April 1919.

94   *Nottingham Guardian*, 25 March 1919.

95   *Mansfield Reporter*, 28 March 1919.

Given what had gone before, the outstanding feature of the strike was its economism. The demands of this strike did not lend themselves to easy generalisation, in the way the two earlier ones had done. The butty system campaign automatically embraced wider issues; the demand for a day wage evolved out of the agitation, allowing the unofficial group to begin to raise its programme. It was an attack on the hierarchies in the pit, and on the Lib-Lab politics which had been sustained by it inside the NMA. The structural vulnerability of the NMA over this issue allowed the URC to lead a campaign which was as much against the anachronisms of the union as against the owners, and in which a rank and file direct actionist current emerged and coalesced. The Crown Farm redundancies exposed the danger of uncontrolled demobilisation to the union, confirming the warnings of the militants that behind the owners' veiled concern for the soldier-miners there lay a hidden agenda of victimisation and union busting. Once on strike, the extension of the demands to include the 30-hour week and £1 day were a logical step.

The demands in the third dispute, however, were purely economistic, and politically barren in the sense that they did not automatically raise issues that divided the unofficial group from the official union, or encourage a debate over the direction and strategy of the MFGB. This could only come from outside, through the explicit intervention of the unofficial group, and this it did not attempt. In fact, at first the reverse was the case, as the strike exhibited a parochialism which, far from challenging, the group actively nurtured. The officials and Council justified their decision to suspend notices on 10 March by arguing that as a constituent member of the MFGB, the national programme must take precedent over the lesser issues which affected Notts. A sectional strike at this stage would be a diversion from the priorities of hours, wages, and nationalisation. Bunfield gave the official line: 'However serious their programme was, the national programme was more serious than local grievances'.[96]

The Notts miners, however, were suspicious that the officials were simply using the national situation to indefinitely delay action over the local demands. 'The county programme looked like getting swallowed up in the Federation's programme. That was the reason for the present drastic action'.[97] This was an understandable suspicion. However, if the unofficial group had been consistent with its previous *modus operandi*, it would have led action over the local issues, whilst simultaneously attempting to introduce the wider questions contained in its Charter. In fact, they counterposed the two. Arthur Thompson, for

---

96   *Nottinghamshire Free Press*, 14 March 1919.
97   *Nottinghamshire Free Press*, 14 March 1919.

example, welcomed the strike on 13 March in Sutton: 'The spirit manifested by the men was a splendid thing. Their case had a priority of right over the Federation's programme'. When at the same meeting Bunfield attempted to raise the question of nationalisation, he was heckled with cries of 'We don't want to hear about that, this a local programme'.[98] The syndicalist emphasis on industrial action no-matter-what led the unofficial group to encourage localism. In so doing, however, they deprived themselves of any possibility of changing the basis of the strike when Sankey's report was issued on 20 March. By the time of the ballot on the Report, the bulk of the Notts miners were enjoying the best pay and conditions ever experienced in the coalfield. The concessions contained within the Report were widely viewed as another step forward in what had been a highly productive three months of union activity. The militants who advocated rejection of the Report now found themselves almost utterly isolated, as the Notts miners voted overwhelmingly, by 30,885 to 1,764, for acceptance.[99]

4   Direct Action in Nottinghamshire: An Assessment

For the direct actionists of the unofficial group, the hours and wages demands of the Charter were a means to an end, rather than an end in themselves. Owen Ford told the Chesterfield Road mass meeting that:

> He did not want to stop at the 20s per day idea ... If you get the six hours and 20s, see what possibilities there will be in the future. You will go on till you get the lot.[100]

It is clear that they felt their ultimate goal of a socialist society lay within their grasp. In a meeting called to discuss the situation in Russia, Goodall said:

> They were in the greatest period of history, and could see the dawn of civilisation. The sermon on the mount must be practised and brought down to earth.[101]

---

98   Ibid.
99   *Mansfield Reporter*, 25 July 1919.
100   *Mansfield and North Notts Advertiser*, 31 January 1919.
101   *Mansfield Reporter*, 25 July 1919.

NOTTINGHAMSHIRE 137

However, interviews with Notts miners in local newspapers tend to contradict the militants' belief that their workmates were moving towards revolutionary direct action. Many supported what one might call the minimum programme of reforms put forward by the unofficial group rather than the maximum one of revolution, and the massive vote for the Sankey Report tends to confirm this. Further confirmation is provided by the pattern of struggle in Notts in 1919; the unofficial group's success was contingent upon the inactivity and obstructiveness of the NMA leadership over local issues. Its influence decreased sharply whenever the latter responded in earnest to the grievances of the members.

Should we conclude from this that direct action in Notts was entirely parochial and economistic? If so, then it is hardly appropriate to use the evocative term 'direct action' at all, when the more commonplace 'strike' will suffice. The evidence suggests, however, that it was not. In the first place, where the union leadership did temporise or obstruct, the miners appear to have been enthusiastic in their support for the militants' alternative strategy, one which explicitly generalised from the local to the national. Secondly, many miners seem to have interpreted the Sankey Commission and the government's acceptance of its Report as a victory for direct action, as a climb-down by the government in the face of threatened workers' power.

In addition, there is considerable evidence that accompanying the industrial militancy of these months was a process of political radicalisation, in which the conventional division between politics and economics – the cornerstone of the British labour movement's reformism – was at least partially broken down. Union leaders acknowledged a dwindling in support for the parliamentary road. Varley complained that 'the men were rapidly losing what faith they had in political action'.[102] The Notts County Council elections in March 1919 were 'very quiet … It was impossible to get the people generally to take an interest in the contests, and the polls were lamentably low'.

Trade union leaders themselves clearly perceived that the industrial militancy contained a political challenge to the Labour Party. Varley, speaking at an Alfreton Labour Party meeting, said:

They were inseparably bound up with constitutional methods: if not, why did they need a Labour Party? They could not get away from this state of Parliament, and whilst direct action could be justified on industrial

---

102 *Nottinghamshire Free Press*, 24 January 1919.

questions, [I urge] the miners to ponder over the grave responsibilities which such action would entail in political questions.[103]

The Labour Party emerged as hegemonic from the post-war upheavals in the labour movement, but it faced a serious and popular challenge from the syndicalists, and the unofficial movement had a considerable impact upon the politics of the NMA. Comparison of the Council Minutes in 1919 and those of previous years demonstrate that subsequent to the delegate elections of January 1919 the Council swung significantly to the left. In the course of the year, it pressed for the removal of all troops from Russia, the abolition of conscription, and the repeal of the Defence of the Realm Act. It voted to take all means necessary to protect members who had been conscientious objectors. It twice voted to withdraw the NMA's investment in War stock, despite strong opposition from the Board of Trustees, in protest at government policy over Russia. It voted for the abolition of income tax for all workers, and for a coalfield wide strike and demonstration each year on 1 May.[104] There was mass support for action to pull the troops out of Russia; at the MFGB Annual Conference in July, Spencer reported that once again the miners were on the verge of striking and highlighted intervention in Russia, the economic blockade and conscription as the major factors behind the unrest. 'They are very strong on these three points'.[105] When the Yorkshire strike provided the catalyst for unofficial action in Nottinghamshire shortly afterwards, local newspapers reported that these were amongst the demands of the strikers.[106]

None of this amounts to support for the revolutionary ideas of the unofficial group. But the generalised radicalism is clear, and the fact that the mass of Nottinghamshire miners were prepared to follow a group of revolutionaries is significant. There is a demonstrative link between industrial militancy and political radicalism in the direct action movement here.

However, there is a problem in proceeding beyond generalities of this sort, and attempting to establish to what extent the unofficial leaders provided political leadership in the struggles of 1919, or characterising the political content of the direct action movement. The problem is not one of unavailability of resources, but rather the ambiguous and often contradictory political ideas of that leadership.

---

103 *Mansfield and North Notts Advertiser*, 5 July 1919.
104 NMA Council Minute Book 1919.
105 MFGB Minute Book 1919.
106 *Mansfield and North Notts Advertiser*, 25 July 1919.

As we have seen, the unofficial group functioned by way of an agreement to engage in joint action around a specific set of immediate goals, but beyond this consensus was impossible. The most coherent and easily identifiable set of ideas was provided by the SLP in the form of industrial unionism. They were tireless propagandists, constantly holding street meetings and distributing literature. One leaflet in particular, entitled 'Get ready for the revolution', led to at least one arrest under DORA in Mansfield, and was discussed in the House of Commons.[107] The main thrust of the SLP's message was that the trade union movement as it stood was incapable of fighting effectively:

> They needed to be organised thoroughly, and until they were they would have their strikes and get what they asked for, but each time they would shortly afterwards be in the same position as before. So long as that went on they were going to be wage slaves forever.[108]

The answer was a combination of militancy and organisation. Pamphlets like *Fifty Points on Industrial Unionism*, advocated tactics like the lightning strike, sit-ins and sabotage. The SLP members saw themselves as facilitators of industrial struggle rather than as a political leadership. In orthodox syndicalist fashion they consciously spurned political leadership in favour of organisational solutions. Owen Ford said: 'When the workers were organised in the International Industrial Union of Workers they would not be sending telegrams for somebody to lead them'.[109]

The SLP had an influence way beyond its numbers in 1919, but its rejection of leadership placed restrictions on how far that influence could go. The other identifiable grouping among the unofficial leadership was that associated with Herbert Booth's committee, and they were members of the ILP. In the context of a high level of industrial struggle in 1919, it was profoundly influenced by syndicalist and revolutionary ideas. Reports of speeches by Deakin, Norris, and Owen, for example, reveal that they attached a great deal of importance to industrial unionism.[110] Indeed most of the speeches made by unofficial leaders at the time are concerned with harnessing the industrial power of the miners, and are shot through with syndicalism. However, although they leaned towards revolutionary ideas, they retained a commitment to the Labour Party's political pro-

---

107   *Nottingham Guardian*, 19 August 1919. Daniel Lazarus was the man arrested.
108   *Mansfield and North Notts Advertiser*, 5 July 1919.
109   *Nottingham Guardian*, 23 July 1919.
110   See *Mansfield Reporter*, 5 January 1919, for good examples of speeches by Norris and Deakin, *Nottinghamshire Free Press*, 11 April 1919 for Owen.

ject. Norris was an ILP councillor, and in April 1919 Arthur Thompson, Andrew Clarke and Deakin stood as Labour candidates in the Urban District Council elections in Sutton, as did Ben Smith. Although there was a low poll, Labour did well, with Smith, Deakin and Thompson being elected amongst eight successful candidates. Labour's campaign was described as 'the first serious, organised attempt' in local elections in Nottinghamshire, and as 'a remarkable success'.[111]

The simultaneous advocacy of syndicalist ideas with an enduring commitment to Labour amounted to neither opportunism nor hypocrisy. What it suggests instead is that the unofficial group was not immune to the political confusion of the left, whose ideas were in a state of flux at this critical historical juncture. It was not until the summer of 1920 that the various strands coalesced or fell apart, with the formation of the CPGB and the refusal of the majority of the membership of the ILP to join it. In 1919, the leadership of the unofficial movement in Notts represented a political hybrid; its ideas straddled both reform and revolution. Some were out for 'revolution and even bloodshed', whilst others described themselves as peaceful revolutionaries.[112] Some, like those associated with the SLP, were convinced that parliament was useless, whilst others were unwilling to abandon it entirely. And some held contradictory ideas. Henry Hicken, who would later become leader of the Derbyshire miners, described himself as 'a revolutionist', and 'an industrial unionist', whilst concluding that '80% of the electorate were workers, and they could send to parliament 100% of their own class, who could alter the present state of things'.[113] Walter Owen, who described himself as 'a freelance, ready to fight any official in Notts or Derbyshire', often used a revolutionary vernacular in his speeches, speaking of the need to overthrow capitalism. Yet the hazy nature of his politics meant he could be swayed by the Sankey Commission, in the aftermath of which he declared, 'Industrial life and conditions could be altered only by evolution and not revolution'.[114]

The search for a political characterisation of direct action in Notts concludes that it was not underpinned by any coherent theory. One finds instead an ideological diversity bordering on formlessness, semi-developed ideas and contradictions borne along by the explosive militancy of the coalfield – hence the amorphous title of 'direct action'. Beyond this lowest common denominator the militants themselves could not go.

---

111   *Nottinghamshire Free Press*, 11 April 1919.
112   *Nottinghamshire Free Press*, 17 January 1919; Mansfield and North Notts Advertiser 4 April 1919.
113   *Nottinghamshire Free Press*, 17 July 1919.
114   *Nottinghamshire Free Press*, 28 March 1919.

The organisation of the unofficial group matched its political incoherence, divided as it was between supporters of the SLP and the ILP. Although they were often able to sustain an operational cohesion around immediate and definable goals, sometimes their differences were felt. There was not, for example, a cohesive intervention in the critical mass meeting in the field off the Chesterfield Road on 25 January and this allowed the officials to salvage something from the situation when all had seemed lost.

The priority of the militant miners was to improve the organisation of the trade union, rather than create a separate rank and file organisation. Although at the extreme end of its political spectrum it had revolutionary goals, the unofficial group was not a revolutionary body. In fact they were, in the tradition of *The Miner's Next Step*, a ginger group, albeit a very militant one. Norris, for example, said: 'It was time that the trade union movement was quickened up and made to represent the rank and file'. Goodall 'hoped there would be a gingering up of Notts officials to give sanction to the movement' for the unofficial charter. Norris again: 'The unofficial movement was out to quicken the pace, and improve the machinery of trade unionism'.[115] As such they saw no need for any formal organisation, revolutionary or otherwise. It was enough to formulate a set of demands around which to agitate, keep all the militants in touch with each other, and push for action wherever possible.

The militant miners stood at the head of a genuine mass rank and file movement in Notts in 1919. In the first three months of the year alone there were over half a million strike days. There existed a vibrant culture of rank and file-ism, which threw up its own leaders and agenda outside the structures of the NMA, and often eclipsing the latter. Unofficial mass meetings became a new feature of coalfield life, challenging the Executive and the Council for the position of governing body of the miners. Again and again the unofficial leaders were able to pack out the large venues; the Sutton Town Hall, and King's Palace Theatre, the Mansfield Victoria Hall and Grand Theatre, and the Market Places of both towns, these became the debating and organising centres of the movement, often usurping the authority of the Council Chamber at Basford.

The unofficial group achieved many things. Its victories in terms of local pay and conditions have been dealt with in depth. These helped to win it a mass base of support for its unofficial Charter, and posed a serious challenge to the officials. However, we have seen that at the high points of its influence in 1919, the aspirations of the unofficial leaders were almost limitless. They believed that the day of reckoning had arrived, that in the immediate future

---

115 *Nottinghamshire Free Press*, 31 January, 28 March, 4 April 1919.

they would have greatly increased pay and leisure hours, better housing, and workers' control of the mines. In short they felt that the final defeat of coal capitalism was on the agenda. In the sense that this was their ultimate goal in 1919, they failed. The confusion of their political ideas, the organisational flaccidity, and the parochialism, were some of the factors which ensured that they were incapable of fulfilling their most ambitious project.

CHAPTER 5

South Wales

They are proceeding calmly and with restraint. They are not rushing into 'Direct Action'. They are proceeding cautiously and constitutionally, looking first of all to Parliament to bring about reforms which are desired by the workers.[1]

The MFGB was going to ask the Prime Minister to put a Bill to Parliament reducing miners' hours from eight to six, and if this didn't happen, a day would be fixed upon which all miners would leave the mine after six hours. (Loud Applause) Do you realise what this means? It means that we are going to challenge the coalowners' right to control their own mines and it is going to be done.[2]

∵

1   The Heartland of Syndicalism

The predominance of coal in South Wales meant that changes wrought in the industry by the war were felt in the community as a whole more keenly than by any other mining community in Britain. Until 1914, Liberalism had shown a strength and resilience that suggested its hegemony would remain intact for a considerable time to come.[3] In fact hindsight allows us to see that its heyday had already passed in the years either side of the 1906 election victory. Profound changes had taken place in South Wales society. Mass immigration from England into the mining areas had diluted religious evangelism,[4] and the rising curve of industrial conflict had exposed the contradictions in the assumed cross-class community of the Progressive Alliance.

---

1 Vernon Hartshorn speaking about the South Wales miners, *South Wales News*, 18 January 1919.
2 Noah Ablett to Ton Pentre miners, *Western Mail*, 21 January 1919.
3 Morgan 1973.
4 Lewis 1980, pp. 15–16. Between 1901 and 1911 there was a 40 percent increase in the mining workforce, and by the latter date 24 percent of the adult male population, and 19 percent of the adult female population, were non-Welsh immigrants.

Nonetheless, this alliance continued to dominate politics, remaining 'the most significant political tradition in South Wales down to 1914'.[5] In the three General Elections of 1906 and 1910, Labour could win no more than four seats. Of these, three were held by Mabon, William Brace and John Williams, old school Lib-Lab miners' leaders who, following the 1908 affiliation vote, had taken the Labour whip without any noticeable alteration of their politics. The fourth, Keir Hardie, held the second seat in the double-barrelled Merthyr Tydfil constituency in these elections (and in the Khaki election of 1900), but he too owed his success to the fact that he was seen as a candidate who stood on the radical wing of the Progressive alliance, and was challenged by the Liberals on only one occasion. The three miners' leaders who stood as socialist candidates for Labour prior to 1914, James Winstone, Vernon Hartshorn and C.B. Stanton, all met with failure.[6]

The war accelerated and made more apparent the erosion of Liberalism. In particular, the perceived inequality of sacrifice by mine owners and mine workers, summed up in the charge of 'profiteering', elicited an increasingly militant response. The first and most spectacular episode was the coalfield strike of 1915, but the rising curve of pit and district level strikes from 1917 onwards were of equal importance in highlighting the class division in South Wales society. This militancy helped socialists to shake off what Hartshorn had described in 1910 as 'the deadweight of generations of Liberal tradition and prejudice'.[7] K.O. Morgan brings us up to date:

> At the conclusion of the First World War, the politics of Wales were on the verge of massive and dramatic changes. The ascendancy of the Liberal Party, the dominant feature of Welsh political history for two generations from the time of the 1868 general election down to the First World War, was to be shattered for ever.

As evidence, Morgan gives the results of the 1922 elections, in which Labour won all 14 coalfield seats.[8] However, both Morgan and Stead (in a less direct way), whilst convincingly arguing against an interpretation of the development of South Wales politics which sees in the pre-1914 period any proof of an inevitable rise of Labour, have a tendency to accept it as inevitable from

5   Stead 1973, pp. 329–53.
6   Stead 1973, pp. 332–41. Gregory 1968, pp. 119–44.
7   Stead 1973, p. 339.
8   Morgan 1988, p. 99. In fact, Morgan gives figures for the whole of Wales, in which Labour won 18 out of 36 seats.

1918. Thus the war is seen as the dividing line in a neat pre-1914 Liberal/post-1918 Labour dichotomy. However, an examination of South Wales society in the immediate post-war period causes question marks to be raised against this simple schematic.

A methodological approach in which general elections provide the main evidence for shifts in political opinions has some weaknesses. In the first place it encourages accounts which tend to play down the significance of the inroads made by the left into the power of the Lib-Labs in the SWMF leadership after the Cambrian Combine defeat in 1911. The left, which breached the dam of Lib-Labism in the SWMF, contained a strong syndicalist current. In addition to the ILPers Barker, Hartshorn and Winstone, the SWMF Executive contained by 1912 a number of URC supporters.[9] As there was no general election in the eight years following December 1910, this shift in industrial politics had no opportunity to register in electoral terms. A second and more important objection is that their electoral focus means that inevitably neither Stead nor Morgan fully takes account of the importance of syndicalism per se, a philosophy which, in its pure form, eschewed parliamentary action.

It is no accident that of all the coalfields, syndicalist ideas first found a following in South Wales, and the tradition which was established here, from *The Miners' Next Step* onwards, bore the hallmarks of the coalfield society which produced it. It was the striking physical configuration of the coalfield which instantly set it apart from all others in Britain, and which provided the basis for the unique place occupied by the SWMF in mining trade unionism.[10] Firstly, there was its sheer size. Fifty miles long and eighteen miles wide at its broadest point, it stretched from Pontypool in the east to Kidwellty in the west, spanning the whole of West Monmouthshire, most of Glamorgan, the southern part of Breconshire, and the greater part of East Carmarthenshire. Mining was by far the most important industry in South Wales, providing the foundations upon which the transport and metal industries had been built, and within the coalfield area it was, with few exceptions, the only source of employment. By our period it was the most intensely mined area in Britain, with over 200,000

---

9   These were Noah Ablett, Frank Hodges, Ted Gill, Noah Rees and Tom Smith. See Dai Smith 1993, p. 74. Bellamy and Saville 1972–84, vol. 1, pp. 37, 51–3, 150–2, 350; vol. 2, p. 1.

10   Several accounts of the South Wales mining industry stress this. For a contemporary account, see Commission of Enquiry into Industrial Unrest, No. 7 Division, Report of the Commissioners for Wales and Monmouthshire, 1917, (Cd. 8668), *passim*. See also the opening chapter of Francis and Smith 1980. For a recent account, see Gilbert 1992, pp. 55–139.

mine workers and their families crammed into the deep and narrow valleys
carved by the streams and rivers which made their way southwards from the
Brecon Beacons.

The supreme position of the mining industry provided the cornerstone for
South Wales syndicalism. In the first place, it gave the miners a strong sense
of their own industrial power, one heavily underscored by the victory over the
government in 1915. Secondly, it seemed to provide an ideal terrain for the class
struggle, a point made by the Industrial Unrest Commissioners and reiterated
by the *Merthyr Pioneer* in 1919:

> The class war has manifested itself more strongly in the coalfield (because
> of) the geographical concentration of the colliery worker into comparat-
> ively small and dense communities in the coal areas, and his employment
> in huge masses, making for a discipline and unity not possible in many
> industrial occupations.[11]

The rising curve of industrial conflict from around 1907 onwards was also due
to the difficult geological conditions of the coalfield, with its uneven seams
and frequent faulting. In the labour-intensive mines (there were almost as
many repairers as colliers), productivity was low and falling, placing enormous pressure both on the owners to hold down wages, and on the colliers to
increase piece rates.[12] The cold winds of industrial relations in the coalfield
filled the syndicalists' sails. Furthermore, struggles over the control of the production process were an integral part of industrial conflict. The Cambrian
Combine strike was, according to David Egan, at root an act of 'resistance
to an attempt by owners to seize control over their mines and break the
miners' customs and practices'.[13] This concern to maintain control over their
working places was an important factor in propelling miners to trade unionism:

> Consequently ... the SWMF became an institutionalisation of employees'
> attitudes to work and thereby took the form of organisations which were
> much more than means of improving wages and conditions of employ-
> ment.[14]

11   *Merthyr Pioneer*, 28 June 1919.
12   Williams 1973; Ablett 1915.
13   Egan 1986, p. 21.
14   David Egan 1988, p. 52.

The poverty of civic life in the valleys, bemoaned by the Industrial Unrest Commissioners as one factor behind the unrest, also pushed the mining communities towards creating their own social institutions and amenities, from the Miners' Institutes and their libraries, to rugby clubs and choral groups.[15] As D.K. Davies has observed, by the end of the first decade of the twentieth century:

> The structures of social activity throughout the coalfield had been formed mainly through the communal efforts of the miners themselves rather than being the product of paternalistic influence.[16]

There was, therefore, both at work and in the community 'a reliance upon and a growing consciousness of the need for organisations that could be controlled'.[17] Paramount amongst these was the local miners' lodges. As Liberalism and Lib-Labism declined, so too did the chapel, and in a world increasingly dominated by the SWMF, the local lodge became the dominant social institution.

> Embedded in the lives of the communities, the SWMF lodges were powerful instruments in generating and sustaining solidarity. The talk in pit, pub and home was not only of rugby, soccer and boxing but of lodge politics, of the miners battles ...[18]

An essay entitled *The Miners' Progress*, written by a CLC student in 1920, showed why many young militants after the war continued to view the SWMF as Ablett and Co. had done before it.

> Questions relating to rent, housing accommodation, local profiteering in food, clothing, etc; in short, all the circumstances of the miners' social life in addition to those purely economic and those exclusively pertaining to the mining organisation and industry, are discussed at Lodge meetings, and policies are there framed to meet the given situations ... In

---

15   Francis and Smith 1980, pp. 1, 8; Commission of Enquiry into Industrial Unrest, No. 7 Division, Report of the Commissioners for Wales and Monmouthshire, 1917, (Cd. 8668), *passim*.

16   Davies 1991.

17   Francis and Smith 1980, p. 1.

18   McIlroy et al. 2009, p. 142.

fact, the local Miners' Lodge ... is now a very powerful social organisation in its district.[19]

It was only to be expected that many would see their own trade union, the most powerful organisation both at work and in the community, as the most realistic and readily available mechanism for social advance. Parliament, by contrast, could seem remote and irrelevant. URC activists felt that if they could perfect Federation organisation, and wrestle control of it away from the moderate leaders, then they would have a powerful weapon that could not only defend miners' conditions, but also beat a path towards revolutionary trade unionism. By erasing the word 'conciliation' from the union lexicon, they could make incremental inroads into the profits and control of the owners, 'so erecting the new society within the lap of the old'.[20]

As a formula for social emancipation this was, for many, more in tune with the realities of coalfield life than the long parliamentary road to distant Westminster. In the first place, it was more practical; as Ablett was always fond of asking, 'Why cross the river to fill the bucket when you can fill it this side?'[21] Secondly, in a society in which the social and economic lives of its inhabitants took place in the shadow of the pithead, the democratisation of the colliery was a very compelling goal. Commenting on the government's post-war promises, Ablett said:

It was no use using brave words like 'liberty', 'freedom' and so on, without having their substance, and man must control the means by which he lived before he could have that substance.[22]

Returning to the objections raised earlier against Stead and Morgan, social and industrial conditions in the coalfield gave rise not only to the Labour Party, but also to a vibrant strain of home-grown syndicalism. At the end of the war this had, in a mono-industrial society dominated more than ever by the SWMF, both a well-established tradition stretching back to 1910–11, and an immediate

---

19   'The Miners' Progress: Five Years in the South Wales Coalfields', by 'A Labour College Student', printed as a supplement to the Workers' Dreadnought, 24 April 1920.

20   The quotation is from an article by Ablett in Plebs Magazine, vol. 9, no. 3, April 1917; TMNS, pp. 16–31; see also discussions of the pamphlet in Francis and Smith 1980, chapter 1; Egan 1978, pp. 64–80; Woodhouse 1969, chapter 2; Dai Smith 1993, pp. 72–3; Gilbert 1992, pp. 78–82.

21   Interview with D.J. Davies, 3 November 1972, SWML, DJD/16/6.

22   Merthyr Pioneer, 20 December 1919.

vitality which can, if we allow it, be obscured by the rise of Labour. There is no intention to marginalise Labour here; indeed, the fact that nine Labour MPs were elected in South Wales in the December 1918 general election is demonstrable evidence that its rise had already begun. The point is that this rise was contested. There was significant competition from influential activists to the left of the Labour Party who held that progress to a socialist society lay along the path of 'Direct Action'.

Peter Stead gave a salutary warning to those seeking an inevitable rise of Labour pre-1914: 'Surely it is a mistake in any period to pick winning sides long before the battle is resolved?'[23] The warning is equally applicable to 1919, when South Wales was still poised between two epochs. The old world of the Liberals was vanishing fast, but the new world had not yet taken shape. Undoubtedly labour, to all intents and purposes miners, would play the key role in moulding it, but on the threshold of the campaign for the Miners' Charter, there seemed no guarantees that those miners would be constrained by the gradualism of the Labour Party.

2   The South Wales Miners' Federation: Organisation, Leadership and Politics in 1919

The SWMF was, as its title suggests, 'a Federation within a Federation'.[24] Mining trade unionism had developed in the form of separate organisations within rather than across valleys, conforming to the geographical contours of the coalfield, and at the formation of the SWMF in 1898, these district unions had retained a high degree of autonomy. In this regard, little had changed over the intervening decades, and in our period the Federation was composed of nineteen constituent district organisations, each with its own funds, internal government structures of executives and conferences, and a full-time agent elected on a permanent basis by ballot vote of the membership. In the larger districts, or where the agent had become an MP, a full-time sub-agent was also elected, of which there were eight in 1919.[25]

The locus of power within the Federation thus lay to a large extent in the districts. These varied considerably in size and tradition, and the amount of

---

23   Stead 1973, p. 341.
24   Jevons 1915, p. 131.
25   Arnot 1967, pp. 74–6. Davies 1991, pp. 70–2. Adams 1988, pp. 283, 324. SWMF Minutes 1919–21.

power they wielded varied accordingly. Under the constitution, each district numbering 3,000 was entitled to elect one representative to the Executive Council, with an additional representative for every 6,000 additional members, an arrangement which guaranteed that the large towered over the small in the governing body of the SWMF. The giant Rhondda No. 1, for example, had six representatives on the EC, whilst tiny Blaina had only one. Although it was not an automatic right, the universal tradition was for districts to elect the agent as their SWMF EC representative, and their sub-agent where they were entitled to more than one place. Because of this, the EC was dominated by full-time officials. In 1919, of the 37 EC members, 27 were agents or sub-agents, four were full-time officers of the Federation, and only six were lay members, four of these coming from Rhondda No. 1. The federal structure of the SWMF constitution therefore entailed the creation of a very large union bureaucracy at the head of the organisation, which concentrated enormous power in its hands both at a district and a federal level.

As a counterweight to this there was an additional governing body in the shape of the SWMF delegate conference. This was, in theory at least, the sovereign body of the Federation, which had the power of decision-making upon the more important matters affecting the coalfield, including drawing up demands, sanctioning deals negotiated by the EC, or initiating strike ballots. The delegate conference also had responsibility for deciding on rule changes, and for electing the President, Vice-President, General Secretary and Treasurer, the four full-time Federation officials. According to the constitution, these were subject to annual re-election, although the custom was to allow the positions to go unchallenged. In instances where the EC was reluctant to call a delegate conference, it was often the case that the membership could exert pressure via the lodges and districts and force them to do so, and occasionally the bigger districts even threatened to withhold funds from the Federation unless the EC complied. Each lodge was permitted to elect two delegates to the SWMF conferences, and given that there were over 350 lodges at this time, these were important events in the coalfield calendar and opportunities for large numbers of the activists to come together.[26]

Whilst the delegate conference allowed some devolution of power to the lodge officials and the rank and file, the union nonetheless remained very top-heavy. This and the 'extreme federalism' of the SWMF were, according to the URC, the main barriers that stood in the way of it becoming the vehicle by which

---

26   South Wales Miners' Library (SWML) Interview with Will Coldrick, WC/106/22; *Merthyr Pioneer*, 1917–20; Arnot 1975, p. 13; SWMF Minutes 1919–21.

workers' control could be achieved.[27] Two concerted efforts were made, in 1912 and 1917, to bring about the constitutional remedies advocated by *The Miners' Next Step*, but both were unsuccessful.[28] By 1919, little or no headway had been made in transforming the character of the SWMF leadership beyond that of the 1911–12 period. Whilst at that time a sea change had taken place, the gains made by the left were not so sweeping as to prevent the survival of strong currents from the Lib-Lab era.

In the first place, William Brace, Tom Richards and Alfred Onions all survived the challenge and remained important figures within the SWMF leadership in 1919. Brace had returned to the MFGB EC in 1912, and had remained there ever since, except for three years during the war when he served in Coalition governments. In 1919, he was MP for Abertillery in his native West Monmouthshire, and President of the SWMF. Richard's career was also far from over. Whilst he had to wait until 1921 to be re-elected to the MFGB EC, he remained there until his death ten years later, serving as vice-President and President. In 1919, he was MP for Ebbw Vale and General Secretary of the SWMF. Onions was never returned to the MFGB EC, but he had retained his position as Treasurer of the SWMF and he was, from 1918, MP for Caerphilly. In 1919, therefore, three out of the four full-time officials of the SWMF were old Lib-Labs, who had been in their posts since the foundation of the Federation in 1898.[29] The other full-time official was James Winstone, who whilst remaining on the left throughout his life, was hardly a young militant in 1919, he too having been a founder member of the SWMF EC.[30] In the 1918 general election he had been defeated at Merthyr by the Coalition Liberal Sir Edgar Jones.

Furthermore, as Dai Smith has emphasised, while the breakthroughs of 1911–12 represented 'real incursions into power ... they were *not* a triumph of the extreme left'.[31] Whilst Hartshorn, Barker and Winstone were supporters of the Plebs League, and associated themselves with the mid-Rhondda militants led by Ablett during the Cambrian Combine dispute, their involvement on both these fronts had more to do with a desire to finish off 'Mabonism' than any commitment to syndicalist doctrine. Already by 1912, Hartshorn in particu-

---

27   Interview with R.P. Arnot, 6 March 1973, SWML, RPA/53/17.

28   Francis and Smith 1980, pp. 16–22; Davies 1991, pp. 70–2; Adams 1988, p. 307.

29   Bellamy and Saville 1972–7, vol. 1, pp. 51–3, 285–7, 259–60. In fact Brace had been vice-President from 1898 to 1912. Richards and Onions were General Secretary and Treasurer respectively throughout their careers from 1898 onwards.

30   Bellamy and Saville 1972–7, pp. 350–1. He had been the first EC member from Tredegar District, and agent for Eastern Valleys District from 1891.

31   D. Smith 1993, p. 74 emphasis in the original.

lar had emphatically declared his opposition to syndicalist aims and methods, and by 1919 he had become one of the leading opponents of the URC. During its agitation against the comb-out in 1918, Hartshorn had launched a strong attack on those Maesteg Lodges who had sent delegates to a URC conference.

> They are a set of nincompoops without the intelligence of a tom-tit ... There was no room for such a body in the Federation. It was causing disunity. The unofficial body had to be fought to the death and he was out to fight them.[32]

Whilst Barker and Winstone retained more cordial relations with the militants, neither of them showed any tolerance of unconstitutional trade union action, as their attitude to the unofficial agitation against the Sankey Report would show. K.O. Morgan has written of these ILPers that 'almost in spite of themselves they had become symbols of the official processes of collective bargaining, of constitutionalism, [and] social cohesion'.[33]

Both Dai Smith and Peter Stead also stress the strong elements of continuity in the values and priorities of the SWMF leadership, despite the personnel changes. According to Smith:

> One sort of leadership did not then replace another sort as a result of this pre-war activity. Old loyalties, the re-grouping of interests, the capacity of older leaders to shift their allegiances all blurred the outcome.[34]

It should be added that the younger leaders displayed the same capacity. Stead defines the situation as one in which older and younger elements merged in the SWMF leadership to create a new industrial consensus. The syndicalists who were from now onwards a regular feature of the EC were a part of this blend, along with the Lib-Labs and ILPers. Although there were tensions and hostilities, their positions as part of the SWMF bureaucracy meant that they

---

32   *Merthyr Pioneer*, 3 April 1918.
33   Morgan 1973, p. 302. In his desire to highlight the extent to which there was continuity in the SWMF leadership from the Lib-Lab period beyond 1910/11, Morgan overstates the case when he says they became symbols of the Progressive Alliance. The imperatives of the SWMF leadership after this date were not identical to those of the Progressive Alliance – indeed, the rise of the left within the union was due above all to a growing recognition that such an alliance was incompatible with effective trade unionism in changed conditions.
34   D. Smith 1993, p. 75.

operated within the constraints of the new consensus. To its dismay, the URC found that prior involvement in the unofficial movement did not inoculate its members against the moderating influences of trade union office. Already by 1914 they were working well within the parameters of the collective bargaining which *The Miners' Next Step* had vowed to destroy, co-operating with the EC in demanding higher price lists rather than rejecting them in their entirety and demanding a universal day-wage. This led to a blistering attack on Ablett, Rees and co. by Will Hay and C.L. Gibbons:

> They were pledged to abstain from supporting reactionary policies; they were to keep revolutionary policies and militant programmes to the fore; they were to force the EC to take action along lines laid down by the militant section in the coalfield. Have they done this? Unhesitatingly we answer 'No'. They have ceased to be revolutionary except in words.[35]

Although in theory the URC rejected all leadership, believing that the power held by leaders inevitably corrupted even the best of them, in practice many of its most able and well-known members ended up as office holders within the Federation.[36] By 1919 in fact, most of the original leaders of the URC were on the EC and moving to the right.[37] In an article entitled 'What has become of the Unofficial Reform Committee?', which appeared in the *Merthyr Pioneer* in July 1918, these men were chastised by Dolling and Watkins, who were part of the younger generation of militants:

> Today there are those in the socialist ranks who, having grown respectable and law-abiding, act the part of the puppy dogs of capitalism. These we expect to bark like any capitalist mongrel because we attempt to live up to the faith that is in us. But from you we expect better things.[38]

The contradiction between the theory and practice of the URC flowed from its general strategy which, as a series of articles in the *Merthyr Pioneer* in the spring

---

35 *South Wales Worker*, 13 June 1914, cited in Woodhouse 1969, p. 112.
36 See TMNS, p. 21. 'All leaders become corrupt, in spite of their own good intentions. No man was ever good enough, or strong enough, to have such power at his disposal, as real leadership implies'; see Woodhouse 1969, p. 147, for how the movement of Plebs Leaguers which eventually established the URC was known as the 'No Leader Movement'.
37 Woodhouse 1969, p. 147.
38 *Merthyr Pioneer*, 13 July 1918.

and summer of 1918 showed, had remained unchanged since *The Miners' Next Step*. Dolling and Watkins characterised the URC as:

> a Ginger Group, constantly attempting to galvanise the EC into life, and focussing their efforts in the direction of our programme, using any and every legitimate means that the circumstances seem to justify.[39]

The whole strategy of the URC was to work inside the existing structures of the SWMF, notably within the lodge and the delegate conferences, in order to pressurise the EC from below, rather than, in the manner of the engineering shop stewards, to develop an alternative leadership structure that could directly compete with it. These different types of rank and file movements reflected the different conditions in which the South Wales miners and the Glasgow engineers operated. The latter were faced with a high degree of sectionalism and craftism, and union branches which were organised on a residential rather than a factory basis, in an urban sprawl where there was no automatic connection between the two. In this situation, the shop stewards found that they needed a structure which was based upon the workplace, and entirely independent of the competing official unions.

By contrast, in South Wales there was no great division between colliery, lodge, and community, and whilst sectionalism between different grades did exist, the vast majority of mineworkers were in the same union. Consequently, there were not the same pressures to seek independence from the official machine. In fact, it seemed far more realistic and practical to concentrate efforts upon reshaping the SWMF. Hence the word 'reform' in 'Unofficial Reform Committee', and the centrality of the idea that to transform social and economic conditions in the coalfield, one had merely to transform the union which bestrode it.

However, in face of its failures to secure significant changes to the SWMF constitution, and given its hostility to leadership per se, the URC was left making vague appeals to miners:

> The day has gone when workers should wait for rusty 'leaders' to give a 'lead'. The day has come when the rank and file should 'take the lead' by developing its own initiative.[40]

---

39   Ibid. At other times, the role which they ascribed to the URC was to 'police' or 'supervise' the EC members, and 'to urge them along when they display tendencies to loiter or impede the progress of the workers'; *Merthyr Pioneer*, 3 August 1918.
40   *Merthyr Pioneer*, 3 August 1918.

In this situation, the URC's practical role became the equally hazy one of acting as 'the educator and interpreter of the rank and file itself, giving lucidity and sequence to the aspirations of the inchoate mass of workers'.[41]

For the most active URC members, this was an ultimately unsatisfactory resolution of the problem of praxis. There was an almost irresistible logic, if control over the EC was the objective, for them to seek election to the EC itself. There was an additional pressure; these men tended to have passed through the CLC classes and courses, either in London as full-time students, or in the local classes that were run by Plebs Leaguers in the valleys. As a consequence, as well as being experienced trade unionists, they were often amongst the most articulate and well-educated men of their communities. In a society where a Social Darwinist idea of leadership was strong, they often perfectly fit the requirements. As Edmund Stonelake said, upon his return from the CLC in London: 'I had greatness thrust upon me'. This combination of factors meant that for most that had the chance, the gravitational pull of office was too strong to overcome.

3   The Re-Emergence of the Unofficial Reform Committee

The SWMF that emerged from the war was, then, in many respects identical to the one that went in to it, remaining federal and top-heavy. The inroads that the URC had made into the leadership had in fact turned out to be unwanted defections to the bureaucracy, which had successfully assimilated militants and moderates in a working partnership. Despite these obvious problems, the syndicalist left remained buoyant in 1919, optimistic that the SWMF could be captured for its purposes. In part, this was due to the one success of the 1917 Rules Conference, where an amendment had been passed that made it an objective of the SWMF to organise all miners in the coalfield 'with a view to the complete abolition of capitalism'.[42] In 1920, a SWSS supporter at the CLC wrote:

> Probably there is no other powerful union in the country the constitution of which is so revolutionary. And the inclusion of this programme in the rules is by no means mere window dressing. Nearly all the activities of the miners' organisation are influenced by the degree to which they bring the overthrow of capitalism nearer.[43]

41   *Merthyr Pioneer*, 13 July 1918.

42   Arnot 1975, p. 143.

43   Supplement to the *Workers Dreadnought*, 24 April 1920.

This rule change reassured the URC that the SWMF's form of organisation was well suited to syndicalist purposes. Will Hay, for example, wrote before the war that the 'student of industrial unionism' would find the SWMF 'the organisation most nearly approaching his ideal'.[44] Apart from the fact that the union organised the vast majority of the workers in the industry, this belief rested upon the power devolved to the lodge meeting and delegate conference. With a sufficiently well organised rank and file movement operating through these channels, the federal and bureaucratic distortions of the SWMF could be overcome. The 1915 strike was seen as a resounding vindication of this theory. In the aftermath of the strike, which took place against the almost total opposition of officials and EC, the URC's members and supporters emphasised how, by organising in the lodges, they had been able to capture a majority at the delegate conference of 12 July, and launch the strike. By the same means, coupled with the requirement in the constitution for a ballot before any deal could be agreed, the rank and file were able to retain control of the strike and prevent a sell-out.[45] It was this experience which led Noah Rees to claim that: 'In no other union is the opinion of the rank and file so fully felt and so quickly exercised and registered in any agitation'.[46]

The other ingredient needed for a successful repeat of the 1915 experience was a militant rank and file, and from 1917 onwards there was ample and growing evidence of this. Particularly encouraging for the URC was the fact that the industrial unrest was overwhelmingly unofficial. In his evidence to the IUC, submitted on behalf of the SWMF EC, Frank Hodges wrote that the inequality of sacrifice between the classes in the war effort, and the rising cost of living meant that the average miner:

> was losing his traditional respect for authority, even that of the institution nearest to him, his trade union, which appears incapable of rendering him any real assistance … He is now more readily inclined to enter upon unauthorised strikes both locally, at the separate collieries, and as a coalfield. It has been with the greatest difficulty that the men's leaders have restrained their men from showing their resentment by means of a strike.[47]

---

44   Cited in Davies 1991, p. 46.

45   See report of meeting chaired by Ablett, *Llais Llafur*, 9 October 1915.

46   *Justice*, 12 August 1915, cited in Adams 1988, p. 324.

47   Evidence of Frank Hodges on behalf of the South Wales Miners' Federation to the Industrial Unrest Commission, EC Box 7, Edgar L. Chappell Collection, National Library of Wales (hereafter NLW).

He concluded with a warning that unless 'some real measure of relief' was given, the industrial unrest would increase to a point where 'the men will be in open revolt'. Hodges had drawn attention to a definite and growing trend in the industrial relations of the coalfield. In its Trade Report and Review of the Year, the *Western Mail* complained that 1917 was probably 'the most troublesome ever experienced in the Welsh coal trade so far as labour questions are concerned', with 61 unofficial strikes between July and November.[48] The situation worsened over the next two years, with the number of recorded wildcat strikes climbing to 174 in 1918, and 242 in 1919, most of which took place between January and August. In 1921, following defeat in the lockout, the number plummeted to 25.[49]

The problem was that faced with a constantly rising cost of living, the Conciliation Board proved slow and cumbersome, and confidence in its ability to deliver real benefits reached an all-time low.[50] Miners turned in increasing numbers to pit and district level direct action methods to defend their living and working conditions. These were often highly effective. A series of strikes, and mob actions against stores suspected of hoarding food took place at towns and villages in several valleys from the summer of 1917, forcing the Food Controller to increase the supply of meat to South Wales in February 1918.[51] In the summer of 1918, the refusal of the Tredegar Iron and Coal Company to recognise the newly-formed Tredegar Combine prompted 'a locally led, and sustained, revolt that spread, through sympathetic action, to other valleys'.[52] At the high-point of the three week strike, 50,000 miners were involved in the unofficial action which forced the Coal Controller to intervene and compel the Company to meet the Combine officials, amongst whose number was, incidentally, the young Nye Bevan. It was, as Smith has noted, another victory 'won against all cautious officialdom'.[53]

---

48   *Western Mail*, 1 January 1918.

49   Monmouthshire and South Wales Coal-Owners Association Records (hereafter MSWCOA), NLW, Record of Stoppages (Without Notice), Prosecutions, Damages and Indemnities, vol. 1. The tonnage lost through such strikes was 282,713 in 1918, leaping up to 723,357 in 1919, before, again, falling away dramatically after 1921. *Western Mail*, Trade Report and Review of the Year 1919, 1 January 1920.

50   Evidence submitted by miners' organisations to the IUC in South Wales contains more references to the failure of the Conciliation Board than to any other factor as the major cause of the unrest. NLW, Edgar L. Chappell Collection, Box 7.

51   *Western Mail*, 25, 27 August 1917; *Western Mail*, 14, 19 January 1918; *Western Mail*, 6, 16 February 1918.

52   Dai Smith 1993, pp. 198–99.

53   Ibid.

The other most important unofficial agitation of 1918 began amongst the traffic men in the Afan valley.[54] In the face of the SWMF EC's refusal to call a delegate conference to discuss their long-standing wage demand, an unofficial committee was established here, and in the Anthracite and Western Districts, in the summer of 1918. By November, the committee had established itself in Maesteg, Aberdare and Dowlais, and it held an unofficial conference in Perth in order to link up with the URC in the Rhondda and West Monmouthshire valleys.[55] This grade of workers was responsible for a high proportion of the unrest towards the end of 1918 and into 1919, fighting a series of local battles whilst the SWMF leaders were concerning themselves with the developments surrounding the Miners' Charter.[56]

In January and February 1919, as the unrest reached epidemic proportions, the owners' representatives repeatedly threatened to make increased use of the courts to curb unofficial strikes, but were persuaded by Brace to allow the SWMF leaders an opportunity to reassert their authority over their members. In his regular column in the *Western Mail*, seen as the mouthpiece of the coalowners, he warned the miners that unofficial strikes spelled 'Death to Trade Unionism', and pleaded with them for a return to disciplined and well-ordered collective bargaining methods.[57] His exhortations had no discernible impact whatsoever, and shortly afterwards he told the owners that the leaders had run out of ideas, 'and think the only course is for the owners to proceed with prosecutions'. Significantly, the owners only felt confident to take such action on 14 occasions in 1919, almost all of these occurring in the second half of the year, after the campaign for the Charter had dwindled away. In 1920, as the level of militancy dropped, and the owners began to prepare their counter-offensive, the number rose to 43, reflecting an already changing balance of forces in the coalfield.[58]

The problem for the owners was that in the first half of 1919 the general level of militancy was such that any attempt to use the courts against miners was likely to backfire, and provoke more unrest. The prosecution of six hauliers from Tirydail colliery in the Amman valley in February was a case in point. These six were among 18 men, some of whom were ex-servicemen, who had stopped work to attend the funeral of a workmate, Private R. Jones, who had died of wounds sustained in France. The six were fined, and the Anthracite

54   This grade included hauliers, riders, trammers and shacklers, i.e. men engaged in the transport of coal along the roadways in the mines.
55   *Merthyr Pioneer*, 7 September and 9 November 1918.
56   MSWCOA, NLW, Record of Stoppages etc., vol. 1.
57   *Western Mail*, 12 February 1919.
58   MSWCOA, NLW, Joint Standing Disputes Committee Resumes, vol. 2, (1918–47).

Miners' District Association was informed by the owners' lawyer that were the miners to agree to a six month period with no unofficial strikes, the penalties would not be enforced. Against the strenuous opposition of the agents, J.D. Morgan and John James, the Tirydail men immediately struck, and over the weekend pulled out all 8,000 Amman valley miners, who then set up a strike committee with one delegate from each of the 21 pits involved. The strike lasted a week, and came very close to forcing a walkout of the whole Anthracite District. Whilst unsuccessful in forcing the owners to withdraw the fines, it was a highly effective warning of the miners' disdain for the legal sanctions of the owners.[59]

It was the industrial unrest, and its unofficial nature, which provided the basis for the re-grouping of the URC from the beginning of 1918, and its subsequent expansion beyond the Rhondda. In relation to this second point it is important to emphasise the geographical spread of the strike wave, which washed to all four corners of the coalfield throughout 1918 and the first seven or eight months of 1919.[60] *The Merthyr Pioneer*, which was closely associated with the reconsolidation and growth of the URC, carried an article in April 1918 celebrating the 'Renaissance' of the unofficial movement, and attributing it to:

> the growing tendency to break down the barriers of a 'constitutionalism' which is fitted only for the maintenance of present-day conditions. [This tendency] does more than spasmodically object to the present. It breaks open the ground in order to lay the foundation of the organisation of the near future.[61]

The URC had, since its inception, trained much of its fire upon the Conciliation Board, and by 1919 events seemed to be carrying the SWMF towards the policy of 'open hostility' which *The Miners' Next Step* had advocated.[62] The spread and scale of the unrest was such that established conciliation procedures were at risk of total breakdown, and neither appeals of union leaders nor legal sanc-

---

59   *Cambria Daily Leader*, 3 February 1919; *South Wales News*, 4 February 1919; *Llais Llafur*, 1, 8, 15, 22 February 1919.

60   The most serious and lengthy of the strikes in 1919 as a whole took place in the Dowlais, Ebbw Vale, Tredegar, Anthracite, and Eastern Valleys Districts, as well as in the Rhondda. All were unofficial. MSWCOA, NLW, Record of stoppages, etc., vol. 1 and JSDC Resumes, vol. 2.

61   *Merthyr Pioneer*, 20 April 1918.

62   TMNS pp. 17–19 and 27.

tions of the owners were able to restore them to working order. Significantly, it was during the sittings of the Sankey Commission that conciliation procedures were at their most ineffective, with wildcat strikes at pit, Combine and District levels reaching an all-time high. At the weekly meeting of the JSDC on 30 April, for example, there were no fewer than two dozen such strikes on the agenda.[63] We will return to some of these in the context of the campaign for the Miners' Charter.

The unofficial nature of the unrest, the declining authority of the Conciliation Board and, along with it, the SWMF leadership, were encouraging developments for the URC. Some other characteristics of the unrest were equally heartening. For instance, whilst there was an overall decline in the number of strikes over non-unionism (due to a government brokered agreement that all workers should be in a recognised trade union), there was a marked increase in insistence that miners join the SWMF.[64] In 1917, the main sectional threat to industrial unionism in the coalfield, the Enginemen and Stokers Association, had disbanded and joined the Federation, and in 1919 there were strikes at many pits to force members of other small, sectional unions to do the same.[65] In June, 11,000 Rhondda miners struck for a number of days in support of a demand from the clerks at several collieries that they be recognised as members of the SWMF.[66] Will Picton, interviewed by Francis and Smith in 1973, singled out 1919 as the year in which they achieved their goal of 100 percent membership of the Mardy SWMF lodge.[67] Districts reported the highest ever membership levels. In Aberdare, the union was 'better organised and united than it had ever been', with over 2,000 new recruits in the first six months of the year.[68] According to *The Miners' Progress*:

> Some idea of the progress made by the SWMF during the last few years is to be got by noticing the complete disappearance of non-unionism in the industry. At the present moment we can say that practically every man in the industry in South Wales is organised. Moreover, the SWMF embraces,

---

63   MSWCOA, NLW, JSDC Resumes, vol. 2, 30 April 1919.

64   MSWCOA, NLW, Notices to Terminate Contracts and Strikes on the Non-Unionist Question. The agreement referred to lasted from 18 April 1916 until 14 July 1921.

65   Adams 1988, p. 279.

66   *South Wales Echo*, 19, 21, 26 June 1919. MSWCOA Minutes of Special Meeting, 25 June 1919. The clerks got as far as making a draft agreement with the SWMF EC, before the owners undercut the move by recognising the National Union of Clerks.

67   SWML, Interview with Will Picton 18 May 1973, W.Pi/3.

68   *Aberdare Leader*, 19 July 1919.

with some few exceptions which must speedily disappear, the whole of these men.[69]

This was one of the conditions 'which are necessarily required for the control of the industry by the miners'. As we have seen, struggles over the control of the production process had been, since the turn of the century, a feature of trade unionism in the South Wales coalfield, and had formed an elemental building block in the development of the URC's strategy of encroaching control. Such struggles were so prominent in the unrest at the end of the war as to encourage a belief amongst those associated with the URC that the strategy was becoming a practical reality. 'The Miners' Progress' argued that the degree of control that the miners held in the pits in this period meant that if ever a joint control scheme were introduced, it would only be legalising what in reality already existed. The pit committees attached to each lodge held considerable sway over questions of safety, hiring and firing and working practices, and even reserved the right to demand the removal of inefficient or unpopular managers. So effective was this control that 'Victimisation, which was once a very frequent and brutal practice of the coal owners, has been fought and, at least in its naked character, thrown aside'.[70]

Writing in the *Workers' Dreadnought* in March 1919, Frank Phippen made similar observations:

It is astonishing to think what extent the miners control their industry ... In most pits today, if a manager requires overtime labour ... he has to consult the Pit Committee and lay all the facts before them. Before any new arrangements in regard to the working of the mines are made, the men must be consulted with a view to having their permission, and if any official is disliked by the workers, steps are taken to effect his removal.[71]

When the *South Wales Echo* sent a reporter into the valleys to talk to miners about the causes of wildcat strikes in 1919, he was told that 'they nip in the bud any tendency on the part of the management to take advantage of the individual workmen',[72] and newspapers and coalfield records are littered with examples of strikes to defend men victimised for refusing to carry out instruc-

---

69   Supplement to the *Worker's Dreadnought*, 24 April 1920.
70   Ibid.
71   *Workers Dreadnought*, 22 March 1919.
72   *South Wales Echo*, 24 June 1919.

tions. Often these were cases of refusals to work overtime in the interests of getting soldier-miners into the pits.[73]

These were some of the main features of the industrial context in which the URC nourished old roots and sank fresh ones in the 1918–19 period. Furthermore, the unrest was accompanied by a process of political radicalisation within the mining communities which also assisted the URC revival. The unofficial nature of the unrest, as we have seen, was due to the unswerving commitment of the SWMF leadership to the Conciliation Board. This commitment was underscored, until November 1918, by its concern to maintain the support of the Federation for the war effort, both in terms of maintaining production and providing the army with a source from which it could replenish lost fighting stock. The unrest thus confronted, implicitly at first but increasingly explicitly, the political as well as industrial priorities of the leadership, and led large numbers to question the assumptions upon which their support for the war effort had been based.[74] D.D. Evans, interviewed by Francis and Smith in 1972 remembered how:

the war itself, the shortages and everything else that occurred during the war, the things that took place had a tremendous effect upon the people. It was a period in which now the young men, or people that were prepared to shed old ideas, had opportunities of doing so. There were ideas floating around, ripe around everywhere see ... So I would say that the period of the First World War was a great dynamic period of thought in the working class communities.[75]

---

73   A few examples: at Raglan colliery in Garw District, there was a strike over a rider who had refused to carry out management instructions as it involved work which went beyond his lodge's definition of his job as 'roping, unroping and riding his journey'. At Nixon's Navigation collieries at Mountain Ash and Merthyr Vale in the Aberdare and Taff and Cynon Districts, all the men struck for over three weeks when two men were sacked for refusing to unload a tram of rubbish in their working place. At the Lewis Merthyr collieries in the Rhondda, 5,500 struck for over a week over the sacking of a roadman who had refused to return from the face to the main haulage road to do some repairs because he did not want to be subject to the temperature change which this would have involved. See *Western Mail*, 16, 28 July, and SWMF Minutes of Council Meeting 21 July 1919.

74   There are numerous references to this in the evidence submitted by miners' and other labour organisations to the IUC in 1917. Edmund Stonelake, for example, wrote: 'On all sides they see increased profits to the master, increased food prices to the workmen, with no corresponding increase in wages'; Edgar L. Chappell Collection, NLW, EC Box 7.

75   SWML, Interview with D.D. Evans, 5 December 1972.

From 1917/18, this enabled socialists to break out of the far-left ghetto to which the war had confined them and, for the first time in years, to reach a substantial audience. Until this point the URC, though still formally in existence, had been in reality little more than individual groupings of militants operating at lodge level, mainly in the Rhondda. The comb out of miners for the armed forces in the summer of 1917, and its extension in January 1918, provided the issue around which the unofficial movement began to reorganise in earnest. Prior to this, anti-war activity had been confined to the ILP's pacifist campaigning, but now the URC began an attempt to 'direct ILP influenced miners away from pacifism and towards industrial action against the war'.[76] From here on in, the *Merthyr Pioneer*, which had previously carried an ILP pacifist line, gave active support and encouragement to the URC militants, providing them with a much needed mouthpiece in the coalfield. Although they were unable to win enough support to force a strike against conscription, and despite considerable police harassment, they nonetheless made important organisational gains in these campaigns.[77] In the first quarter of 1918, the *Merthyr Pioneer* reported that the URC was holding regular meetings and establishing new branches, and two unofficial conferences open to delegates from across the coalfield were held in Cardiff. In April, the URC held its biggest conference, with nearly 200 delegates present. By now it was clear that, apart from the Rhondda, the URC had developed a sizeable base in the Aberdare and Monmouthshire Western valleys, and that there was a smaller, though real, organised URC presence in the Eastern valleys, Garw, East Glamorgan and Ogmore and Gilfach districts.[78] In June, *The Times* wrote:

From nearly every part of the South Wales coalfield evidence is accumulating of a new and almost unprecedented activity on the part of the labour extremists. The extension of the ramifications of the notorious 'URC', with its extreme propaganda of industrial unionism and syn-

---

76 Woodhouse 1969, p. 134.

77 The URC conferences in January and March were both raided by the police, and a number of leading URC members were prosecuted for their anti-war activities. A.J. Cook and George Dolling were imprisoned for three months for arguing that 'the only solution to the food problem was a revolution'; *Merthyr Pioneer*, 5 January, 30 March, 20 April 1918.

78 See *Western Mail*, 3 April 1918, for a report of this conference. See also Woodhouse 1969, p. 144. Otherwise, evidence for the spread of the URC in this period has been gleaned from the regular reports in the *Merthyr Pioneer*.

dicalism, the evidence of new discontent as seen in sporadic strikes ... are causing considerable misgivings in the minds of responsible persons.[79]

In addition to this, as we have seen, in the winter of 1918 the traffic men's unofficial organisation established a network of militants in contact with the URC in the west of the coalfield, in the Afan valley, Anthracite and Western Districts.

The growth and geographical spread of the URC was underpinned by the enormous expansion in the number of CLC classes in the last year of the war. R. Lewis's study of independent working class education in South Wales has shown how important these classes were, arguing that in this period they 'created the world view of a whole generation of leaders among the workers of the valleys'.[80] Moreover, these classes provided local nuclei for the re-emergent URC. Prior to 1914, 'CLC classes came very clearly to function as the local units of the unofficial movement', and this continued to be the case at the end of the war.[81] It is no coincidence that the main areas of strength of the URC in 1919, the Rhondda, Aberdare and Monmouthshire Western valleys Districts, were also the main areas of the CLC expansion.

There was one further factor that fuelled the reorganisation of the URC and helped to popularise the idea of direct action in South Wales. This was the Russian revolution, or revolutions, both of which had a considerable impact on the political life of the valleys. The February overthrow was universally welcomed. Nye Bevan told the 1951 Labour Party conference that he remembered miners 'rushing to meet each other in the streets with tears streaming down their cheeks, shaking hands and saying "at last it has happened"'.[82]

Of course, support for a democratic revolution against a foreign tyranny did not make one a revolutionary at home in a parliamentary democracy, as the misgivings of much of the British left to the methods of the October revolution would show. However, in South Wales the Bolshevik revolution was still widely supported, despite hysterical press reports of men buried alive at 'musical executions', the 'mass slaughter' of clerics, and the 'nationalisation of women'.[83] At the Rink in Merthyr, meetings about the revolution regularly attracted capa-

---

79   *The Times*, 10 June 1918.

80   Lewis 1980, p. 282 and *passim*, esp. pp. 207–93.

81   Woodhouse 1969, pp. 136–8.

82   Cited in Coates and Topham 1970, p. 95.

83   *Cambria Daily Leader*, 18 February 1919; *Llais Llafur*, 15 March 1919; *South Wales Echo*, 17 January 1919.

city audiences during 1919. At the end of February, for example, the veteran BSP propagandist Albert Ward defended the dictatorship of the proletariat to the 'loud cheers' of around 3,000 people.[84] A commemoration of the murdered German revolutionaries Rosa Luxemburg and Karl Liebknecht attracted a similar number.[85] Edward Soermus, a Russian entertainer and Bolshevik, whose performances combined violin recitals with speeches about the revolution, gained celebrity status in the mining communities before his arrest and deportation in February 1919.[86]

The 'Hands off Russia' campaign seems to have attracted wider support here than anywhere else in Britain, and on 21 July 1919, when the rest of the British labour movement was holding protest meetings over intervention, in excess of 50,000 South Wales miners took unofficial strike action for the day. The largest contingent came from the mid-Rhondda, but they were by no means unsupported, as collieries struck in the Tredegar, Anthracite, Western, Western valleys, Dowlais and Merthyr districts. At Clydach near Swansea, for example, all the collieries came out on strike, and in the evening, the Public Hall was densely packed with 'a most determined and enthusiastic audience, who loudly applauded the revolutionary speeches delivered by the local speakers'.[87]

As Francis and Smith have noted, enthusiasm for the Russian Revolution, industrial militancy, and hatred of conscription 'contributed to an internationalist perspective rooted in the direct and indirect experience of the South Wales miners'. In July 1921, this internationalism would lead the SWMF conference to vote for affiliation to the Red International of Labour Unions (RILU), 'one of the very few instances of a British trade union, of any size, indicating its adherence to "revolutionary" trade unionism'.[88]

As detailed information about the revolution was hard to come by, and many of Lenin's writings were not available in Britain until later in the 1920s, understanding of the Bolsheviks' politics was hazy at best. The tendency on the left was to take from what they knew of the revolution those aspects that seemed to best fit their experience as activists in South Wales. Inevitably, this

---

84   *Merthyr Pioneer*, 22 February 1919.
85   *Merthyr Pioneer*, 11 January 1919.
86   *Merthyr Pioneer*, 15 February 1919; Soermus was deported to Russia in February 1919.
87   *Merthyr Pioneer*, 26 July 1919; MSWCOA, NLW, Record of stoppages etc., vol. 1 lists all
      collieries which took strike action.
88   Francis and Smith 1980, pp. 29–30. In fact, the resolution at SWMF conference mistakenly
      referred to the RILU as the Third International.

involved considerable distortion. Some, for example, understood the soviets as analogous to Urban District Councils.[89]

For the syndicalists, the main attraction of the revolution was the fact that the workplace was the foundation stone of the new political and economic system. In the first place it was felt that if such a society were transplanted to South Wales, idle coalowners and royalty holders would be relegated to social positions which reflected their parasitic worthlessness. As S.O. Davies told a meeting of miners in December 1918:

> The working class dictatorship of Russia said that the man who did not work in a useful and legitimate manner should be ostracised so far as having any voice in the life of Russia was concerned.[90]

Secondly, and more importantly, the example of the industrial power of workers as the lynchpin both of the destruction of the old system and the construction of the new, struck obvious chords with the syndicalists. W.W. Craik, in his account of the CLC, reckoned that it was this example above all which influenced the post-war CLC students from South Wales.[91] Beyond a doubt it was also responsible for the rekindling of a revolutionary Noah Ablett, who saw Russia as 'the one burning political and industrial question of our day'.[92] In early 1919, he told miners at Maesteg:

> I rejoice in Bolshevism. I am proud of it. We can do the same in this country, and now that the war is over, I wish good luck to the German Bolshevist as well. Good luck to any of the working class where the germ of Bolshevism reigns.[93]

Ablett felt that the Bolshevik method was so applicable to South Wales that anyone who was against it 'required a doctor and an operation'.[94] Again, the detailed theories of Bolshevism were not well understood. The *Merthyr Pioneer's* description of it as 'a more democratic, more intelligent, more determined trade unionism than we have had in the past' was fairly typical.[95] For Ablett:

---

89   For example, Albert Ward at the Rink, *Merthyr Pioneer*, 22 February 1919.

90   *Merthyr Pioneer*, 26 December 1918.

91   Craik 1969, p. 122.

92   *Merthyr Pioneer*, 5 July 1919.

93   Cited in Davies 1991, p. 189.

94   *South Wales News*, 22 July 1919.

95   *Merthyr Pioneer*, 16 August 1919.

A Bolshevik is a scientific trade unionist – one who has recognised that the method of production today has gone beyond political government and must be based on the new idea of common work. Bolshevism is socialism with working clothes on. Bolshevism is the system of the future, and the scientific development of the working class constitution.[96]

W.H. Mainwaring expanded on this theme. He too was 'proud to be a Bolshevik', the meaning of which was:

that the workers of the country, organised upon the basis of the industries they worked in, should take full control and administration of such industries in the interests of the community.[97]

In 1919, before the formation of the CPGB began to redefine what being a Bolshevik meant, syndicalists were able to claim the Revolution as a vindication of their own ideas, as the successful application of the basic method of *The Miner's Next Step* on a national, multi-industrial scale. *Industrial Democracy for Miners* should be seen as an update, or development, of the 1912 pamphlet in the light of the Russian experiment, a bold assertion of what the militancy of the miners could achieve if properly channelled. The Revolution sharply increased the value of syndicalist coinage, and swelled the authority of the URC, especially when viewed from the vantage point of the industrial unrest with its elements of workers' control. Many of the leading lights of the URC in 1919 would go on to play important roles in the Minority Movement and the Communist Party in South Wales in the 1920s, though, reflecting the diversity of ideology within syndicalism, others took different trajectories. Ablett, for example, distanced himself from both the CP and the MM, W.H. Mainwaring became a CP member for a brief time before resigning when it passed him over for A.J. Cook as its candidate for MFGB president, whilst S.O. Davies saw no conflict between support for the CP and loyalty to the Labour Party. Communist Arthur Horner famously went on to lead both the SWMF and MFGB, while his colleague in the URC, Nye Bevan, gained even greater notoriety in the Labour Party.[98]

But all this lay in the future. In 1919, the nature and extent of the industrial unrest, the growing hostility to the war amongst a considerable section of

---

96   *Merthyr Pioneer*, 28 April 1919.
97   *Rhondda Leader*, 15 February 1919.
98   J. McIlroy et al. 2009, pp. 142–4.

miners, the expansion of the CLC, the Russian Revolution, and the 'ideological ferment' which accompanied all of these, provided the soil in which the URC could sink fresh roots.[99] Its resurgence began in 1917 and 1918, blossoming in February 1919 at the height of the industrial unrest to produce the South Wales Socialist Society, which reportedly had branches in every district of the Federation.[100] By the time of its formation, the SWSS activists had already popularised a brand of direct action with a strong revolutionary syndicalist subtext, one which was highly influential amongst the activist left.

Many ILPers in South Wales found the syndicalist case compelling. At the ILP Divisional Conference in February 1919, an amendment was passed which deemed it 'futile' to call on the government to socialise the means of production, and called instead for the 'absolute overthrow of the capitalist system'. According to the Pioneer, the speeches for the amendment 'espoused industrial unionism', and Morgan Jones of Bargoed protested that it was 'the reflex of the fervour of the Continent', warning that the door was being left open 'for the introduction of some new, unconstitutional methods into the activities of the ILP'.[101] Looking back over 1919, the Pioneer noted how:

> the intense industrial activity of the past year has ... augmented to a considerable extent the industrial faction of the ILP rank and file, who are drawing inspiration from Russia and the theoretical system underlying the Bolshevik faith and action.[102]

W. Harris, a political organiser for the SWMF whose task was to build up the Labour Party in South Wales, complained in July 1919 that 'the direct actionists seem to be increasing in leaps and bound ... My greatest difficulty as an organiser is in the districts where the direct actionists are strongest'.[103]

The growth of the SWSS, and the popularisation of direct action, was a serious concern for defenders of the status quo in South Wales. In the interstices between two epochs there was considerable anxiety that, given the waning influence of Liberalism and non-conformism, traditional mechanisms of social control might prove incapable of containing unrest within manageable bounds, and that 'extremists' might become dangerously influential. Major

99   The phrase comes from an interview with Will Coldrick, 24 September 1973, SWML, WC/106/9.
100   *Merthyr Pioneer*, 22 February 1919.
101   Ibid.
102   *Merthyr Pioneer*, 3 April 1920.
103   *South Wales News*, 28 July 1919.

James, head of Merthyr's Watch Committee, described his town in February 1919 as 'the headquarters of one of the most seditious parts of the country, where extremists, Bolsheviks, Pacifists and pro-Germans spoke ... almost every Sunday'.[104] Around the same time, the Home Office received a report from Captain Lionel Lindsay and the army authorities at Chester warning that:

> Ablett and his kind are undoubtedly gaining ground amongst the working classes of South Wales, especially in those districts that have the reputation of being socialistic and 'storm centres' of agitation, disputes, and stoppages of work.[105]

Lindsay concluded that 'these centres badly need some counterpropaganda', an opinion shared by Tom Jones.[106] It was with this in mind that Sir Edgar Jones, coalowner and Liberal MP for Merthyr, established in January 1919 the Welsh Democracy League. An essentially middle-class organisation, representing middle-class values and interests, its aim was to counteract 'the activities of revolutionary bodies and to pursue a vigorous campaign against doctrines and influences with Bolshevik and Syndicalist tendencies'.[107]

In particular, it was concerned about 'Bolshevik secret classes' (almost certainly a reference to CLC classes) which were 'spreading the gospel of direct action'.[108] C.B. Stanton, who before the war had been a prominent URC member, but had since become ultra-patriotic MP for Merthyr (1915), and Aberdare (1918), believed that propaganda was insufficient means with which to fight the far left. He demonstrated what K.O. Morgan has called his 'quasi-fascist' tendencies when he argued in July 1919, during a series of violent clashes in South Wales between supporters of his British Empire Union and supporters of the Russian Revolution, for the creation of a 'Patriotic Fighting Brigade', to 'put down the terrorism of these people'.[109]

Some of those who hankered after a return to the order and values of traditional Welsh society clearly felt embattled in the volatile atmosphere of the valleys at the end of the war. While some sought to mobilise middle-class and right-wing working class elements against the left, others sought refuge in unorthodox spiritualism as a revivalist fervour swept through a number of min-

---

104   *Western Mail*, 11 February 1919; *Merthyr Pioneer*, 15 February 1919.
105   Cited in Hopkins 1974, p. 137.
106   The Thomas Jones, C.H., Collection, Diary, NLW/Z/1919, p. 36.
107   *Merthyr Pioneer*, 15 February 1919.
108   *Western Mail*, 28 January 1919.
109   *Western Mail*, 22 July 1919.

ing communities in 1919. Newspapers reported 'remarkable scenes' at meetings which lasted all day and into the night, and the emotional outpourings at 'huge congregations' in Dowlais and Aberdare evoked comparisons with the Evan Roberts revival of 1904–5. Led by Pastor Stephen Jeffreys, himself personally converted by Roberts, there were frequent reports of congregations preparing themselves for the imminent second coming, of the divine healing of soldiers' wounds, and of other assorted miracles.[110] The revival of 1919, like most of those of the nineteenth century, took place in a period of 'acute tension and unrest'.[111] Although it did not reach the same level as the 1904–5 revival, there is a sense in which Francis and Smith's explanation of the latter as 'the expression of anguish' caused by the 'denial of a previous pattern of life' is applicable here.[112] The feverish evangelism that affected parts of the coalfield in 1919 was, perhaps, the last gasp of a rapidly vanishing society, anxious as to what might take its place.

The left, by contrast, was optimistic and ebullient, confident that the future lay with the working class. The meetings at Merthyr's Rink over the New Year period became, almost automatically it seems, celebrations of the miners' industrial power, and the disappointing election result seemed hardly to matter at all. James Winstone told a post-election rally:

> We are by no means dismayed. There is a force rising up in this country that all the forces of darkness will never stem. This crowd, this coalition crowd of capitalists and landlords and their henchmen will be snuffed out as a bit of snowflake before the sun.[113]

Against the groundswell of industrial militancy and political radicalism, Hartshorn's confident and frequent assertions of the hegemony of gradualism and parliamentarianism seemed almost hopelessly out of step. S.O. Davies's opinion that the 'trade union weapon was the only weapon to contend with the new government' seemed much more in tune with the mood in the coalfield.[114] Furthermore, as the showdown over the Miners' Charter approached, there was anticipation on both left and right that direct action might develop into some kind of revolutionary movement. S.O. Davies told a mass meeting in

---

110   *South Wales Echo*, 14 March 1919; *Aberdare Leader*, 13, 27 December 1919; *Western Mail*, 2 January 1920.
111   Hobsbawm 1973, p. 279.
112   Francis and Smith 1980, p. 7.
113   *Merthyr Pioneer*, 4 January 1919.
114   Ibid.

Merthyr on Christmas Day 1918 that the Russian Revolution was sweeping westwards, and that 'this country would be involved in a revolution no less thorough in the immediate future'.[115] Veteran agitator Tom Mann told the same meeting:

> Coming much more rapidly than I would even have dared to dream ten years ago are events of a drastic industrial and social character, to which the only term that is fittingly applicable is revolution. The revolution is right here. It is coming in this country. Of that I am profoundly convinced. Parliaments remained the same in every country – in the hands of the ruling classes. Trim your lamps, and be ready for the hour when you will be called upon to demonstrate your attitude to Parliament and industry.[116]

For the syndicalist left, at the height of its influence and for the first time boasting a coalfield-wide organisation, the prospects for transforming economic and social life by direct industrial action had never seemed so good. As D.K. Davies has observed:

> Indeed, one interpretation of the whole issue of syndicalism/industrial unionism in the South Wales coalfield might be that the pre-war period witnessed the laying of the theoretical groundwork, while events during and after the war seemed to offer the opportunity to test its validity in practice.[117]

4   Struggles over Sankey

For that section of the SWMF which saw direct action as the way forward in 1919, and for the SWSS in particular, the challenge was whether the energy of the widespread but often localised wildcat strikes could be harnessed and hitched to the wagon of the Miners Charter.

The first major test of strength between the unofficial movement and SWMF officials over the Miners' Charter came in the shape of the demobilisation crisis. This was of no less critical importance in South Wales than in other coalfields,

---

115   *Merthyr Pioneer*, 26 December 1918.
116   Ibid.
117   Davies 1991, p. 157.

and presented challenges and opportunities for both left and right.[118] On the one hand stood C.B. Stanton and the British Empire Union, who wanted the 'cowards and traitors' who had entered the mines after August 1914 to pay the price of the reabsorption of demobilised miners. On the other were the militant socialists for whom the only solution to the crisis was a new social order in the mines, in particular a six-hour day and nationalisation with workers' control.

Both left and right found constituencies of support within the Federation. However, it was the right-wing position that laid the basis for the scheme proposed by the Joint Demobilisation Committee (JDC), consisting of representatives of owners and miners. The scheme also proposed the reintroduction of the double shift system, which had long been scrapped in most of the coalfield.[119]

The left accused the SWMF leaders of collaborating with the owners in scapegoating the post-1914 men for the crisis, encouraging divisions amongst the workforce, and paving the way for the creation of a large pool of unemployed labour which would undermine union organisation. More generally, opposition derived from the fact that acceptance of the scheme would mean acceptance of working conditions far below pre-war standards. There was also widespread and vehement criticism, on safety grounds, of the proposed reintroduction of the double shift system.[120] The situation was further complicated by the case of returning soldier-miners who were being re-employed at lower grades than they had been on before they went to war, often as labourers, for as much as 10s a shift less than they could earn as a skilled miner.[121]

The situation quickly reached crisis point at pits all over the coalfield. The Coed Ely colliery at Tonyrefail was typical, and helps to illustrate a common problem. Several categories of miner were, under the agreement in operation,

---

118   SWCA, South Wales Joint Demobilisation Committee (SWJDC) Minutes 30 January 1919. The total number of South Wales miners who had gone into the armed forces was 67,182, a figure representing almost 30 percent of the workforce.

119   *Western Mail*, 21 January 1919; SWMF Conference 25 January 1919.

120   Boyns 1986–7, p. 155; SWCA, E242, SWMF Statistical Department, Accidents in Coalmines in 1919. The difficult geological conditions in the coalfield, and the dry and fiery nature of the coal, meant that injuries and deaths through falls and explosions were consistently higher in South Wales than elsewhere in Britain. In 1919 alone there were to be 268 deaths and over 25,000 injuries in the coalfield. Safety was therefore a supremely important issue, and a key standard by which any post-war industrial order would be judged. In these terms, double-shifting was anathema, for it did not allow the pit to cool and settle before being worked again, thus increasing the chances of falls and slides.

121   *Merthyr Pioneer*, 11 January 1919.

eligible to 'go on the coal', where the highest earnings could be made. These were pit-boys who had turned eighteen, returning soldiers who had been either hauliers or colliers, and the hauliers and colliers who had replaced those who had gone off to fight. At a mass meeting on 19 January, many argued that all ex-soldiers over eighteen should have the right to go on the coal. However, given the restricted number of jobs, it was completely impossible for all those who claimed eligibility to do so. The situation could easily have collapsed into sectional infighting, and the demand to get rid of the post-1914 men was indeed raised. However, urged on by a strong and well organised left, the meeting voted instead to scrap the existing agreement, and to adopt the principle of equal rights for all. To make this a workable policy, and free up enough places on the coalface, the meeting demanded the immediate introduction of the six-hour day.[122]

The Coed Ely response was part of an increasing trend in the coalfield, as miners sought a solution to the demobilisation crisis in the demands of the Miners' Charter. This trend had been set in motion by the Western Valleys District Council as early as 16 December 1918, when it had called for a special MFGB conference to start the fight for the six-hour day.[123] In the context of the demobilisation crisis, the demand for nationalisation with workers' control became a logical corollary of the demand for shorter hours, which many felt could not, on its own, provide a long-term solution. The extensive development of the mines and the opening up of new workings which alone could guarantee real job security for all miners could not be trusted to the owners.[124] It was this combination of shorter hours and nationalisation with workers' control which S.O. Davies had in mind when he told miners that demobilised men 'could only be brought back into the coalfield by nothing short of a revolution in the South Wales coalfield'.[125]

The JDC was soon complaining that men were refusing to work double shifts, and were refusing to identify the post-1914 men to whom notices should be given.[126] In many pits miners refused to work overtime, or operated a 'stint', to force management to take on more men.[127] By the beginning of Janu-

---

122   *Merthyr Pioneer*, 25 January 1919.

123   *Merthyr Pioneer*, 21 December 1918.

124   See for example the interview with Walter Lewis, Rhymney Valley agent, in *South Wales News*, 2 January 1919.

125   *Merthyr Pioneer*, 26 December 1918.

126   SWCA, SWJDC Minutes 30 January 1919.

127   MSWCOA, NLW, Joint Standing Disputes Committee Resumes, 12 March 1919, 3 June 1919, 28 July 1919.

ary, as unemployment became more acute, and skilled men were increasingly forced into labourers' jobs, strikes began to break out. On New Year's Day, the Glamorgan Colliery at Llwynypia was shut down for the day in protest.[128] In other places, strikes occurred as owners fought to resist the incipient forms of workers' control by which miners were seeking to get more men into the pits.[129] On Monday 3 February, 3,000 miners at the Llanbradach Colliery in the Rhymney valley, who had earlier voted for the sacking of the post-1914 men, now reversed their position and struck for the employment of 200 ex-soldiers with no redundancies. At the Killan Colliery near Swansea, 700 miners also struck.[130]

Unofficial action of this kind at pit and Combine level soon made itself felt in the Districts, and pressure upon the SWMF leaders to pull out of the JDC's scheme grew. On 10 January, after meetings of all the lodges, the Monmouthshire Western Valleys District Council demanded a special SWMF conference, by the middle of January, to begin the fight for the six-hour day, or the whole District would strike. Matters were brought to a head by the miners in S.O. Davies' Dowlais District. There were two main grievances involved in the dispute. Firstly, workers at the washeries of Guest, Keen and Nettlefold's had joined the SWMF but were being refused surface worker's union rates. The most immediate grievance, however, was the dismissal of nine colliers to make way for returning soldiers. During the week ending 25 January, the Dowlais miners held a mass meeting, at which they insisted upon the reinstatement of the nine colliers as well as the employment of all ex-soldiers. The meeting pointed out to the employers 'the extreme state of disrepair' of the collieries, which alone required the employment of a large number of workers, and again, nationalisation with workers' control was agreed to be the only long-term solution.[131] It was resolved that unless the demands were met immediately there would be a strike. The following Monday the whole Dowlais district, numbering about 8,000 miners, carried out its threat. Within three days, the owners had given way to all the demands. The Pioneer heaped praise upon S.O. Davies for supporting the unconstitutional action, and reported that '[W]e have had a short,

---

128   *Western Mail,* 2 January 1919.

129   In late January, for example, thousands of miners at the Ocean Coal Company's collieries in the Rhondda struck over bonus payments which had been withheld when a ban on overtime had been enforced by the Combine Committee; *South Wales News,* 22 January 1919.

130   *South Wales News,* 4 February 1919.

131   *Merthyr Pioneer,* 25 January 1919; *Western Mail,* 23 January 1919.

sharp and incisive fight in the Dowlais District, in which the miners have won a decisive victory over the Company'.[132]

The Dowlais strike forced reluctant officials to convene a special conference of the Federation where the JDC scheme was overturned by 1,647 to 1,098. Nobody bothered to even suggest an amended version of the scheme in its place. Instead, the vote reflected the growing conviction that the only solution to the crisis lay with the struggle for the Miners' Charter. The mood was urgent; demobilisation meant too much was at stake to tolerate protracted negotiations over the Southport Programme drawn up at the previous week's MFGB conference, and it was resolved to communicate this to the MFGB EC. A resolution was carried naming the first week in March as the date when the talking must stop and the strike start.[133]

The demobilisation crisis in South Wales was critical in two respects; firstly, as in all the coalfields, it threw the issues facing the miners at the war's end into the sharpest focus, one in which the Charter became an immediate need, not a distant desire. In the second place, the manner in which the JDC's scheme was overturned once again seemed to confirm the soundness of the URC's method. In the opening skirmish of 1919 it had, as in 1915, captured the delegate conference and enforced its own policy on the SWMF against the will of the officials. In the weeks following the SWMF conference of 25 January, the pressure for an immediate fight for the Charter continued to build as the problems caused by demobilisation, and the accompanying industrial activity, continued to escalate. This pressure was reflected at the series of MFGB conferences which took place in February and March in the lead up to the Sankey Commission and the issuing of its Interim Report. The South Wales delegation was one of those that overturned the MFGB EC's decision not to recommend a national strike in the ballot, and several delegates later led the opposition from the conference floor against acceptance of the Sankey Commission. Even Hartshorn was initially doubtful about the wisdom of recommending acceptance of the Commission. In the days before the MFGB EC finally decided to do so, he said:

> This movement of miners had now reached a stage where it was very uncertain whether it was possible to defer the matter for ... even one day. The movement sprang from the miners themselves and not from their leaders.[134]

---

132   *Merthyr Pioneer*, 1 February 1919.

133   SWMF Special Conference 25 January 1919; *Western Mail*, 27 January 1919.

134   *Western Mail*, 22 February 1919.

Antagonism between the militant section of the rank and file and the more moderate officials was sharpened during this period by the unbending opposition of the latter to the idea of direct action. William Brace and Tom Richards participated in an almost hysterical campaign for a 'no' vote in the *Western Mail* which, in the days prior to the ballot on 19 February, was almost entirely given over to articles that prophesied economic and social ruin in the event of the Charter's triumph. In a dramatic breach of customary trade union discipline, Tom Richards, in his capacity as General Secretary, declared:

> Most emphatically I say they should vote against the strike. I consider that the decision of the conference in submitting such grave issues to a ballot without proper time being secured for their consideration to be altogether unwarranted. For this reason I have no hesitation in advising every miner who attaches any importance to my advice to vote against it.[135]

Brace's weekly column was titled simply 'No Support for a Strike'. On the day of the ballot, the newspaper carried the banner headline 'Leaders Oppose Strike' and claimed, mendaciously, that an SWMF EC majority supported this view.[136]

It is hard to measure the precise impact of this intervention by the two most senior SWMF officials. It may help to explain why the proportion of miners who voted for a strike policy was considerably lower than the national average in South Wales. An 80 percent turnout resulted in a majority of 117,302 to 38,261, a ratio of only three to one, compared to six to one nationally, ten to one in Yorkshire, and twelve to one in Notts.[137] This vote provides an important corrective to the tendency to characterise the coalfield as a straightforward 'cockpit of militancy'.[138] H.S. Jevons rightly differentiated between the militant younger miners, often without family responsibilities, and the older generation of 'temperate, industrious and thrifty' miners.[139] Richards and Brace obviously still had a significant bedrock of support amongst the generation of Lib-Labs from which they came. There were also large discrepancies between the Districts. Dowlais, where S.O. Davies had led a vigorous campaign, secured an eight to one majority, whilst Blaina and the Eastern valleys, where the more moder-

---

135   *Western Mail*, 19 February 1919.
136   Ibid.
137   MFGB Result of Ballot on National Strike 1919. The small size of the majority has been
        pointed out by Arnot 1975, p. 164.
138   Carter 1915, p. 455.
139   Jevons 1915, p. 125.

ate agents James Manning and A. Jenkins presided, only just got the two-thirds majority required.

By breaking ranks at such a critical moment, Brace and Richards no doubt widened the gap between themselves and all but the more moderate miners. Richards's action in particular was described as 'despicable' in a resolution passed by a 'monster demonstration' in the upper Rhondda, which called for his resignation, a call taken up by many lodges and several Districts.[140] More interesting is the response of officials and EC members. Even Noah Ablett, whilst denouncing Richards's actions as hypocritical, defended Richards against rank and file demands for his head.[141] Left and right joined forces on the SWMF Council in a unanimous resolution which rejected calls for Richards's resignation, because 'his services ... [were] better known to the Council than to the rank and file'.[142]

Brace saw the strike weapon per se as 'at best a terrible instrument', the prospective national strike as 'a far reaching disaster', and went so far as to claim that, despite the overwhelming majority in the MFGB ballot, 'it would be a misreading of the miners' desire to conclude that they desire to stop work'.[143] At the crucial MFGB conference which discussed the offer of the Sankey Commission, Brace, Richards and Onions did 'their utmost to secure the endorsement by the conference' of the government's offer. According to the *Western Mail*, they were decisive in swinging the Welsh delegation behind the MFGB EC.[144] Clearly, however, the vote within the delegation was a close run thing, as several South Wales delegates, including Ablett, S.O. Davies and rank and file SWSS members, angrily demanded that the EC act upon the strike mandate. S.O. Davies made their position clear:

> As long as I have a clear mandate, especially from the rank and file of the MFGB, I hold no allegiance to a conference of this kind, nor to the Executive, in submitting a recommendation before us in the face of the clear and definite mandate which the rank and file has given.[145]

140   NLW, Tom Jones, CH Collection, CS/55 Report no. 3189, 4 March 1919, by J.R. Murray.
       Western Mail, 25 and 26 February 1919. SWMF Council Meeting 5 April 1919. 'Several letters were received protesting against the action of the General Secretary in relation to the ballot for tendering notices and asking that he tender his resignation'.
141   *Merthyr Pioneer*, 2 March 1919.
142   SWMF Council Meeting 17 March 1919.
143   *Western Mail*, 22 and 25 February 1919.
144   *Western Mail*, 28 February 1919.
145   MFGB Conference 27 February 1919, p. 30.

Neither S.O. Davies nor the SWSS were able to initiate action against the Sankey Commission in South Wales, most miners preferring to await the outcome of the hearings. However, they did conspire to lead a highly important strike in Dowlais while the Commission sat in the House of Lords. The strike was a sequel to the January dispute. Having returned to work with the assurance that the washery workers would be recognised as SWMF members, the owners then reneged on the agreement. The whole Dowlais District again came out unofficially on 3 March, and *The Pioneer* reported:

> The story of broken faith on the part of the coal owners is one of the most sordid in recent industrial history. The turning down of the washery workers and the consequent betrayal of their promise to the men by the Conciliation Board raises the issue far above that of a small strike in one District.[146]

The SWSS mounted a 'Conciliation Board Must Go' campaign, with 'missionaries' from Dowlais and SWSS members distributing leaflets throughout the coalfield calling for a SWMF conference.[147] Despite his recent defence of Brace and Richards, Noah Ablett took an active part in the campaign. On 14 March, he told a mass meeting at Merthyr:

> He would take the officials and agents off the Executive except in so far as they acted in an advisory capacity, raise the lodges to the position of governors in the coalfield, meeting in monthly conferences, the agents to carry out instructions and the Executive to function as it should – to execute under command. The Dowlais strike was 'starting the business'. The miners were coming to a point where they must reshape their organisation.[148]

The Dowlais miners were still on strike when the MFGB conference met on 21 and 26 March, and voted to recommend acceptance of the Interim Sankey Report. The SWSS now looked to use the Dowlais strike as a platform from which to launch a wildcat action against the Sankey Report. On Sunday 23 March, following the MFGB's provisional acceptance of the Report, the SWSS held meetings in several Districts to drum up support for its strategy.[149] In their

---

146   *Merthyr Pioneer*, 8 March 1919.
147   *Merthyr Pioneer*, 22 March 1919.
148   Ibid.
149   *Workers' Dreadnought*, 5 April 1919.

strongholds in the Rhondda, large numbers turned up to vote for action, and on Monday morning many collieries in the Rhondda Fach came out on strike. After pithead meetings, the Lewis Merthyr pits at Trehafod in the Pontypridd area also struck despite the previous evening having voted 'unbounded confidence' in the leaders on the recommendation of the agent, Ben Davies.[150]

Considerable confusion surrounded the strike. The *South Wales Echo* reported that it was the beginning of an attempt at all-out action, whilst the *Western Mail* saw it as no more than a one-day protest stoppage. Concerned agents hurried back to the Rhondda from London, and made 'strenuous efforts to get the men to return to work, there being throughout the day a fear that the stoppage might spread to other parts of the coalfield'.[151]

At a mass meeting that evening in Perth, Will John and D. Watts Morgan, the Rhondda agents, struggled long and hard to persuade the men to observe the conference decision, but they voted to stay out at least until the reconvened conference on the Wednesday, in order to pressurise the MFGB leaders to reject the Sankey Report. The main protests were that there had been too much secrecy in negotiations, that the concessions won were too vague, and that the delay in starting the strike was playing into the hands of the government. Generally, however, the picture remained confused, as 2,000 men on the late shift at the Naval Colliery in Penygraig and two collieries at Treherbert in the Upper Rhondda voted at pithead meetings to return to work, and the agents reported that the men were calming down.

Tuesday saw a continuation of the strike, which spread within the Rhondda to Clydach Vale where SWSS members, with the help of colleagues from Lewis Merthyr, were able to transform an existing strike over a shortage of timber into one for the Miners' Charter. In the rest of the Rhondda, the strike remained patchy, with half a dozen afternoon shifts voting to descend the pits, but by the evening there were still about 20,000 on strike. With the action looking set to continue, the SWMF EC convened an emergency meeting in Cardiff, and passed a resolution which strongly condemned the strikers and called on them to work on day to day contracts until the MFGB's position on the final proposed terms of settlement was known.[152]

---

150   *Western Mail*, 24, 25 March 1919. The information with which the following account of the strike has been constructed is taken mainly from the *Western Mail*, the *South Wales News*, and the *South Wales Echo* during the week ending 29 March. Where other sources have been used, these will be identified.

151   *Western Mail*, 25 March 1919.

152   SWMF Special Council Meeting 25 March 1919.

Up until the MFGB conference the following day, the strike was largely confined to the powerbase of the SWSS in certain areas of the Rhondda. However, as news of the decision to recommend acceptance of the Report in a ballot came through, the strike took on 'alarming dimensions'.[153] Inside the Rhondda, over two-thirds of the collieries now joined the unofficial action, making the total number of miners on strike here about 30,000. More ominously for the officials, large numbers of miners outside the Rhondda began to follow suit. On the evening of 26 March, mass meetings at Abertillery and Crumlin in the Western valleys of Monmouthshire voted to strike for the full Charter. Town criers were cheered as they paraded the streets announcing the decision, and the following morning 17,000 miners came out. Simultaneously in the neighbouring Eastern valleys, 7,000 struck in the Pontypool area, to be joined by a further 4,000 from Varteg and Griffithstown the following morning. The *South Wales News* reported that Rhondda SWSS members (referred to as 'the Unofficial Party') had been in the Monmouthshire valleys since the first evening of the strike, 'to help their counterparts here to get action off the ground'.[154]

In Aberdare, where there were also SWSS bases, undisclosed thousands struck. In the Swansea, Neath and Port Talbot regions of the Anthracite, Western and Afan valley Districts, where the unofficial traffic men's committee had originated, 10,000 joined the growing strike wave. In Tredegar, Nye Bevan was amongst those members of the Combine Committee which pulled out 6,000 at Risca, Ynysddu and Nine Mile Point. Before the week was out, collieries in the Rhymney valley, East Glamorgan, Merthyr and Ebbw Vale Districts had also joined the movement, which at its peak involved from 75,000–85,000 miners, or half the SWMF membership. Only the smallest Districts of Maesteg, Ogmore and Gilfach, Garw and Blaina seem to have remained entirely unaffected.

The action was in many ways erratic and uneven in form, often exploding in one locale, fizzling out, then exploding again somewhere else. Some only came out to make a short protest, whilst others looked to strike until the Charter was conceded in its entirety. Spontaneity was a hallmark of the strike; in a sense it was an almost cathartic episode, a mechanism for venting the pent up frustration and emotion which had accumulated during the war, and which had been invested in the campaign for the Miners' Charter, which now seemed to be in such grave danger. However, alongside this spontaneity the hand of the SWSS was also visible. For a brief moment, the elemental anger and emotion expressed in the strikes became harnessed to a network of activists

---

153 *Monmouthshire Evening Post*, 27 March 1919.
154 *South Wales News*, 28 March 1919.

and once again, as in Scotland and Nottingham, the game of sabre rattling, negotiation and compromise being played between Lloyd George and Smillie was disrupted.

Having failed to nip the strike in the bud, the SWMF leadership was forced to call a delegate conference for Saturday 29 March, to allow an opportunity for the reconsideration of the MFGB's recommendation that the Sankey Report be accepted. Here was another opportunity for the SWSS to use its support in the rank and file to overturn the leadership. At mass meetings held in all Districts prior to the conference, its members argued that the strike should continue, and that the ballot should be boycotted, in order to undermine the compromise and push the MFGB into a national strike.[155] Unofficial conferences were held in several places in the days beforehand to ensure the selection of the maximum possible number of militant delegates.[156]

On the day, things went badly for the officials. Despite their misgivings, Winstone and Barker joined forces with their MFGB EC colleague Hartshorn to argue for acceptance of the Sankey Report in the interests of MFGB unity, but '[t]hese arguments seemed to have very little effect on much of the conference which was in no mood to compromise'.[157] There followed a strong challenge from the floor led by Nye Bevan of Tredegar who had been mandated to demand that no ballot should be held in South Wales whatsoever. After a struggle William Brace eventually managed to rule this amendment out of order, but he could do nothing to prevent the conference from voting, by 168 to 102, that the members be 'strongly recommended' to vote against the Sankey Report in the ballot.[158]

This was the high-water mark of the unofficial movement in the first stage of the campaign for the Miners' Charter. The MFGB's endorsement of the Sankey Report had succeeded in bringing the simmering suspicion of the rank and file to the boil, and had driven large numbers into open identification with the SWSS. At this point the MFGB was in danger of losing control of the most important area in the Federation. In the space of a week, the SWSS had managed to initiate a strike of over half the SWMF's membership and in so doing had created the atmosphere in which its revolutionaries, syndicalists and socialists could capture a conference majority against MFGB policy. In the process, the idea of direct action for the original Charter was resurrected at the eleventh

---

155   *Monmouthshire Evening Post*, 27 March 1919; *Western Mail*, 28 March 1919; *Western Mail*, 29 March 1919.

156   *South Wales News*, 8 April 1919.

157   *Western Mail*, 1 April 1919.

158   SWMF Special Conference 29 and 31 March 1919; *Western Mail*, 1 April 1919.

hour. For the unofficial leaders, much more was at stake than a shilling and an hour; the great industrial upheaval, the manifestation of working class power in Britain, was slipping away. In the opinion of Frank Phippen, writing in the *Workers' Dreadnought*, the concessions contained in the Interim Report were 'trivial details':

> The Miners' Charter is dissolved in a maze of small points ... Unless the rank and file take things into their own hands, the great Triple Alliance will subside as ignominiously as though it were a little union in a half-organised industry.[159]

Once again, however, as happened in other coalfields, at the very moment that a genuine challenge to the official strategy appeared to be gaining momentum, the forces behind the unofficial movement melted away. In overwhelming numbers the South Wales miners ignored the recommendation of the SWMF conference and voted to accept the Interim Report. In a large turnout the SWSS was swamped, even though it reportedly campaigned hard across the coalfield.[160] Every District of the Federation produced a majority for acceptance and in some collieries there were, remarkably enough, completely unanimous votes.[161] The final result, 142,558 to 19,429, was, according to the *Western Mail*, a record majority in the history of the SWMF.[162]

Given the size of the movement in the final week in March, and the fact that the sovereign body of the Federation had recommended rejection, this was a startling reversal of recent trends in the coalfield, and a shock for the militants. There is no doubt that prior to the ballot result, many from across the political spectrum took seriously the revolutionary predictions of the SWSS and its fellow travellers. The South Wales miners had been widely considered to be the shock troops of the direct action movement, who might undo the carefully constructed mediation of the MFGB leadership and drag miners and government onto the battlefield.

The fact that these same miners then abandoned the SWSS in droves, and voted by seven to one for compromise rather than confrontation, requires careful explanation. Arguably the most important factor – the political rhetoric surrounding the publication of the Interim Report of the Sankey Commission –

---

159   *Workers' Dreadnought*, 29 March 1919.

160   *Western Mail*, 12 April 1919.

161   *South Wales News*, 11 April 1919. The collieries in question were at Troedrhiwyair in Tredegar, and at Wernavon and Maesmelyn in the Afan valley.

162   MFGB EC 15 April 1919, p. 2; *Western Mail*, 12 April 1919.

is discussed in the concluding section to Part One. This made itself felt across all the coalfields. For the moment we shall remain in South Wales, to re-examine the nature of both official and unofficial leaderships in the light of the events of the early months of 1919, to suggest why the miners decided, when push came to shove, to follow the former and not the latter.

In the first place, the SWSS's ability to compete effectively with the SWMF machine was hampered by its own organisational shortcomings. These had plagued the unofficial movement since its foundation in 1910–11. In fact, it is not strictly accurate to classify the SWSS as a clearly defined organisation, with recognised rules, aims, constitution and membership. It was not a political party, but rather a network of union activists who could link up with each other when there was occasion to do so.[163] At times of increased industrial activity, local reform committees might take on a more structured form with weekly branch meetings, as happened in parts of the Rhondda and Aberdare valleys from 1918 onwards, but even here the resemblance was closer to the informal CLC discussion groups from which the URC had originated than to the branches of a permanent and stable organisation.[164] In 1920, a SWSS activist named Hewlett would explain to engineers at a Shop Stewards and Workers' Committee Movement conference that they had a misconception of the nature of the miners' unofficial movement:

> I know there is an idea abroad that South Wales is covered by a network of unofficial committees. This is not so. In fact, there is no permanent unofficial organisation in the coalfield. What does happen when it is necessary is that the advanced or rebel element does meet and discuss matters, arrive at decisions, then goes back to their respective pit committees and lodges, put their own views forward, have them thoroughly discussed, and if their opinions are accepted, the delegates to the Councils and conferences are instructed accordingly.[165]

This operational method did not require permanent, self-contained organisational structures, and indeed, it was often the case that such structures as there were lapsed completely. This point was made by Morgan Jones of Bargoed at the founding conference of the SWSS in February 1919, when he described the

---

163 Woodhouse 1969, *passim.*
164 See the regular reports in the *Merthyr Pioneer* in 1918, e.g. 16 March 1918 for the Aberdare URC, 9 November 1918 for the Tonyrefail branch in the Rhondda.
165 *The Worker*, 4 September 1920.

'tendency of the unofficial to die periodically for no discernible reason'.[166] If the aim was, as Dolling and Watkins said, 'to galvanise the Executive Council into life', then conversely, when the EC responded, the natural reaction would be to decrease the pressure.[167] The level of rank and file activity can, therefore, be gauged in inverse proportion to the level of official activity, so long as the latter was 'progressive'. So, for example, during the negotiations over the minimum wage in 1912 the URC stopped sales of *The Miners' Next Step* and agitation for union reform lest it disrupt the unity of the Federation, and it 'remained in a state of suspended animation' during the national strike.[168]

It seems that in early 1919, acceptance by the MFGB of much of the URC's Charter also induced, if not outright passivity, then at least complacency.[169] The timing of the foundation of the SWSS is a case in point. Although the steady escalation of unofficial activity since 1917 had done enough to make an impact on SWMF and MFGB policy, no effort was made to consolidate and build upon these successes by establishing a coalfield-wide organisation, or at least not until the MFGB was already locked into negotiations with the government. Only a month was to pass before this fledgling organisation was forced to square up to the combined organisational might of the SWMF and MFGB official structures. Competing with these called for a well-rooted independent opposition force capable of acting with cohesion in all areas of the coalfield. The unofficial movement, in the guise of the SWSS, did not allow itself sufficient time to approach this state of affairs before it was called upon to act.

Moreover, when the SWSS belatedly woke up to the fact that compromise had become official policy, its 'ginger group' status led it to pursue a flawed strategy. The strike it launched on 24 March was not an attempt to bypass the officials with independent rank and file action, rather it was conceived of as a 'police' action, designed to stiffen the resolve of the most militant agents and EC members, in the hope that they would stand up to the compromisers led by Brace, Richards, Onions and Hartshorn. 'Syndic' wrote in the *Workers' Dreadnought*:

> It is plainly seen that there are two elements among the leaders, one vacillating and anxious to compromise, wishing to be regarded as 'sane and reasonable', in a word politically-minded men who are affrighted of the wrath of the usual capitalist bogey – the public: and the modern men

166   *Merthyr Pioneer*, 22 February 1919.
167   *Merthyr Pioneer*, 13 July 1918.
168   Woodhouse 1969, p. 86.
169   Woodhouse 1969 p. 151.

who know we have to wring any advance on, or even retention of, our present status by our own organised industrial strength. It is to back this latter element, although it will probably incur their condemnation, which this sporadic strike has broken out.[170]

'Syndic' was undoubtedly correct in identifying a split inside the SWMF leadership over the Sankey Report. He was no less correct, however, to suggest that the left officials and agents would oppose the strike. The SWSS's intention was to exploit the split in the leadership, but by steering the rank and file behind the more militant (or less moderate) leaders, it ended up tailing the left wing of a body of officials and agents whose ultimate loyalty was to the official apparatus of which they were a part. Whenever this group perceived a threat from below during the campaign for the Miners' Charter, it tended to close ranks with the officials to its right. The Brace and Richards affair during the national ballot showed this to be true of even the most militant amongst them. The support of men like Ablett, S.O. Davies, Ted Williams and Noah Rees were crucial in protecting them when there was a real possibility that Richards in particular might be forced out.[171]

The rank and file movement against the Interim Report in March elicited a similar response. Winstone and Barker were prepared to argue against acceptance inside the MFGB EC, but having been defeated there they threw their efforts into halting the strike wave.[172] In fact, not a single member of the EC would go as far as supporting the strike, and the Special SWMF EC meeting on 25 March that condemned it was unanimous.[173] Even those agents most closely identified with the SWSS, Ablett and S.O. Davies, revealed that there was a limit beyond which their rank and file-ism would not go. They too campaigned hard against the strike in their own Districts, and would go no further than endorsing the SWMF conference recommendation to vote against acceptance of the Report in the ballot.[174]

In terms of the SWSS's strategy of making full use of the SWMF constitution, the capturing of the SWMF conference on 31 March should have been a great success. It was, in fact, a pyrrhic victory. In the first place, the conference decision meant that the SWSS abandoned its attempts to continue the strike,

170   *Workers' Dreadnought*, 29 March 1919.
171   The SWMF EC meeting which decided to drop the matter did so unanimously.
172   *Western Mail*, 27 March 1919; *Llais Llafur*, 5 April 1919.
173   SWMF EC Special Meeting 25 March 1919.
174   *Western Mail*, 5, 12 April 1919.

participation in which had actively united those who were opposed to a compromise. Once the strike was over, the opposition forces lost this focus, making
defeat in the ballot more likely. The sudden loss of momentum can be seen in
Merthyr, where Ablett subsequently admitted 'such was the spirit of discontent manifested' that he had had 'the greatest difficulty' in stopping the strike.
At the mass meeting which reluctantly agreed to work on day to day contracts
pending the ballot result, 3,000 miners had unanimously voted to reject the
Sankey Report.[175] Yet the intervening fortnight saw a dramatic change in attitude; although the minority against acceptance was proportionately bigger in
Merthyr than in any other district (except for S.O. Davies's Dowlais), they were
still out-voted by two to one.[176] In the light of this, the *Workers' Dreadnought*'s
explanation of the collapse of rank and file opposition to compromise is per-
suasive: 'after having marched to the top of the hill in sight of victory, the miners
were asked to retrace their steps, and this action caused chaos in the ranks, and
demobilisation resulted'.[177]

When one considers that in the period between the end of the strike and the
ballot, the railway and transport workers settled their disputes, and early balloting in other areas of the MFGB had already produced big majorities in favour
of the Sankey Report, the loss of momentum which the strike had provided
indeed seems an important factor. With the sense of forward motion gone, the
concessions contained in the Sankey Report appeared more attractive. Ablett
reckoned that in the run up to the ballot, 'a fear ... gained a grip that the back pay
promised from January 9th would be lost ... in the event of an adverse majority'.[178] Most agents played up these fears. Owen Powell from Aberdare said that
the gains made in the Sankey Report were:

> greater than anything the workmen had ever had before. If we embark
> on a strike and lose, we lose everything. Take and secure what has been
> granted and then fight for the rest of our programme again.[179]

Following the 'demobilisation' of the strike, the recommendation of the conference was swamped by a well-orchestrated campaign by the SWMF leadership
for acceptance of the Interim Report. Apart from Ablett and S.O. Davies, only

175   *Western Mail*, 12 April 1919.
176   *Western Mail*, 10 April 1919.
177   *Workers' Dreadnought*, 26 April 1919.
178   *Western Mail*, 12 April 1919.
179   *Western Mail*, 5 April 1919.

Ted Williams of Garw and Arthur Jenkins of the Eastern Valleys held to the conference mandate. The rest threw their energy and resources into preventing
South Wales from becoming a rogue area inside the MFGB. There was nothing
half-hearted or defensive about this campaign, even though it flew in the face of
what was supposedly the sovereign body of the Federation. Council members
toured the valleys warning of dire consequences if South Wales went out on
a limb, and many of them followed Brace and Richards's earlier example and
participated in the *Western Mail*'s campaign for acceptance. *The Mail* carried
full lists of all officials, agents, and many checkweighers and lodge secretaries
who were recommending a 'yes' vote.[180]

In order to appreciate why this campaign achieved such dramatic results,
it is vital to understand that the SWMF leaders were, at one and the same
time, leaders of the union and of the wider mining community. As Bill Paynter
succinctly put it:

> A single union operating in single-industry communities, this was the
> unique environment which moulded the 'Fed' into an exceptional kind of
> trade union. The branch chairman and secretaries were much more than
> representatives of miners dealing with problems of wages and conditions
> of employment ... They were village elders to whom the people went with
> their worries and woes. They were the guides, philosophers and friends
> to a community as well as the trade union leaders dealing with the pit
> boss.[181]

The extent to which this was true can be seen in the fact that of the 37 SWMF EC
members in 1919, seven were MPs, and another 14 were JPs.[182] As Roy Jenkins has
pointed out with respect to his father, Arthur Jenkins, agents not only ascended
vertically through the SWMF:

> but spread their activities horizontally into politics and other forms of
> public service so that they were very often well known figures with an
> influence well beyond the members of their union.[183]

By 1919, long before he became an MP, Arthur Jenkins was not only agent for
the Eastern Valleys, but was also a district and county councillor for the Labour

180   *Western Mail*, 5, 9 April 1919.
181   Cited in Gilbert 1992, p. 54.
182   SWMF Records 1919.
183   Jenkins 1991, p. 10.

Party. This 'tandem of miners' leader and local politician' also existed below the level of the agents; many a lodge secretary and checkweigher was also a councillor or representative on the local Board of Guardians.[184] Working as they did within the received institutions of South Walian society, these men stood overwhelmingly for incremental social advance. Given the revolutionary tone of much of the political discourse which accompanied the campaign for the Charter in South Wales, and the widespread rumours of army activity in the area, it should not come as a surprise that leaders of this type threw their weight behind the Sankey compromise.[185] When they chose to override the SWMF Conference decision, they did so not just as union leaders concerned to maintain the unity of the MFGB, but as the representatives and guardians of the whole community.

Under certain conditions the unofficial movement was able to function as an alternative leadership in industrial matters. The delegates to the SWMF conference on 31 March had been elected by relatively small numbers of miners who were, by and large, activists of some description or other. These were under the sway of the SWSS to a great extent. However, whilst being much more than a marginal force, the collective of SWSS members did not represent a community leadership in the same way that the official SWMF leadership did. *The Miners' Next Step* had pointed to ways of combating a moderate union bureaucracy within the structures of the SWMF, but its syndicalism meant that it provided no critique of the wider political role of that leadership in the community context. Whilst the SWSS could compete with the official SWMF leadership in the environment of the workplace and lodge, when it came to major political issues that concerned the future of the whole community, it was far less well equipped. When the working men of that community voted in March 1919, they followed the advice not of the syndicalist militants, but of the village and community leaders and elders who promised the security of steady progress without the risks of confrontation. By the time it became clear that much of this promise had been made in vain, the post-war militancy was already on the ebb.

---

184   Dai Smith 1993, p. 76.
185   *Workers' Dreadnought*, 29 March 1919.

CHAPTER 6

Selling Sankey

I think I need not remind this conference that this is a fateful day in the history of our country.[1]

There could be no doubt that not only a dangerous, but perhaps the most critical moment in the history of the country has been passed. There is no doubt as to what would have happened if an industrial dispute had taken place.[2]

∴

The acceptance of the Sankey Report by the miners was one of the most important moments in British industrial politics in the interwar years. If not yet an irreversible turning point in the fortunes of direct action, it was a crucial milestone along that road, and a critical victory in the MFGB leadership's quest for a post-war settlement without confrontation. Many historians have made the point that acceptance of the Report changed the industrial environment, but a focus upon events at the purely national level has tended to produce accounts that see the Sankey compromise as a relatively straightforward process.[3] The regional studies of militancy in the first quarter of 1919 reveal that the strategy pursued by MFGB officials was far more vulnerable to a challenge from below than has hitherto been recognised. In coalfield after coalfield, genuine rank and file direct action movements emerged, led by groups of syndicalists, revolutionaries and socialists, which threatened the drift towards conciliation, and made a national strike against the wishes of the EC a real possibility.

---

1 Smillie at MFGB Special Conference convened to discuss the Interim Sankey Report, 21 March 1919, p. 2.
2 Jimmy Thomas, following the decision of the reconvened MFGB Conference on 26 March to accept the Sankey Report, *The Times*, 28 March 1919.
3 See for example Clegg 1985, pp. 283, 286–7; Morgan 1979, pp. 62–4; Mowat 1955, pp. 30–6; Middlemas 1979, pp. 142–5; Arnot 1953, p. 201. Miliband 1964, pp. 66–7 acknowledges that the MFGB leaders had difficulty in getting delegates to accept the Sankey Commission, but in his chapter 'Parliamentarism versus Direct Action', the entire episode takes up only half a page.

High expectations generated by government promises of reward for years of wartime sacrifice and restraint clashed with the high cost and low standard of living to produce unprecedented militancy over local wages and conditions, militancy which conventional collective bargaining procedures proved unable to contain. When the element of demobilisation was introduced to this already volatile mixture, the result was often explosive. The unemployment (or employment on lowly grades) of thousands of returning soldier-miners was both a threat to union organisation, and an insult to men who had risked their lives for indifferent and ungrateful rulers. With 'homes fit for heroes' nowhere in sight, sporadic strikes, often over local grievances, were forged together and escalated. Although they remained coalfield rather than national affairs, under the influence of the unofficial pretenders they took on national demands, and represented serious attempts by sizeable sections of the rank and file to force an immediate struggle for the Miners' Charter.

For a variety of reasons, none of these revolts proved sustainable. Organisational shortcomings and ambiguous attitudes to the nature of the trade union bureaucracy, for example, have already been discussed in relation to the individual areas in question. Another common factor seems to have been the relatively restricted demographic prepared to give allegiance to the unofficial leaderships. The evidence suggests that they had most influence among younger, often single, miners, and those who were activists of some description, even if that only meant attendance at lodge and branch meetings. Older miners with families whose union activity was confined to paying subs and voting in ballots provide an ideal type of those less likely to follow the impetuous strategies of rank and file rebels, and they provided officials with an important anchorage.

In seeking out the weaknesses of the challenge from below, primacy must be given to the episodic and localised nature of the militancy. Although in all coalfields under study the strikes threw up common, national demands, they remained ultimately parochial affairs. In the first place, the ability of the unofficial leaderships to create and hold together movements for national demands was to a large degree determined and constrained by local factors. In Scotland, for example, the unofficial revolt came very early, its timing decided not autonomously by the Fife and Lanarkshire URCs with an eye to the situation elsewhere in the MFGB, but by workers in the engineering industry on the Clyde. Hopes that a general strike of Scottish workers over demobilisation would provide the catalyst for a movement south of the border did not materialise, and when the engineers' strike collapsed, the miners' strike collapsed along with it. By the time the unofficial movement in South Wales was at its peak two months later, the movement in Fife and Lanarkshire was in a trough.

In Nottinghamshire, the surge of militancy in the New Year saw the butty system, which had long buttressed the moderate leadership, collapse like a pack of cards, albeit temporarily. A small, poorly organised group of militants attained considerable respect for its part in this victory, and was able for a brief time to fill the vacuum left by the dislocated power structure in the NMA, leading miners in Notts and Derbyshire in a strike for the Miners' Charter. However, the election of a more combative and left-wing NMA Council which was more prepared to fight over local issues closed down the space in which the militants had operated. The generalisation of demands seen in the strike which originated in the Crown Farm demobilisation crisis was not repeated in the March strike over fork-loading.

The South Wales experience was also local and specific. The relative longevity of the syndicalist tradition in South Wales meant that here the perspectives of the direct actionists, both economic and political, were more ambitious than anywhere else. 'Revolution' was an accepted part of the discourse of industrial politics in South Wales in 1919, and could even set the terms of the debate, especially prior to the acceptance of the Interim Report. However, the mono-industrial society which gave rise to revolutionary syndicalism also produced by far the largest trade union bureaucracy of all the federated areas, one which was looked to by the wider mining community for political and social, as well as industrial, guidance. The mobilisation of this all-pervasive leadership proved too much for the SWSS. At the very moment when revolutionary trade unionism appeared most feasible, the SWMF showed itself to be ultimately a force for social stability and gradual change.

Arguably, in terms of the unofficial leaderships' goal of a national strike against the Sankey process, the differences between these regional direct action movements were not decisive. Common demands, especially that for the six-hour day, shows that there was a firm basis upon which parochialism might have been overcome. So too did common action. In the days following the MFGB's acceptance of the Sankey Report on 26 March, there were unofficial strike movements in several other coalfields aside from South Wales. In Derbyshire, 'conversations between miners on their way to work, and railwaymen at the Chesterfield stations led to an unofficial strike of about 8,000 miners ... against the MFGB's decision'.[4] Further south in the Midlands there was also a significant level of action; in North Staffordshire, Cannock Chase, and South Staffordshire, something in the region of 15,000 miners, 'influenced in

---

4   Williams 1962, p. 618.

their decision by the South Wales strike', also struck in protest.[5] So too did between 20–30,000 in the Sheffield area of the South Yorkshire coalfield.[6] Altogether, leaving out the strike which was underway in Nottinghamshire at this juncture, over 100,000 miners were involved in this brief but threatening outburst.

Had there existed a national organisation linking up the rank and file movements in the various coalfields, it is surely possible that the unofficial strikes of late March might have been forged into an effective assault on the compromise agreed by the MFGB. However, according to J.D. MacDougall, no more than 'tentative efforts' had been made to establish a National Miners' Reform Committee – 'save a few meetings of representatives, nothing came of it'.[7] Contact between the two strongest areas of the Reform Movement, Scotland and South Wales, was important in popularising *The Miners' Next Step*, and in rejuvenating the Committees at the war's end, but congress of this kind was occasional and informal, and involved little or no strategic co-ordination.[8] In March and April 1919, William Paul and John Maclean toured South Wales, 'to see if it was possible to link the rebels here with the rebels on the Clyde'.[9] However, in May, Walton Newbold, a future Communist MP, lamented:

> In recent industrial troubles what has been conspicuously and regrettably noticeable has been the lack of understanding and cooperation between the workers of one area and those in others. The greatest misapprehensions are to be found in South Wales about events on the Clyde, and on the Clyde about the intensity and purpose of industrial and political tendencies in South Wales.[10]

The unofficial movement was ill-equipped to weld local outbreaks of militancy into a force of sufficient power to decisively affect national developments. Just as there were no national organisation or leadership structures, nor was there a national newspaper or broadsheet. The comparative wealth of information about the SWSS in this period is due only to the support of the *Merthyr Pioneer*, which carried no information about events elsewhere. In Scotland,

---

5   *Staffordshire Sentinel,* 27 and 28 March 1919; *Staffordshire Chronicle,* 29 March 1919; *Dudley Chronicle,* 29 March 1919.

6   *Sheffield Independent,* 28 March 1919.

7   MacDougall 1927, p. 774.

8   MacDougall 1927, pp. 767, 774.

9   *Merthyr Pioneer,* 26 March 1919.

10   *Forward,* 10 May 1919.

The Worker contained occasional articles about the Fife and Lanarkshire URCs, but was to all intents and purposes the organ of the Clydeside shop stewards, whilst the unofficial group in Nottinghamshire was unsupported by any regular publication. The movements in each area were influenced or supported to varying degrees by revolutionary political groups and parties; the Socialist Labour Party and British Socialist Party in Scotland, the Workers' Socialist Federation in South Wales, and the SLP and Industrial Workers of the World in Nottinghamshire, but none of these constituted an effective national force which could bridge gaps left by the unofficial movement.

Prior to the establishment of the Minority Movement in the 1920s one cannot talk about a national unofficial movement at all. Rather there were a number of separate, area-based movements. The local nature of their activities did not preclude shared national or even international perspectives in 1919, but their organisational reach thwarted their more vaunted economic and political aspirations. All attempts to co-ordinate agitation and action stopped, by and large, at the coalfield boundaries. Whilst the national leadership was troubled at various times by loose cannons on the decks of the MFGB, it was able to lash fast each in turn, and remain at all times in overall control of the ship.

It is worth returning briefly to re-evaluate why Smillie and the EC chose to participate in the Sankey Commission. As we have attempted to establish, right from the start of the campaign they were fighting shy of a confrontation with the government. Two days after the meeting between the EC and Lloyd George, Sidney Webb visited Smillie to encourage him to accept the Prime Minister's offer. Smillie was 'depressed with a cold and the feeling of responsibility'. He personally was willing to go along with a Commission, but felt that the rank and file 'were out for a fight' and would brook no compromise.[11]

Smillie's anxiety was the product of a number of factors to which reference has already been made, not least experience of union fragility in Lanarkshire. He was sensitive to the dangers which a full-scale confrontation with the government held for stable organisation. More than this, however, he was alert to the dangers of state violence in a clash that must raise the question of power. Like Jimmy Thomas, he took seriously government threats to 'use all the resources of the state without the smallest hesitation … to win and to win quickly'.[12] In this situation, defeat for the miners and an employers' counter-

11   Cole (ed.) 1952, pp. 150–1.
12   Bonar Law, 113 HC Deb.5s, c.2348, 20 March 1919.

offensive might be accompanied by bloodshed and the destruction of working class communities. At a Triple Alliance conference later in the year, he said:

> There can be no doubt at all in the mind of any unbiased person that the Government were making preparations to crush out any movement initiated by the Transport Workers, by the National Union of Railwaymen, and by the Miners' Federation to ultimately use their trade unions in the interests of the members and that the Government were prepared to meet that by calling out the military to shoot down our people.[13]

Nye Bevan's suggestion that Smillie and the MFGB EC were at least as nervous of victory as defeat also appears plausible. After all, the size of the initial strike vote in the national ballot, and the nature of the militancy on show in coalfields of varying traditions, suggest victory was a strong possibility had the leadership sanctioned action. Certainly most of the labour movement expected it. Even as late as 22 March, *The Herald* felt that the miners would strike, and their victory would establish a beachhead for the rest of the working class.[14] Robert Williams was just one of many left leaders who felt that 'the hour of capitalist domination has struck. We are on the eve of the Proletarian Revolution'.[15]

Lloyd George was not alone in forecasting dire political consequences should the miners gain victory in battle in the political, social and economic conditions of 1919. His concerns were echoed with increasing urgency by key labour figures as the confrontation loomed closer. The Webbs pondered: 'Blockading the miners will be a difficult and dangerous task. The railwaymen and the transport workers might be drawn in; the army might refuse to act. And then?'[16] Macdonald felt that a strike for even fairly limited goals would quickly escalate; it would become 'difficult to keep discipline in the attacking army of Labour', and as the country polarised, the government would get reinforcements from the rest of society. 'Then both armies and issues will be completely changed, and what was a strike for an apparently definite and comparatively small change in reality becomes a revolution'.[17] In a widely reported speech at Watford in the lead up to the anticipated strike of the Triple Alliance, Jimmy Thomas appealed for reason in the name of fear:

---

13   MFGB Report of Proceedings at a Conference of the Industrial Triple Alliance 23 July 1919, p. 8.
14   *Herald*, 22 March 1919.
15   Ibid.
16   Mackenzie (ed.) 1984, p. 336.
17   MacDonald 1919, p. 77.

> Victory for either side may be purchased at too great a price ... The next three weeks might determine whether the country, having emerged successfully from a world war, could avoid an industrial dispute which, in its consequences, might be almost as dangerous as defeat by Germany. It was therefore essential that all classes should endeavour to understand the situation, and, without class prejudice or bias, should recognise that they had responsibilities which they could not escape.[18]

Miliband has observed that it was not weakness in the ranks that frightened the trade union leaders in 1919, it was strength:

> The militancy of their followers did not make them feel stronger: it struck them with apprehension. They held a formidable instrument in their hands. Much of their energy was devoted to persuading their members it should not be used.[19]

This was as true for left-wingers like Smillie as for right-wingers like Thomas. In the political context of the first quarter of 1919, none of the possible scenarios flowing from a national strike appeared appetising. Defeat would spell disaster. Victory might signal the beginning of an industrial uprising, to which not one of them would subscribe. Far from encouraging them, talk of revolution, and the miners' role in initiating it, unnerved them. Their aim was to bank the militancy of their members, and cash it in for concessions, not to unleash it in a bid for a decisive victory against the government. To say that the government 'sensed' the leaders' fear is a gross understatement. The leaders were shouting it at the government from the rooftops. It took no particular wizardry to understand that, should the government offer the miners' leaders a way out, they would take it. Lloyd George duly offered a Commission on terms favourable enough to the MFGB that Smillie might sell it to the membership and retain control of the union. The Sankey Commission was nothing more than a stage prop in a mutual act of escapology from a conflict which neither the MFGB leadership nor the government wanted.

For the syndicalists, the acceptance of the Sankey Report represented a calamitous defeat. To them the Miners' Charter was not a negotiable set of separate demands, but the Magna Carta of the direct action movement. Syd Jones of the SWSS wrote an article in *Solidarity* entitled 'The Miners' Charter in the Torture Chamber' in which he asked:

---

18   *Labour Leader*, 8 March 1919.
19   Miliband 1964, p. 65.

Whenever in the history of the class struggle had we clinched the argument so tightly? When was there an appeal which would attract the workers to the banner of freedom? When did our leaders beat the big drums with so much vehemence?[20]

With the Triple Alliance lined up behind the miners, the entire Sankey charade represented 'the shameful betrayal of the working class' and a victory for 'the counter-revolution'. 'The slaves are asked to fondle their chains, to remain and acquiesce in their slavery, and work on for the preservation of their masters' existence and luxurious ease'.[21]

However, the size of the ballot majority for acceptance of the Interim Report showed how marginal such a view had become inside the MFGB, despite the promise of the first quarter of 1919. Given the importance of the unofficial movements in the campaign for the Miners' Charter up to this point, their sudden loss of influence requires explanation. Organisational parochialism accounts for the failure to convert local direct action movements into a national one prior to the ballot, but cannot account for the small number of votes against the Report within each area of the Federation.

The most common explanation is that the revolutionary perspectives of the unofficial groups were entirely out of step with the bulk of the miners, for whom the goal was not revolution but substantial reform. The marginalisation of the unofficial groups at the end of March suggests that the support they had received until now was subject to serious qualifications and limitations. Participation in local direct action movements had been an indication not of revolutionary aspirations or convictions, but rather was the temporary transference of loyalty from procrastinating officials to the only other available leadership, in order to further the struggle for reforms. Once these had been secured in the guise of the Interim Report, the vast majority of miners switched their allegiance back to the officials, both locally and nationally. The revolutionaries in the direct action movement had been left high and dry, their grandiose plans shipwrecked on the rocks of the rank and file miners' mundane economic concerns.[22]

Whilst there are obvious truths contained here, this explanation fails to explain why and how the rank and file challenge melted away in March 1919. It seems feasible when viewed at the level of national politics, government and

---

20   *Solidarity*, April 1919 (the newspaper of the National Shop Stewards and Workers Committee Movement).

21   Ibid.

22   Woodhouse 1969, p. 166 tends to this view.

union leadership, but it does not square quite so easily with the picture that has been uncovered in the coalfields. Undoubtedly, participation in local direct action movements did not entail automatic identification with revolutionary syndicalism, yet neither was it symptomatic of a narrow and politically disinterested economism. There was beyond doubt a collective aspiration for some kind of radically new economic and social order amongst the miners, and a collective instinct that the occasion for its realisation had come, despite the fact that the size of the Coalition majority had slammed shut the parliamentary road. Higher wages and shorter hours were an important part of such an order, but a part nonetheless. The miners' vision of reconstruction, although often inchoate and ill-defined, also incorporated wider themes of social equality, unfettered democracy, social ownership and workers' control, full employment, international peace, and civil liberties. There were differences about both means and ends, but to the extent that many miners felt their aspirations could only be realised by the use of their industrial power, there was a widespread identification with direct action as a political philosophy. But direct action as an industrial strategy requires momentum and feeds on action. Deprived of these, it quickly starves. Failure to go forward meant that it could only go backward.

The acceptance of the Sankey Commission's Interim Report signalled the beginning of a collective belief that some route to post-war reconstruction other than direct action had been opened up. The scale of concessions contained in the Report itself were an important sweetener. Hartshorn persuasively told Maesteg miners:

> The minimum today was 76s9d per week (compared with 29s thirteen years ago) and the hauliers' wages had been increased from 25s3d per week to 85s9d today. In regard to hours they had secured a reduction which was equivalent to four months holiday with full pay. They had had £30,000,000 per annum added to their wages during the last few weeks, and a reduction in hours which were equivalent to £13,000,000 more. The profits of the owners declared for the whole country was only £39,000.000. The miners had £4,000,000 more than that, so there was no profit at all for the owners. (Laughter).[23]

These gains were impressive and unprecedented. For generations of trade unionists in the mines, this was the stuff of dreams. But whilst these gains were

23   *South Wales News*, 6 May 1919.

vital in inducing the miners to vote for acceptance of the Report, they cannot on their own explain the enthusiasm with which it was embraced. Given the elevated horizons of the miners in 1919, it is highly questionable whether 2s and an hour would have been enough, on their own, to allow the leadership to call off the national strike. The key to the success of the Sankey Commission was its presentation, by those opposed to direct action, as the vehicle by which the wider vision of a reconstructed post-war order could be peacefully – and relatively rapidly – attained. The economic gains of the Interim Report were offered as milestones on the road to this goal, and helped to secure a belief amongst the miners that the Sankey Commission represented a feasible political alternative to direct action. The public presentation of the Sankey Commission by politicians, union leaders and the press – especially the socialist press – was such that by the time it came to the ballot on the Report, only the most hardened sceptic could doubt that it was merely the *hors d'oeuvre* before the main feast.

The proceedings of the Sankey Commission themselves went a long way to dispel fears that it was simply a delaying mechanism. The miners' representatives knew they could not allow the Commission to become a sterile exercise in the amassing of dry statistics; rather it had to give the appearance of a historic head to head tussle with the coalowners. Those of a more questioning disposition might have seen it as little more than an elaborate piece of political theatre designed to draw the miners' attention away from direct action ... but politicians and press played ball.

Unlike the National Industrial Conference (NIC), which opened on the same day, the Sankey Commission was not tainted by the brush of Whitleyism and class collaboration. In his opening speech to the NIC, Lloyd George appealed for 'fair play for all classes' and co-operation between employers and union leaders who should see themselves as 'trustees for the whole country'.[24] By contrast, the Labour representatives on the Sankey Commission made no attempt to find common ground, and instead adopted an unashamedly partisan approach. Smillie's riposte to owners' complaints about the cost of the miners' demands was typical of their blunt and uncompromising attitude: 'We say that the landlords and the capitalists will have to do with less wealth produced in future, and the workers will have to receive more'.[25]

A well thought out division of labour on the miners' side enabled them to pursue the owners in a pincer movement. Smillie, Hodges and Smith stayed on home ground, and confronted the owners using their extensive knowledge

24   *Manchester Guardian*, 5 March 1919.
25   CIC 1919, vol. i, p. 80; Smillie to A.G. Hobson.

of the mining industry. The Labour intellectuals, Webb, Chiozza-Money and R.H. Tawney, intervened whenever possible to use the information extracted by the miners' officials to broaden the issue beyond wages and hours, into the realm of public policy and nationalisation. This method of cross-examination landed punches from the start. The first witness, Arthur Lowes-Dickinson, a chartered accountant with Price Waterhouse and adviser to the Coal Controller, was called upon to produce figures which revealed for the first time the full extent of wartime 'profiteering'. In the five years up to 1913, average profits after the deduction of royalties had been £13,000,000, a figure that had tripled to £39,000,000 by 1918. Closely questioned for over a day, by Webb and Chiozza-Money in particular, Lowes-Dickinson explained that the large increase in profits was due to wartime price increases which had been necessary to ensure production continued at less profitable collieries. He conceded during the course of his testimony that theoretically, in a unified industry with all profits pooled, such price increases would not have been necessary.[26]

The inference, which Webb and Chiozza-Money drew out, confirmed what a million miners had known for several years; the owners had taken advantage of the national crisis during the war to line their own pockets. In his *Facts from the Coal Commission*, Page Arnot wrote the following of Lowes-Dickinson's evidence:

> [t]he revelations disclosed by him of the profiteering in coal during the war caused an immediate revulsion of public feeling in favour of the miners and against the coalowners.[27]

The *Manchester Guardian*, which was not in favour of public ownership, described the evidence as 'sensational' and 'a striking illustration of the miners' case for nationalisation'.[28] The testimony of Richard Redmayne, Chief Inspector of Coal Mines, was also damaging to the owners. Although he was ambiguous about nationalisation, he was hostile to multiple private ownership, which he described as extravagant and wasteful.[29]

Sidney was a star performer, but Beatrice Webb displayed no bias when she wrote in her diary at Sankey's half way stage:

---

26   CIC 1919 vol. i, p. 80 and vol. iii, Appendices 5–11, pp. 7–17.
27   Arnot 1919, p. 7.
28   *Manchester Guardian*, 5 March 1919.
29   CIC 1919, vol. i, p. 214.

The ostensible business of the Commission is to examine and report on the miners' claim for a rise in wages and a reduction of hours; but owing to the superior skill of the miners' representatives it has become a state trial of the coal owners and royalty owners conducted on behalf of the producers and consumers of the product, culminating in the question 'Why not nationalise the industry?'[30]

Webb captures the essence of the miners' success at the Sankey Commission. Attuned to the mood in the coalfields, and sensitive to the widespread suspicion that the whole episode was a diversion, they sought to use the critical opening sessions to not only demonstrate the soundness of the argument for nationalisation, but to put the coalowners in the dock. As one observer neatly wrote, past crimes against the miners 'fluttered into the King's Robing Room like the forgotten wives of a bigamist'.[31] The imagery of the Commission and the accusatory tone of the miners' advocates injected a note of class struggle into the proceedings, and it was in this form that Sankey entered the popular consciousness. *The Daily News* wrote:

No one who attends its proceedings can help coming away with the impression that it is the mine-owners and not the miners whose case is on trial. So skilfully have Mr. Smillie and his colleagues managed the proceedings that they have become virtually a labour tribunal, before which the coal owners and magnates from other industries have to plead their cause. More than once, especially when Mr. Smillie or Mr. Webb has let himself go, I have been reminded of reports of the proceedings of revolutionary tribunals in France or Russia.[32]

Smillie in particular proved adept at utilising the latitude allowed by Justice Sankey in cross-examination to voice the class hatred of the miners towards the mineowners and their supporters. Lowes-Dickinson's careful neutrality was pierced with the question: 'Would you prefer to be a miner to a chartered accountant?' Benjamin Talbot, a Sheffield steel magnate, was told that a man of his wealth should be 'ashamed' of trying to stop the miners from improving their conditions. Thomas Mottram, Divisional Inspector of Mines for Yorkshire, was reminded of a colliery where electrical haulage equipment had only been

---

30   Cole (ed.) 1952, p. 152.
31   Gleason 1920, p. 39.
32   Cited in Gleason 1920, p. 48.

installed because the fierce temperatures underground were killing pit ponies at the rate of forty a month: 'Isn't it remarkable that men and boys should continue to work eight and nine hours a day in conditions under which it is impossible for horses to live?'[33]

Commentators described Smillie as an unconventional Commissioner, but his gruff, no nonsense approach, combined with his eagerness to acquaint industrialists and civil servants with the harsh realities of the miners' lives, threw the coalowners on the defensive, and guaranteed sensational press coverage. Ramsay MacDonald felt that Smillie had manipulated the proceedings with 'genius'.[34] His greatest coup lay in summoning the titled royalty owners, Lords Durham, Dunraven, Dynevor, Londonderry, Tredegar, Bute and Northumberland. *Ways and Means*, E.J. Benn's organ of conciliation backed by progressive employers, described the scene:

> Peer after peer has been made to confess that he is the owner of a fortune by reason of the foresight of an ancestor three or four hundred years ago. Lord Durham, for example, is drawing an income of a thousand a week out of ancient land, most of which was acquired by various means by his ancestors in a long past century. Lord Dunraven is a more interesting case. He is drawing an income from coal secured under common land; the surface appears to belong to the public and the mines to Lord Dunraven.[35]

Smillie questioned the peers' royalties on moral and legal grounds, demanding to see title deeds which they could not produce, and claiming that, just as their ancestors had received property for services rendered in war, now the miners claimed the same.[36] These sessions were the most dramatic scenes in the unfolding drama of Sankey. *Forward*, the Scottish ILP's paper, wrote that 'the story of the origin of Lord Bute's title deeds was making wild revolutionaries of tame grocers'.[37] Arthur Gleason, who gives over a whole chapter to the Commission in his impressionistic but occasionally insightful study of British labour in 1919, described the impact outside the Commission:

> The fact that they [the peers] had to come was a moral victory. And the word of it ran through Britain. Smillie was the lord high executioner, the

---

33   CIC 1919 vol. i, pp. 31, 133, 107.
34   *Forward*, 29 March 1919.
35   Gleason 1920, p. 52.
36   CIC 1919 vol. ii, p. 651.
37   *Forward*, 25 May 1919.

judge, the people's man, and in the name of the people he had issued orders to the privileged class, which they unwillingly but humbly obeyed.[38]

The image of Smillie as tribune of the oppressed was carried widely in the press, particularly on the left. *The Daily Herald* wrote:

> When he speaks it is as if the inarticulate millions speak through him. He insists not on the profit or loss of high wages, but on the shame of not paying them. He does not argue, he states, and each statement stabs like a sword point. He asks no mercy and shows none. I think his eyes have always before them the sordid lives and heart breaking labour of those men in the dark underground who breathe the foetid air in which horses may not live and men must.[39]

Nobody denied that the miners' side outperformed the owners. Smillie's pugnacious moral humanism in combination with the economic and public policy expertise of Webb, Tawney and Chiozza-Money was too much for their ill-prepared adversaries. Beatrice Webb wrote:

> The other side are absurdly outclassed: the three mine-owners are narrow-minded profit makers with less technical knowledge than the miners' officials, or, at any rate, less power of displaying it, without the remotest inkling of the wider political and economic issues which are always being raised by the miners' advocates.[40]

The Times concurred:

> There will be no difference of opinion among dispassionate readers on one point, which is that of the three parties concerned the miners came out far the best. Their case was better presented, but it was also a better case than that of the Government or the mine owners. We do not say that the miners' demands are justified in full, but the Coal Controller's department and the mine owners cut a sorry figure.[41]

---

38   Gleason 1920, p. 49.
39   *Daily Herald*, 15 March 1919.
40   Cole (ed.) 1952, p. 153.
41   *The Times*, 18 March 1919.

The left revelled in the triumph of the miners' Commissioners. *Forward* wrote of the owners being 'publicly flayed' before being 'carried out feet first'.[42] After having watched an industrialist 'trampled to the flatness of his own shadow', Arthur Gleason described the witness chair as having taken on 'the atmosphere of the electric chair at Sing Sing'. Gleason likened the miners' Commissioners to pieces of heavy artillery, with Smillie the 16-inch field gun, Smith the howitzer, Webb the machine gun, and Chiozza-Money the snipers' rifle. Witnesses for the owners were 'carried away on a stretcher ... They had not expected to attend a slaughter'.[43]

Miners were captivated. In the coalfields, entire communities turned out to hear checkweighers, lodge secretaries and MFGB officials recount the Commission's proceedings and explain its findings. In South Wales, for example, 'vast audiences' were 'spellbound' by reports of the hearings.[44] The discrepancy between the profits of the owners and the conditions of the miners was not news to the mining communities, but their daily exposure in the national press, their domination of national political life, was deeply satisfying. The humiliation of the owners by Smillie's team was a source of pride and a cause for celebration. Protests over housing conditions, which had hitherto been largely ignored outside trade union and labour circles, were now being echoed by the Chairman of a Commission appointed by a special Act of Parliament and sitting in the House of Lords. In the iconography of the left, the Commission ceased to be scorned as a red herring, and became a Trojan horse from which, behind enemy lines, a bold attack on capitalism had been launched. Smillie's duels with the coalowners were portrayed as a historic showdown between labour and capital. According to the *Labour Leader*:

> What is really happening in the King's Robing Room is, on a smaller scale, what is happening in Germany at this moment – a struggle between the two conceptions, 'Socialisation' and 'Private Enterprise'.[45]

Comparison with Germany, where armed workers' councils were fighting a civil war against the Freikorps, was farcical, but this is typical of the way in which a heroic mythology was created around the Commission. Page-Arnot's *Facts* had a similar tone; here William Straker's evidence on nationalisation becomes 'the open expression of the conflict between the opposing principles of autocracy

---

42   *Forward*, 24 May 1919.
43   Gleason 1920, pp. 34, 38.
44   *Merthyr Pioneer*, 29 March 1919.
45   *Labour Leader*, 13 March 1919.

in industry and industrial democracy'.[46] Lit in this way, it was as if the Robing Room had become the stage upon which the actual class struggle was being fought – and won – by proxy. The doors to the citadels of power, forever closed to the working class, now seemed to be giving way under the battering ram of the Miners' Charter. *The Daily Herald*, initially highly sceptical of participation in the Commission, now felt it possible that 'the industrial history of England will be changed for ever' by its disclosures.[47] The *Labour Leader* went further:

> The Revolution is evolving. Labour has already got as far as the King's Robing Room ... The rugged figure of Robert Smillie, the miners' intrepid leader, in that gilded setting is symbolical of the day that is yet to dawn – but will dawn – when Labour shall enter its full inheritance.[48]

Decked with symbolism of this kind, Sankey seemed much more than a royal commission of inquiry. As scepticism receded, the way opened up for those in the MFGB leadership who had only weeks before clung defensively to the Commission as a foil against direct action to present it as a positive political alternative to it. The actual concessions contained in the Interim Report hardly featured in Smillie's speech to the MFGB conference on 21 March in which he recommended acceptance. Rather, Smillie concentrated on the Commission's future potential. He invested it with quasi-legislative powers, and assured delegates that not only would nationalisation and joint control 'inevitably follow', but that the Commission would continue to sit over the following year, issuing regular reports which would transform the mining communities.[49] It is worth quoting one particular section of his speech in full, to demonstrate what miners felt they were voting for when they accepted the Interim Report. Referring to Justice Sankey, Smillie said:

> He is anxious; his mission is to improve the miners' standard of life ... The Chairman of the Commission, if it continues its sittings, has an idea that it has extraordinary powers. They have the power to make people produce almost anything ... The Chairman is anxious, if the Commission sits again, to take up mining social questions one by one, and that we should devote ourselves for a month or two months ... in finding out not merely the state of housing, and where they are worst, but in drafting and drawing plans

---

46   Arnot 1919 p. 37.
47   *Daily Herald*, 15 March 1919.
48   *Labour Leader*, 13 March 1919.
49   MFGB Special Conference 21 March 1919, p. 16.

for the building of houses, not on the lines of the present colliery houses, which have been built by the owners, but in the shape of little villages out in the country, where they would have room to breathe, and room for the children to live and breathe fresh air. He is anxious that the Commission should also have some method as to how the money is to be secured, and how the plans of these towns are to be made, and that a beginning should be made at the earliest possible time. He wants parliament to confer on the Commission the right, having decided a thing, to have the power to enforce it, where enforcement was necessary.[50]

Smillie envisaged that the Commission would continue to sit for a year or more, and, in addition to housing, identified pithead baths, transport, and underground safety as areas in which it would intervene. At the meeting between government and MFGB representatives on 25 March, when the amendments to the Interim Report which the miners had asked for were turned down, Bonar Law sugared the pill by stressing that what Sankey proposed was 'certainly something new in Commissions. It really suggests what is in effect executive action'. He then went on to promise that the reports, which would be issued 'on subject after subject', would not be thrown in the wastepaper basket as might ordinarily be the case, but would, as with the Interim Report, be observed by the government 'in the spirit and the letter'.[51] The protests of delegates from the unofficial movements that Bonar Law was an unscrupulous Tory politician, and that the Report did not 'contain one principle whatsoever', were drowned out at the MFGB conferences on 21 and 26 March. The reality was, of course, that the Sankey Commission, which Smillie now began to refer to as the 'Miners' Industrial Commission',[52] was legally only empowered to make recommendations to parliament, but this was entirely overlooked. W. Whitefield, the aged leader of the Bristol miners, told delegates that 'this is the best day in my life and that makes my heart rejoice'. For years he had been hoping that:

> the day would come in Britain when in place of waiting years for the House of Commons to pass these things, there would be used a greater power than the House of Commons, and I thank the Almighty that we are going to do it now.[53]

---

50   MFGB Special Conference 21 March 1919, p. 5.
51   Report of Proceedings between the Government and the MFGB 25 March 1919, p. 31.
52   MFGB Adjourned Conference 26 March 1919, p. 3.
53   MFGB Special Conference 21 March 1919, p. 8.

Delegate after delegate, all of them officials, took the floor to echo Whitefield. For John Robertson, it was 'the greatest leap forward ever taken in connection with the mining industry'.[54] James Brown predicted that the miners' Commissioners 'will be looked upon as being the founders of the greatest Charter ever gained for the colliery workers as a whole'.[55] Even the legendary Mabon was moved to come briefly out of retirement, to proclaim the Commission as 'the greatest industrial victory in history'.[56] Presented as it was, few could disagree with Brown that it would be 'sheer madness' to turn down such a historic opportunity.[57] Executive members repeatedly stressed that if the miners were to strike and lose, they would lose everything that the Commission promised to bring them.[58] The appeal written by Smillie and Hodges and distributed in the coalfields in the run up to the ballot was unambiguous:

> The Coal Industry Commission has scarcely commenced its work. Its greatest and most important tasks lie before it. Nationalisation, effective control by the producers whether by hand or brain, economies in production, elimination of waste, maximum economic value of the product for the community, with maximum social amenities for the men who produce the ... coal. The choice is between definite and systematic progress and the dangers of social disorder.[59]

Labour Party intellectuals and commentators anticipated that the Commission would establish an immediate precedent for the workers in other major industries. For Beatrice Webb, there would follow 'state trials of the organisation of each industry by a court made up half of the prosecuting proletariat, half of the capitalist defendants'.[60] Campaigning in the Swansea by-election a few months later, Smillie and Hodges made similar predictions. In *Forward*, 'Rob Roy' wrote 'this is not the first time the miners have done the work of social pioneers'. The Coal Commission had 'thrown Parliament into the shade. People feel that here we are at close grips with something big that matters ... What is now wanted is a shipping enquiry and a railway enquiry'.[61]

---

54   MFGB Adjourned Conference 26 March 1919, p. 10.
55   MFGB Adjourned Conference 26 March 1919, p. 18.
56   *South Wales News*, 9 April 1919.
57   MFGB Adjourned Conference 26 March 1919, p. 18.
58   See, for example, Smillie's opening speech, MFGB Adjourned Conference 26 March 1919, p. 3.
59   MFGB Official Statement to the Members, 27 March 1919.
60   Cole (ed.) 1952, p. 153.
61   *Forward*, 22 March 1919.

It was the political branding of the Sankey Commission, more than the economic concessions of its Interim Report, which explains the huge impact it
had on labour movement politics. With impressive sleight of hand, Labour and
trade union leaders sold an inquiry that had in fact saved the state from a titanic
confrontation with the miners, as an anti-capitalist tribunal. Between December 1918 and the end of March 1919, parliamentary socialism had suffered a serious problem of credibility, especially amongst trade union activists. The Sankey
Commission, by rekindling hopes in socialist change via the state, provided a
safe harbour in which the battered ship of reformism could take refuge. Ramsay
MacDonald, who had reluctantly argued that the Labour Party had no choice
but to grudgingly tolerate the idea of direct action if it was to avoid a damaging
split, now wrote that the lesson of the Sankey Commission was 'we can win
... quietly, calmly, by political constitutional methods'.[62] MacDonald, Snowden
and other leading figures took this message to South Wales, as the stronghold
of direct action politics, on May Day 1919. In all cases, the central theme of their
speeches was that Sankey had proved the constitutional road to socialism was
alive and well. There was no need for a workers' revolution as 'the Commission
would result in the old order being destroyed'.[63]

Until this point in 1919, the vernacular of revolution had been the preserve
of revolutionaries and syndicalists. Now reformism began to usurp it. In a
keynote article in the *South Wales News*, written at the end of the second
stage of the Commission, Vernon Hartshorn claimed that the miners were
pioneers, leading the whole working class into an 'unexplored realm'. The
world revolutionary process set in motion by the October insurrection took
different forms in different countries, according to their political traditions. In
autocratic nations like Russia, the revolution had been by necessity a violent
one:

> But in countries like this, where democratic institutions have long been
> established, and free discussion has been allowed, the social revolution
> has taken a form so mild and constitutional in comparison with what
> is going on in some other countries that to many of the people the
> changes which are being affected are almost imperceptible. But the world
> revolution is at work here nonetheless.[64]

---

62   *Forward*, 29 March 1919.
63   *South Wales News*, 6 May 1919.
64   *South Wales News*, 30 June 1919.

According to Hartshorn, the miners had, in the form of the Sankey Commission, unearthed what many direct actionists had denied the existence of, namely a constitutional mechanism by which revolutionary change could take place in Britain. Commenting on the proposals of its second Report, Hartshorn said:

> They go to the very roots of the capitalist system. The recommendations comply with all the forms of constitutional procedure, though they foreshadow a change which is truly revolutionary.[65]

By selling the Sankey Commission as an industrial parliament, the MFGB leadership not only secured a huge majority in the ballot, it encouraged the idea that state socialism was a viable option, even without a parliamentary majority, provided trade union power was responsibly exercised. Obviously, in this regard the Commission's impact was only temporary. The mirage vanished in an instant when the government eventually turned down the recommendation of nationalisation in the second Report, once it was confident that the threat of direct action had finally passed.

Yet for Labour, the Sankey Commission was more than just a means of quieting a rebellious rank and file. It was a national platform from which it could advertise its national policy, something it had not struggled to do in the jingoistic atmosphere of the December 1918 elections. The miners' agenda at the Commission – from profiteering, through housing and social amenities, to nationalisation – corresponded closely to Labour's *New Social Order*. The Commission showcased some of Labour's most talented intellectuals and key policies far more effectively than the parliamentary Labour Party had managed to do since the war's end.

The MFGB leaders had justified participation in the Sankey Commission on the grounds that it would throw public opinion behind the miners in the event that a strike should prove necessary. There is no doubt that the miners' cause became genuinely popular during these weeks. *Forward* wrote that:

> The steady day by day exposures of capitalism at the Coal Commission continue, and the massed effect upon the public mind has been revolutionary. Nothing approaching it has ever before happened.[66]

---

65   Ibid.
66   *Forward*, 22 March 1919.

Smillie told the MFGB conference on 21 March that had a strike taken place three weeks before, they would have had 'the vast majority of the general public – the upper classes and the middle class, and some of our own class against us'. Since then, however, the Commission had 'changed absolutely the trend of opinion in the country in our favour'. The Fife and Lanarkshire Reform Committees agreed with Smillie. They congratulated him for 'the masterly way' in which he had used the Commission, ensuring that 'a strike now would be well supported'.[67] Smillie, however, drew different conclusions:

> We are all anxious to avoid a strike with chaos and anarchy in this country. We have kindled a fire which is not going to go out ... I am sure we shall still further draw public sympathy to our cause, and I believe it will enable us to go forward and raise to a higher altitude than has ever been reached before, the conditions of the workers of this country.[68]

Public opinion had now become too important to waste on a strike according to the EC majority. It had forced the government into an acceptable compromise, and, if it could be carefully nurtured, there was every possibility that it could do so again. Furthermore, they had not failed to register the electoral significance of their work at the Commission. In the Durham County Council elections held during the Commission's first stage, Labour swept into office, holding every seat it had previously held, and gaining 28 additional ones, becoming the first County Council ever with an outright Labour majority. *The Labour Leader* wrote that 'the extraordinary sweep of Durham County ... has not been fully appreciated over the country. It is a great triumph. Labour is supreme'.[69] George Lansbury, writing in the *Herald*, also sensed a shift in the political wind. 'This is', he wrote, 'indeed a revolution'.[70]

With Labour impotent in parliament, the campaign for the Miners' Charter had symbolised the contrasting power of the direct action movement, based on the power of workers at the point of production. Now however, the popular appeal of the miners' cause in general and the breakthrough in Durham in particular encouraged Smillie to begin to seek an open reinterpretation of the MFGB's responsibilities within the labour movement, and reposition the Miners' Charter in the bid for postwar socialist reconstruction.

---

67   *The Worker*, 15 March 1919.
68   MFGB Special Conference 21 March 1919, pp. 3, 7.
69   *Labour Leader*, 3 April 1919.
70   *Herald*, 22 March 1919.

We have not been doing merely trade union work, as a matter of fact, the effective work accomplished by the Commission has enabled the Durham miners to sweep the reactionaries from the County Council ... and they have set up in Durham, for the first time in its history, government by the people for the people ... They have swept out the reactionaries. Why? Largely because of the revelation of that Commission, which sent a flame right through the rank and file of our people, and I feel sure that before that time, July 1921 (when nationalisation and the six hour day would be introduced under the provisional recommendations of the Interim Report) we will not have the reactionary Government which we now have in power. It is a duel movement, the political and trade union side working in hand together. I have great hopes for the future, providing we steadily keep our forces together and get through this crisis.[71]

The political presentation of the Sankey Commission held the key to its effectiveness in heading off a national strike. By appearing to show the creative potential of a 'neutral' state, and giving electoral viability to Labour's postwar programme, it helped to restore some much needed credibility to political reformism. In the process, it weakened the appeal of those who argued that the way forward lay with the industrial power of the working class.

In the new political atmosphere, the unofficial groups and committees in the coalfields found their influence waning. The Fife and Lanarkshire URCs admitted that the moment had come for them to 'justify their existence', and that 'failure to act would have a detrimental effect on all other such organisations'.[72] Unable to mount resistance to the acceptance of the Interim Report, demoralisation set in; both the MFGB leadership and the rank and file were blamed for the collapse of direct action.[73] During the Commission's second stage, *The Worker* wrote:

The devilish cunning of the capitalist class in making 'concessions' seems likely to ensure them a steady seat in the saddle for a long time to come. How calculated to appeal to the human weakness of the leaders was this

---

71   Adjourned MFGB Conference 26 March 1919, pp. 20–1.
72   *The Worker*, 22 March 1919.
73   For example, Bob Selkirk wrote a hymn to the miners, entitled 'The Miners' Beatitudes', which began 'Blessed are they who voted for the S(w)ankey Report, for they shall be Swanked'; *The Worker* 5 April 1919.

proposal of a commission. No risk of jail, no trouble of a strike, no danger of martial law, but such an opportunity of flourishing in the public eye as has never happened before.[74]

It then went on to bemoan the political effect that the Commission had had upon the miners:

> Headlines of Smillie and the Dukes. Why it makes one feel that the revolution has already been accomplished ... Only a small minority of the miners have been able to withstand the influence of this soporiphic.

It was not just unaffiliated miners who were seduced by the Sankey Commission. In its South Wales heartland, the unofficial movement itself was split. The founding conference of the swss, in February 1919, had revealed its politically heterogeneous nature. The *Merthyr Pioneer* reported that an argument between 'political activists and industrial unionists' had turned the proceedings into something of a 'bear-garden'. Nye Bevan, at this stage heavily influenced by syndicalist ideas, 'made a five minute speech which left the impression that the real object of the Unofficial was to smash the ILP'. Eventually, a compromise was reached, and it was decided that the swss should be open 'to all who accept the class war theory', a qualification sufficiently vague to allow the participation of syndicalists, industrial unionists, revolutionaries and reformists.

The unofficial movements everywhere contained many revolutionaries, but they were not revolutionary organisations. Will Coldrick, Secretary of the Eastern Valleys swss in 1919, when asked years later whether it had been 'exclusively industrial unionist', was most emphatic: 'Oh no, anyone who was keen on getting these reforms joined the movement'. He and many others were members of the ILP, and 'believed strongly in Parliament' while they were active in the swss. This involved no contradiction whatsoever as far as they were concerned.[75] As Dai Smith has pointed out:

> The divorce between simon-pure syndicalism and conventional politics has, largely, been a subsequent academic postulate which contemporary practice contradicts at each and every stage.[76]

---

74   *The Worker*, 31 May 1919.
75   SWML, Interview with Will Coldrick, WC/106/8.
76   Dai Smith 1993, p. 201.

Doctrinal diversity was a characteristic of the unofficial movement from the time of the formation of the URC, when ILP branches in the Rhondda had provided one of the networks for individuals interested in syndicalism.[77] The other, the Plebs League and CLC classes, also contained a variety of socialist opinion within its ranks. Indeed, in a letter responding to an appeal for ILP/CLC unity towards the end of the war, Will John Edwards, a checkweigher from Aberdare and URC member, said it was wrong to think of the CLC men as anti-political; in fact, he knew of only six in the whole of South Wales.[78]

Throughout its existence, the variegated political make up of the unofficial movement had on important occasions proved a hindrance, not a help. Where the issue was more or less economic in nature, as for example with the campaign for the minimum wage in 1911–12, political differences could be accommodated without much difficulty, but in situations where these differences could not be ignored, disunity and disorientation were the result. Take the outbreak of war. This had demanded a political response, yet it silenced the URC until the opportunity arose for agitation around immediate economic questions like the rising cost of living later in the war.[79] The campaign for the Miners' Charter straddled economics and politics. Whilst SWSS members of different political persuasions could unite over the demands for wages and hours, divisions opened up immediately when it came to the question of nationalisation or workers' control. More importantly, the high stakes of the campaign for the Charter meant that highly contentious political questions such as direct action or parliament, and reform or revolution, intruded at every stage.

Perhaps the most illuminating example of how the Commission affected the left in South Wales is provided by the *Merthyr Pioneer*, which, whilst being an independent paper, had informal links with both the ILP and the SWSS. In 1919, its columns carried the debates that were taking place on the left, in particular featuring a long-running discussion on whether parliamentary reform or direct action represented the best hope for British socialism. Its editorials, especially in the early months of the year, reflected the shift to the left amongst activists under the combined influences of foreign revolution and a high level of domestic class struggle. Although the *Merthyr Pioneer* never explicitly turned its back on electoral politics, it had little or no enthusiasm for it at this juncture, when it clearly embraced direct action and looked to

---

77   Woodhouse 1969, p. 41.
78   *Merthyr Pioneer*, 15 June 1918.
79   Woodhouse 1969, chapter 3.

the working class industrial power for political change. At one stage it leaned towards those arguing that the ILP should leave the Labour Party and link up with the emerging Third International.[80]

The Sankey Commission had a major impact on this ongoing debate in the columns of the *Merthyr Pioneer*. It was as if a bucket of cold water and a cup of strong coffee had been administered, causing a sober reflection not witnessed in its columns for months. The Commission, which had 'rung the knell of capitalism', signalled:

> a transmutation towards the ultimate triumph of socialism inconceivable to the mass of the population a few short years ago. The ultimate is thus brought nearer to achievement by the steady, unrelenting and logical dynamic of evolutionary progress.[81]

This measured, moderate tone reflected a restored confidence in reformism on the South Wales left, and the retreat of revolutionary politics, including direct action in its pure syndicalist form at least. The size and scale of the unofficial movement in South Wales meant that this political shift was more marked than elsewhere, but there is no doubt that it was replicated across the MFGB. All the unofficial leadership groups in the coalfields in 1919 contained an important revolutionary current, however there were others, probably larger, and coalescing around the ILP, which saw a more limited role for direct action and the self-activity of the working class. As such the unofficial movements were far from immune when a Royal Commission came bearing gifts. The Sankey Report divided the unofficial movements politically, took the wind out of their sails, and left the revolutionaries amongst their number high and dry.

For a few months between April and July, the unrest in the coalfields subsided considerably, and the MFGB reasserted its control over the union. However, nationalisation was still a live issue, and in what John Maclean called 'the manoeuvring for position' between government and MFGB in the period after the issuing of the second Sankey Report, enthusiasm for the Commission waned sharply, reviving the prospect of direct action once more.[82]

---

80   *Merthyr Pioneer*, 12 April 1919.
81   *Merthyr Pioneer*, 28 June 1919.
82   *The Worker*, 19 July 1919.

PART 2

∵

Introduction: A Background Sketch of the Summer's Crisis

The trouble that was fermenting before the war, which broke upon a nation in which another crisis was rapidly germinating, is now coming fast to a head. A tremendous discontent possesses the soul of the people, and a violent passion to set things suddenly aright.[1]

∴

With the acceptance of the first Sankey Report by the MFGB, the situation facing the government eased considerably, and Lloyd George was able to return to Paris to resume his place at the negotiating table. However, respite was temporary. By the beginning of July, he was back. The Sankey Commission had finished its investigation into nationalisation, and the country was once more entering a period of profound crisis.

The crisis of the summer was composed of several elements. The optimism that had greeted Coalition promises of peace and reconstruction was rapidly dissipating. The transition from a war economy was proving difficult. Lloyd George privately admitted that a 'burgeoning national debt, falling productivity, loss of credit, and endemic strikes signalled Britain's declining position'.[2] Churchill's support for the White armies in Russia was widely viewed as an act of base hypocrisy, undermining the idealistic rhetoric of the Versailles peace treaty. Liberals outside the government were outraged by the treaty itself. C.P. Scott, editor of the *Manchester Guardian*, told Lloyd George that whilst he had 'not the least objection to Coalition in itself', he objected strongly to such 'surrenders of principle' as had taken place at Versailles.[3] Liberal concerns increased tensions within the Coalition, and there was a sense that the government might not last.

The government's problems were compounded by renewed social and industrial unrest. There has perhaps been no more volatile summer in modern

---

1 *South Wales Daily Post*, 21 July 1919.
2 Middlemas 1979, p. 149.
3 Wilson 1970, p. 379.

British history than that of 1919. There was a hard edge of bitterness in what Hartshorn described as the 'profound stirring of the masses' which occurred in July and August.[4] The national peace celebrations, which took place over the weekend of 19 and 20 July, only served to increase working class discontent. For many of the poorest sections of society, the military victory had been pyrrhic. Halfway through 1919, and a land fit for heroes was nowhere to be seen. The cost of living had continued to rise, food was still scarce in some places and of poor quality in many more, there was significant unemployment in some sectors of the economy, and nothing substantial had been achieved in the area of social reform. On top of all this was the continuation of conscription (another broken promise) and Churchill's adventures against the infant Soviet Union. Taken together these created a sense of bitterness and betrayal that was as wide as it was deep. This dry social tinder was ignited by Peace Day riots which erupted in spectacular fashion in Luton, and then spread across Britain.

Tension in Luton had been high since the Discharged Soldiers' and Sailors' Association had announced its withdrawal from official celebrations in protest at the refusal of the local authorities to allow it the use of Wardown Park for a memorial service. On the day of the celebrations, thousands of workers and soldiers gathered outside the Town Hall, hurling abuse at the Mayor. After some jostling with police, a section of the crowd attempted to charge into the building and wreck the Assembly Room where the Peace Banquet was to be held that evening. A pitched battle with the police continued well into the night, with the latter suffering sufficient casualties to force them to relinquish their defence of the Town Hall steps. The building was stormed, ransacked, and red flags were flown from windows and balconies before the whole edifice was torched and destroyed. It was not just Luton Town Hall which went up in smoke; so too did peace celebration images of national unity.[5]

Following Luton's example, there developed serious rioting at Greenwich, Coventry, Edinburgh, Swindon, Hull, Wolverhampton and Liverpool, and numerous smaller clashes between crowds and police in other towns and cities.[6] Further research would be required to make a confident assessment of the political character of this non-industrial social protest in 1919. There is, unfortunately, no equivalent in British historiography of Ray Evans's fascinating account of Australia's 'Red Flag Riots' of the same period, which shows how ex-servicemen's sense of betrayal could find political expression in either left-wing

---

4   *South Wales News*, 27 September 1919.
5   *The Times*, 21, 22 July 1919; *Daily Herald*, 21 July 1919.
6   *Daily Herald*, 1, 7, 23 July 1919; *The Times*, 23 July 1919.

or right-wing terms.[7] However, the pattern of rioting in Britain that summer suggests it too was fuelled by the same sense of having been deceived. Often the disturbances began with a perceived maltreatment of ex-servicemen (arrests for drunkenness being a common example), and culminated in an assault by the rioters on the most obvious local symbol of authority, usually the nearest police station or town hall.

While these eruptions on the street went up and down very quickly and posed no real threat to the state, they were a graphic manifestation of a menacing mood amongst the urban working class and poor. These street protests and riots were accompanied by a revival of industrial militancy, which the *Daily Herald* anticipated as 'the beginning of the fiercest of struggles between Labour and Capital'.[8] All three constituents of the Triple Alliance were gearing up for action. Leaving the miners aside for the moment, the NUR was preparing a new national programme for submission to the Railway Executive Committee, and the Transport Workers Federation was demanding a 10s wage rise.[9] The Bakers' union was also in a national dispute over pay, and called strikes in several districts, causing bread to be scarcer than it had been during the war according to some accounts.[10] The aggressive mood amongst railwaymen was shown when a local strike on the Newcastle to Carlisle section of the North Eastern Railway (NER) against the sacking of a driver who had failed an eyesight test was spread by rank and file strike committees to Leeds and York, involving 10,000 men. This then escalated into an attempt by the Vigilance Committees to overthrow the Executive of the NUR and 'smash the conciliation board'.[11]

An already precarious industrial situation worsened when, on 31 July, a mass meeting of police in London voted for an immediate strike against the Police Bill, which outlawed trade unionism in the force. A ballot of NUPPO members held at the end of May had revealed a huge majority in favour of a national strike, by 44,539 to 4,324. PC James Marston, the union's leading figure, claimed that at least 30,000 would join the strike immediately, and warned the government that any attempt to use coercive measures against them would be met with force by police pickets.[12] The *Daily Herald* described the Police Bill as 'the most definite step made even by our present discredited

7   Evans 1987, chapter 8.
8   *Daily Herald*, 23 July 1919.
9   Ibid.
10   *The Times*, 13 August 1919; *Sheffield Weekly Independent*, 9 August 1919.
11   *Yorkshire Post*, 19 July 1919; *Leeds Mercury*, 19, 21 July 1919.
12   *The Times*, 1 August 1919. For the most recent account of the events leading up to the strike, and the strike itself, see Wrigley 1990, pp. 53–79.

and bullying government to destroy the organised movement of the workers', and called upon other trade unionists to support the strikers.[13] There was some early evidence of solidarity action, especially amongst railwaymen. In London, ASLEF members at the Nine Elms depot of the London and South Western Railway, and on the City and South London underground struck in support of the police, and mass meetings in Liverpool, organised by the Vigilance Committees, voted in favour of a general transport workers' strike in the city. In London, the Electrical Workers Union threatened to cut off the capital's power supply.[14]

The response amongst police to the strike call was very poor in most places, but Merseyside, where 65 percent of the force struck, provided a taste of what might happen elsewhere if the elemental anger of the riots was to fuse with industrial struggle that included police action. Tom Mann returned to Liverpool to try to recreate a postwar version of his 1911 movement after an ad hoc strike committee called a mass meeting to organise solidarity with the police, and 6,000 workers and ex-soldiers voted for a general strike.[15] With the exception of 400 dockers, the strike did not materialise, but week-long rioting on the streets of Liverpool and Birkenhead created an insurrectionary atmosphere on Merseyside. On the night of 2 and 3 August, slum dwellers rioted through the night against the army, with one man shot dead. *The Times* wrote: 'Central Liverpool tonight represents a war zone ... there has been fighting and wounds. Soldiers with steel helmets and fixed bayonets patrol the streets'.[16] In addition to 3,000 troops on the streets of Liverpool and Birkenhead, one battleship and two destroyers were positioned in the Mersey, their guns trained on the city.[17]

The press imagined it saw the red hand of Bolshevism in the revolts. *The Times* ran a headline 'Revolution by Strikes: Plot Financed From Abroad'. *The Leeds Mercury* proclaimed a 'Bolshevist Plan to Seize London. Soviet to Be Set Up', while the *Sheffield Independent* reported that an organisation with Bela Kun at its head lay behind the strikes and riots.[18] Paranoid fantasies in retrospect, these headlines nonetheless point up the near hysteria that social instability was producing in some quarters.

---

13   *Daily Herald*, 1 August 1919.

14   *The Times*, 4–6 August 1919; *Daily Herald*, 4 August 1919.

15   *Daily Herald*, 4 August 1919; *The Times*, 4 August 1919.

16   *The Times*, 4 August 1919.

17   Wrigley 1990, pp. 75–6; *Daily Herald*, 5 August 1919.

18   *The Times*, 6 August 1919; *Leeds Mercury*, 11 August 1919; *Sheffield Independent*, 7 August 1919.

A further element in the crisis of the summer was a groundswell inside the working class for the abolition of conscription and the withdrawal of British troops from Russia. Widespread accusations that the government was guilty of betrayal on these issues were difficult to answer, as 'peace and no conscription' had been a central plank of the coalition's electoral campaign in 1918. Socialists had for months been arguing that allied intervention in Russia was an attack by capitalism on an infant workers' democracy. *The Times* reckoned that this campaign had been so effective 'that it has come to be widely accepted by the rank and file of the labour movement' and warned that 'the intensity of feeling among the workers on the Russia question must not be underrated'.[19] A Cabinet paper reported that 'even mild trade unionists are said to be strongly moved' over intervention in Russia, 'because they think it is the thin end of the wedge towards making compulsory military service a permanent institution'.[20]

Resolutions demanding action poured in to Labour Party and TUC headquarters.[21] Anti-conscription and anti-intervention fitted the radical liberal political culture which still had a place inside the new model Labour Party, but it was pressure from the trade unions, particularly the miners, which led the Annual Conference to pass a resolution instructing both the National Executive and the Parliamentary Committee of the TUC to 'enforce these demands by the unreserved use of their political and industrial power'.[22]

The decision of the conference was interpreted in some quarters as a dramatic shift to the left by the Labour Party. The *South Wales News*, for example, declared:

> The Labour Party yesterday abandoned constitutional methods in favour of revolutionary. It reveals in a sensational way the extent to which organised workers have lost faith in political action.[23]

The Workers' Dreadnought was characteristically bombastic: 'A triumph for direct action!', it declared; 'A blow for Hungary and Russia! A step towards revolution!'[24] In fact, the Labour Party resolution typically committed nobody

---

19   *The Times*, 8 July 1919.

20   Report on Revolutionary Organisations in the UK, Cabinet Paper GT 7916, 30 April 1919, CAB 24-78-387.

21   Labour Party Annual Conference Report 1919. See in particular Henderson's speech, p. 122.

22   Labour Party Annual Conference Report 1919, p. 156.

23   *South Wales News*, 28 June 1919.

24   *Workers' Dreadnought*, 5 July 1919.

to anything, as Frank Hodges pointed out.[25] Nonetheless, it received enormous publicity, and fostered the illusion that direct action had conquered the citadels of the labour movement.

This was further encouraged by the decision of a Triple Alliance delegate conference on 23 July to ballot its constituent unions over direct action 'to compel the Government to abolish conscription, to discontinue military intervention in Russia, and military intervention in trade union disputes at home'.[26] *The Daily Herald* was supremely confident: 'The decision to ballot is a decision to strike, there is no doubt as to the result', and it urged that the Triple Alliance incorporate all key working class demands into the strike, notably the withdrawal of the Police Bill, and the nationalisation of the mines. In a 'Manifesto to the Rank and File', Lansbury wrote:

> The heaviest responsibility in the political history of Britain rests upon the rank and file of the Triple Alliance in the forthcoming ballot. What is really at stake is the whole future of democracy. Are the people to prevail, or are the oligarchs, the militarists and the reactionaries to resume their age-long sway?[27]

There was a general feeling that the struggle to determine the shape of postwar British society was coming to a head. The apparent shift in attitude at the top of the labour movement, in conjunction with the epidemic of strikes and riots, suggested that the direct actionists were in the driving seat as the crisis reached its climax.

Once again, as earlier in the year, the mining industry was at the centre of the renewed unrest. A month-long official strike by the Yorkshire Miners' Association sparked off well-supported rank and file strikes in many other coalfields. These strikes were significant for two reasons. Firstly, they signalled a deep disgruntlement with the practical details of the Interim Report. Secondly, they raised the possibility of a strike over nationalisation. Direct action was once again on the agenda, both in the MFGB, and in the wider labour movement.

The controversy over the nationalisation of the mines dominated British politics in the summer of 1919, and was at the epicentre of its crisis. Once again there was widespread speculation about the possibility of mass strikes

---

25   Labour Party Annual Conference Report 1919, p. 160.
26   MFGB 1919, Report of Proceedings at a Conference of the Industrial Triple Alliance 23 July 1919.
27   *Daily Herald*, 24, 28 July, 6 August 1919.

and revolution. As such, there was more at stake in the nationalisation controversy than the question of ownership of the mines. For the government, it was seen as a matter of survival; government ministers themselves canvassed scenarios of being toppled either from within, by rebellious backbenchers, or from without, by labour's massed ranks. For the labour movement, it further sharpened the debate which had been raging all year. The question of whether reconstruction along socialist lines should be pursued through direct action or by constitutional means was finally pinned down. The answer hinged to a large extent upon how the MFGB would act, and this itself hung in the balance during the strike wave which swept across the coalfields in July and August. These months were the decisive ones in which miners and government manoeuvred for position over nationalisation, and in which the supporters and opponents of direct action within the MFGB vied with each other for ascendancy. The outcome of these tussles was already fairly clear when Lloyd George announced the government's rejection of nationalisation to the House of Commons on 18 August. It is of the utmost significance that the Yorkshire miners' strike had effectively collapsed only days before.

The issues raised above will be dealt with in Part Two in the following manner. Chapter Seven provides a contextual backdrop to the nationalisation controversy and gives an account of the coal output crisis, which provided the platform from which the government made its opening moves against the MFGB. Interwoven with these is an examination of the widening division in the labour movement and the MFGB around the question of direct action. Chapters Eight and Nine deal in turn with the general unrest in the coalfields, and the Yorkshire strike, to assess the strength of the challenge posed to the Federation Executive, weigh up the potential which the strike wave had to become a fight over nationalisation, and offer an explanation as to why this potential remained unfulfilled. In the concluding chapter we seek to extend and develop this theme, and give an analysis of the demise of the direct action movement, and the victory of those within the MFGB who favoured the Mines for the Nation campaign.

CHAPTER 7

Perspectives on Nationalisation in the Period of Manoeuvre

'Why should nationalisation produce any change?' 'Because men would feel that they had control, or some control over their own energies, and that they were not merely at the will and direction of another being. They would be in a better position than the horse that they have to drive, or the machine that they have to attend ... it is that desire that cannot be crushed that ... has given rise to more unrest than anything else.'[1]

∴

1   The Preparation of the Government's Counter-Offensive

The Coalition was presiding over a society that was sliding into increasing unrest. The government was beset by a range of problems to which it seemed to possess no solution, and it projected an image of hesitancy and drift. C.P. Scott, editor of the *Manchester Guardian* and someone well acquainted with supporters of both Lloyd George and Asquith, believed simply that 'this government is doomed'. The only uncertainty, he felt, was 'what is to succeed it?'[2] Central to the government's problems was its delay in coming to a decision on whether or not to nationalise the mines.

The final reports of the Sankey Commission were completed by 23 June. There had been four separate reports issued by the Commissioners. The first, signed by five of the six employers, rejected nationalisation outright and recommended an end to government control and a return to unfettered private ownership. Sir Arthur Duckham, an engineer employed by the Ministry of Munitions during the war to run munitions factories, declined to sign the report, preferring to submit one of his own. He recommended the unification of the min-

---

1   William Straker replying to R.H. Tawney, CIC 1919, vol. iii, p. 963.
2   C.P. Scott to L.T. Hobhouse, 3 September 1919, in Wilson 1970, p. 377.

ing industry under private ownership, by the establishment of district trusts, with the miners represented in a minority on the Boards of Directors.

The Labour representatives submitted their plan for nationalisation with joint control in the form of a Bill to be introduced in parliament. Under their scheme, a National Mining Council would be established, structurally analogous to the British Army Council, with a Minister of Mines at its head. Half of the Council members would be appointed by the MFGB, the other half by the government. This arrangement would then be replicated downwards, on a district and pit basis. Sir John Sankey's report, whilst stopping well short of genuine joint control, was nonetheless in favour of full nationalisation. To maximise pressure on the government, the miners' representatives decided to endorse Sankey's proposals, which thus became the Commission's majority report.[3]

The Sankey Commission had been working to a deadline. One month after the final reports had been issued, the government had still made no comment whatsoever upon them, other than that 'the matter remained under consideration'.[4] Procrastination over nationalisation worsened deteriorating industrial relations, shrouded the government in suspicion, and caused anger to rise in the pits. On 22 July, with the Yorkshire miners' strike already underway and unrest spreading across other coalfields, the *Daily Mail* lamented that there was:

> no decision, and no hint of a decision. It is impossible to conclude otherwise than that the Government is greatly to blame for the present situation. They were confronted with a question which demanded a plain 'Yes' or 'No', and they have allowed matters to drift into a very dangerous position.[5]

In private, senior government figures agreed. After a discussion with Lloyd George about the unrest, Maurice Hankey wrote to Tom Jones: 'The settlement of the nationalisation question is the root question of all our trade and labour difficulties'.[6]

One way or another, the government had to cross the bridge, but both acceptance and rejection of nationalisation held serious consequences. The fate of the Ways and Communications Bill, which in its original form left the door ajar to state ownership of the docks, harbours and railways, signalled the

---

3   Cole 1923, pp. 90–100.
4   Cole 1923, p. 102.
5   *Daily Mail*, 22 July 1919.
6   Hankey to Tom Jones, 12 July 1919. Cited in Wrigley 1990, p. 200.

extent of Tory opposition to nationalisation in parliament. A well-orchestrated backbench revolt openly threatened to split the Coalition if it decided to go ahead with such a scheme.[7] Bonar Law signalled a willingness to concede,[8] but even so the government was twice defeated in the House in the following week. On the second occasion some commentators felt that it would have fallen had it insisted on using whips at the division.[9] By the time the bill was passed into law, it had been 'mutilated by the Government's own supporters, and every reference that seemed to hint at nationalisation was expunged'.[10] The *Daily Herald* commented: 'The first great victory of the anti-reconstructionists has been won'.[11]

The coalowners had been caught off guard during the first sittings of the Sankey Commission. Harry Supple has written that 'faced with a new and alarming context for the discussion of their post-war world', they were 'stunned and temporarily demoralised'.[12] In the aftermath of the first stage of the Sankey Commission, there had been 'an air of profound resignation about the possibility of national ownership', many owners expressing a preference for outright nationalisation (with generous compensation) to joint control. The *Colliery Guardian* urged the owners to 'Take the Gloves Off', arguing that the debate was about more than the future of just one industry. The Mining Association of Great Britain (MAGB), it said, 'merely occupy the front trench' in a battle which if lost would mean 'England will change hands'.[13]

Thus encouraged, the MAGB regained its composure during the second stage of the Sankey Commission. From the witness box coalowners warned that nationalisation of the mines would 'be followed by the nationalisation of all industries ... and consequent ruin to the people'.[14] Leading industrialists came forward in support; Sir John McLaren, President of the Leeds Chamber of Commerce, told the Commissioners that, in his view, the nationalisation of the mines would trigger an offensive by workers in other major industries:

> Miners would probably get benefits which would be resented by other workers who would naturally combine to bring about similar results in

7   Joynson Hicks, 117 H.C. Deb.5s, c.1026, 2 July 1919.
8   *Daily Herald*, 1, 2 July 1919.
9   *Manchester Guardian*, 9 July 1919.
10   Cole 1923, p. 102.
11   *Daily Herald*, 2 July 1919.
12   Supple 1984, p. 219.
13   Supple 1984, pp. 221–4.
14   CIC 1919, vol. ii, p. 1137. Sir Lionel Phillips, Chairman of the Central Mining Company.

their trades. It would lead to an enormous increase of bureaucracy until eventually all were working for the state, individual freedom and liberty would be at an end, and conscription of labour would be the result.[15]

The MAGB began to co-ordinate the support of its own members and other employers' organisations, notably the Federation of British Industries and the National Association of Chambers of Commerce, in a campaign against the nationalisation of the mines. They rallied MPs who represented business interests, and the defeat of the Ways and Communications Bill in its original form was the first taste of victory. Sir Edward Carson wrote to Lord Selborne in late June that previously each section of industrialists in the House had only been stirred to activity when its own interests had been threatened:

> I have been impressing upon them that you must get to the bedrock of principle, and unite forces whenever the principle is assailed. I think they have begun to see this – and coalowners, dock trusts, roads, wagon owners are now uniting on the broad question of nationalisation or individual effort.[16]

The Parliamentary Coal Committee, an organisation of coalowner MPs, tapped into this new found unity, circulating a memo calling for steps to be taken 'to protect our great industries against the organised, revolutionary, and predatory forces of direct action, and against the nationalisation of the mines'. Three hundred and five MPs signed the memo, presenting it to Lloyd George in mid-July.[17]

On the other hand, the government had, 'if', as George Barker observed, 'language has any meaning', promised to implement the recommendations of the Sankey Commission.[18] The *bête noire* of the industrialist class, nationalisation of the mines had become the *cause célèbre* of the labour movement, and through its trade union and political bodies it had bound itself by scores of resolutions to support the MFGB in its quest. It appeared that if the government rejected nationalisation, it would be taking on not just the miners, but also the main battalions of the organised working class.

The government was caught between a rock and a hard place. To concede nationalisation would mean subjecting the brittle Coalition to the severest test.

---

15   CIC 1919, vol. ii, p. 1054.

16   E. Carson to Selborne, 29 June 1919, cited in Wrigley 1990, p. 191.

17   *The Times*, 14 July 1919.

18   *South Wales News*, 30 August 1919.

To refuse it was to invite confrontation with the Triple Alliance, with potentially disastrous results. Robert Munro, the Scottish Secretary, wrote to Lloyd George:

> I greatly fear that to turn down the principle of nationalisation absolutely would inflame working class opinion to a dangerous degree; while on the other hand, to adopt it immediately would not only imperil the Coalition itself but might prove to be a national disaster.[19]

As Maurice Cowling has noted, for the previous eight weeks Robert Home had been warning that:

> the coming clash between unions and employers was likely to make the political system redundant, and turn the House of Commons into a shadowy organisation, validating decisions made elsewhere.[20]

There was constant speculation as to how far the fault line of nationalisation was widening beneath Coalition feet. Lord Rothermere's *Sunday Pictorial* argued forcibly that internal tensions were paralysing the government, threatening the country's stability.[21] C.P. Scott, who felt that liberal principles were being sacrificed upon the altar of the Coalition, continued to forecast its downfall:

> The broad fact that emerges from the increasing alarms and stratagems at Westminster is that, while the bulk of the nation is in the mood for great changes, the majority on which the Government rests is not. A period of indecision and uncertainty has set in. It can end only with a new Parliament.[22]

The *Daily Herald* agreed. 'Coalition Crash Coming', predicted its banner headline; 'It is divided within itself. It cannot stand'.[23]

Behind the government's long 'consideration' of the Sankey Commission's findings lay an absence of any definite policy on nationalisation. Government records reveal no discernible strategy, no preconceived plan of action. Rather, the evidence suggests that it tackled the problem on the hoof, *à la* Lloyd

---

19   Munro to Lloyd George 4 August 1919, LG F/1/7/32.
20   Middlemas 1979, p. 148.
21   *Sunday Pictorial*, 24 July 1919.
22   *Manchester Guardian*, 9 July 1919.
23   *Daily Herald*, 1 July 1919.

George, playing for time, and seeking opportunities to extricate itself from the situation intact. The varied political views of Lloyd George and his cabinet colleagues regarding nationalisation of the mining industry, and the process by which they decided against it, has been described and debated in depth elsewhere, and is not the focus of this study.[24] However, a few observations are necessary, insofar as the government's attitude to nationalisation, and the strategy which it adopted in the summer, had an impact upon the internal politics of the MFGB.

In ideological terms, the Coalition government was a hybrid creature assembled at a moment of national emergency with the successful prosecution of immediate tasks as its pragmatic brief. Whilst Tory members were opposed to nationalisation in theory, in practice they had participated in the greatest ever intervention of the government in the nation's economic life, including state control of the mines ... and they had claimed the credit for it. Moreover, on the other side of the Coalition there was, within liberalism, a strand which favoured nationalisation of some sectors of the economy; Lloyd George, after all, had established his liberal credentials in large part by his advocacy of some form of land nationalisation. Winston Churchill had made a speech during the 1918 election campaign in Dundee, in which he had said in good faith that the Coalition intended, in the interest of national efficiency, to nationalise the railways. Although the specific phrase was cut from the finalised version of the Coalition's manifesto (without Churchill's knowledge), Lloyd George and several of his Cabinet colleagues continued to believe that some variant of nationalisation of the railways would be economically beneficial.[25] The Ways and Communication Bill, as introduced to the Commons by the government, testified to this belief. Furthermore, according to Riddell's diary, Lloyd George felt in April 1919 that nationalisation of the mines was inevitable. He told Bonar Law, 'It has to come. The state will have to shoulder the burden sooner or later'.[26]

Of course, there were important economic considerations which, in government scales, weighed against nationalisation. Middlemas writes of the concern of Ministers on both sides of the Coalition to preserve business confidence, and of constant Treasury pressure to decontrol for financial reasons.[27] However, the

---

24   Middlemas 1979, chapter 5; Kirby 1977, especially pp. 24–45; Armitage 1969, chapter 4; Wrigley 1990, especially chapter 7.

25   Wrigley 1990, p. 189.

26   Riddell 1933, p. 49.

27   Middlemas 1979, pp. 135–6.

economic record of the Coalition shows that, as a whole, it was not *in principle* opposed to nationalisation on economic or ideological grounds. This is important for two reasons. Firstly, it encouraged the MFGB EC to believe that the government really might deliver. Secondly, it suggests that the government's rejection of nationalisation must be explained primarily in practical political terms. The main concerns for the government were its own political survival, and the prevention of a further dangerous escalation of class conflict.

In the prevailing conditions, the government decided that rejection of nationalisation carried less risk. To do otherwise would not only have risked bringing down the coalition, but would have been seen as a victory for the miners' industrial power. Industrialists had stressed this at the Sankey Commission. Bonar Law was equally emphatic in cabinet:

> If the miners threatened to strike on the question of nationalisation and the Government gave way, it would mean abandoning the functions of Government by agreeing to a sectional demand in order to prevent a strike.[28]

The government decided that it would have to make a stand and hold the trench of private ownership. A letter from Winston Churchill to Bonar Law suggests that the government had taken this decision as early as 5 July:

> For practical purposes in the present situation we should say, 'We are going to nationalise the railways, but we are not going to nationalise the mines.' We only lose by appearing ashamed of what we really are going to do.[29]

Lloyd George was not as brazen as Churchill. At a meeting at Criccieth on 17 July with senior ministers and advisers, he formally decided against nationalisation. However, it was stressed that this was a provisional decision.[30] The full Cabinet did not finally decide the issue until 7 August, and did not make its decision public until 18 August.[31]

To summarise then, the government's rejection of nationalisation was rooted not so much in theoretical or ideological objections, as in conjunctural political factors flowing from the class struggle in the summer of 1919. On the

---

28   W.C. (546A) 19 March 1919; CAB 23/15.
29   Cited in Wrigley 1990, p. 189.
30   Wrigley 1990, p. 200.
31   W.C. (607A & 608A) 7 August 1919; CAB 23-15-174/204.

one hand, the determined opposition to nationalisation from coalowners and industrialists which was expressed politically in the rebellion on the government's back benches; on the other, the likely consequences of giving way before the MFGB under the threat of direct action in the prevailing atmosphere of industrial unrest.

An important point emerges here with regard to the existing historiography, much of which has been concerned with the question of mines nationalisation and the Sankey Commission largely because, in Wrigley's words, it is an example of 'one of the more controversial episodes in Lloyd George's political career'.[32] A preoccupation with the question of whether the Sankey Commission was just a delaying tactic and a clever piece of political chicanery on Lloyd George's part has tended to produce one-dimensional accounts of what was after all a two-sided conflict. Deciding against nationalisation was one thing; getting away with it was quite another, when faced with the threat of a strike by the MFGB and possibly the Triple Alliance. This study attempts to redress the historiographical imbalance away from Coalition politics and Cabinet intrigue and towards the politics and strategies of its adversaries in the MFGB and the trade union and labour movement. In particular, it seeks to explain why the forces of direct action failed to carry the day against both Smillie and Lloyd George.

There follows an account of how the situation was transformed in the period between the issuing of the Sankey Report on nationalisation on 22 June, and the government announcement that it was rejecting the report on 18 August. It was during this period that the government and MFGB manoeuvred for position over the issue, and supporters and opponents of direct action vied with each other for ascendancy within the MFGB. It begins with the opening shot of the government's counter-offensive, which came in the shape of a 6s rise in the price of a ton of coal, and ends with the defeat of the Yorkshire miners' strike.

2   Nationalisation and the Output Crisis

The government's counter-offensive consisted of a combination of measures to reverse the tide of public opinion which had coalesced behind the MFGB since the first stage of the Sankey Commission, before announcing its decision on nationalisation. Intermittently, it seems, it put out feelers to test working

---

32   Wrigley 1990, p. 207.

class reaction with rumours that nationalisation had been rejected. These rumours were so widespread and so detailed (for example, that the Duckham scheme was the favoured option) that they must have been leaks, controlled or otherwise.[33] The government was, naturally, extremely concerned about public opinion. The popularity of the miners' cause put limits on its ability to manoeuvre. The successful operation of its contingency plans for dealing with a national strike depended on its ability to at least neutralise a significant section of working class opinion. Middlemas and Wrigley have shown that this had been a prime concern of the government since the first hearings of the Sankey Commission, but at that stage support for the miners had been so strong that Lloyd George had felt nothing could be gained 'by challenging the merits of the miners' case'.[34] Now, in July, with a conflict looming once again, Lloyd George warned the Cabinet: 'There must be no division in the public mind ... If a fight was to come, it must be certain beyond a doubt that the miners were in the wrong'.[35]

On 8 July, the Cabinet agreed on a strategy: to exploit the poor state of public finances to paint the miners' demands as excessive and damaging to the national economy. The mechanism was a 6s rise in the price of a ton of coal. Auckland Geddes expressed the hope that: 'By administering such drastic medicine we might bring the community to a state of sanity', turning consumers against the producers of coal.[36] By way of underlining that blame lay with the miners, the price increase, announced in the House of Commons on 9 July, was timed to come into effect on 15 July, the day before the first Sankey Award became operative. In a speech of unrelieved gloom, Geddes explained to the House that the price rise was needed to pay for the concessions which had been made to the miners, and emphasised that not only would it harm the consumer, but it would also damage industry, reduce exports, and increase unemployment.[37]

---

33   Cole 1923, p. 112. On 24 and 25 July, all major newspapers carried the news that the government had not only decided against nationalisation, but that, accurately as it turned out, it had plumped for Duckham's report.
34   Lloyd George to Bonar Law 20 March 1919; LG F/30/3/32. Wrigley 1990, pp. 148, 161–4. Middlemas 1979, p. 146.
35   41 W.C. (596A), 21 July 1919; CAB 23-15-136/7.
36   W.C. (589), 8 July 1919; CAB 23-11-14/15. In autumn, the price was reduced by 10s. Independent accountants showed that there had been no financial justification for the increase, a fact which tends to confirm MFGB suspicions that it had been politically motivated. See Wrigley 1990, p. 197.
37   117 H.C.Deb.5s, c.1817, 9 July 1919.

Particular weight was given to falling productivity, in line with instructions from Lloyd George in Criccieth to 'rub in the reduction of output'.[38] The latest available figures showed that productivity in the mines was plummeting; evidence placed before the Sankey Commission by the Coal Controller suggested that on the basis of the first twenty weeks of the year, average output per miner was only 16.8 tons per month, compared to 19.8 tons in 1913. The total estimated output for 1919 was 230,606,000 tons, compared with 287,412,000 in 1913, a drastic decrease given that there were between 150,000 and 160,000 more miners working in the pits in the later year.[39] Geddes's lurid picture of Britain's dire economic position reverberated with familiar wartime themes of nationalism and patriotic responsibility:

> If we were to pass safely through the dark and anxious days which lie ahead, we must be animated by the spirit of patriotism which prevailed during the war. Production was the most urgent need of the hour.[40]

Lloyd George had spoken in identical terms in a keynote speech in the House of Commons upon his return from Paris ten days earlier, when he warned that whilst Britain had won the war, it was in danger of losing the peace to its economic competitors:

> There is a tendency to assume that now all will come right without any effort. We have output diminishing and costs of production increasing. That is exactly the opposite rule to the one that leads to prosperity. Let us think together, act together, work together. I beg that we do not demobilise the spirit of patriotism too soon.[41]

Ministers would warm to this theme as the summer drew on, particularly in relation to the Yorkshire miners' strike. Rather than attempt to talk down a national crisis, they sought to play it up and blame it on the miners whose selfishness and sloth were threatening the sacrifices the people had made during the war, undermining the prosperity of industry and consumer, and ruining the prospects of successful reconstruction. One unnamed Coalition

---

38   Lloyd George to Bonar Law, 14 July 1919; LG F/95/5/16.
39   CIC vol. iii (A), p. 210. The actual output for 1919 was 229.8m. See Supple 1987, pp. 8–9, table 1.1.
40   *The Times*, 15 July 1919.
41   *The Times*, 4 July 1919.

Unionist MP neatly condensed the government's message by commenting that the 6s increase was 'the first fruits of Sankeyism, which might be a good thing if it saves us from the last fruits of Smillie-ism'.[42]

One final tactical embellishment during the House of Commons debate on the price rise came in the form of an offer by Bonar Law, to the miners' MPs sitting opposite, to delay the increase for three months if, in that period, the MFGB was prepared to co-operate with the Coal Controller to increase output and agree to a strike moratorium.[43] This was, as the *Daily Herald* admitted, 'a clever move'. If the miners refused, they would lay themselves open to further accusations of selfishness and unpatriotic behaviour, whilst acceptance would imply an admission that, on some level, they were to blame for low output. Lansbury was impressed. Throughout the proceedings of the Coal Commission, the miners had been on the offensive, but: 'Mr. Bonar Law's offer in the House has, in effect, put them on the defensive. It will require the greatest care if the position is to be reversed'.[44]

There was almost unanimous agreement within the MFGB that the 6s increase, which was announced on the eve of polling in the Swansea East by-election, and Bonar Law's offer were part of a Machiavellian stunt designed to turn public opinion against the miners, put a stop to the recent run of by-election defeats, and queer the pitch on nationalisation.[45] However, beyond this the consensus within the union would not stretch. The price rise and the offer led to a sharp division in the MFGB, or to be more accurate, it widened and exposed existing divisions over the question of direct action.

Inside the Federation, the MPs were the most implacable opponents of direct action, and many of them also held executive positions within their district organisations. They met in the aftermath of the Commons debate on the price rise, and, urged on by Brace and Hartshorn in particular, voted to recommend acceptance of Bonar Law's offer of a strike moratorium to the MFGB Annual Conference which opened at Keswick on the following day, 15 July.[46]

---

42   *The Times*, 10 July 1919.

43   118 H.C.Deb.5s. c.175–6, 14 July 1919.

44   *Daily Herald*, 15 July 1919.

45   Swansea East was the fifth by-election of the year, the others being West Derby, Hull, Central Aberdeenshire and West Leyton. The Coalition held on to West Derby, but lost the others to Liberal candidates. Compared to the General Election, the coalition vote in these seats fell by 18,489, whereas that of its opponents rose by 9,407. *The Times*, 12, 15 March, 12 April, 1 May 1919.

46   *Westminster Gazette*, 15 July 1919.

Their decision was couched in terms of the miners' duty in a national emergency. The wartime spirit of ministerial responsibility and statesmanship was rekindled in William Brace, who toured South Wales in the following weeks appealing to the miners for extra exertion. He told Abertillery miners that:

> At the moment, coal was going to determine the future destiny of the race. If the country would not be given coal they would perish, and be swept as a nation ... into broken fragments.[47]

Always, his final appeal was the same: 'Men, for God's sake, give the nation the last ounce of coal it is possible to give'.[48] For Brace, the output crisis was first and foremost a national crisis, one that, as ever, must take precedence over the class struggle.

Vernon Hartshorn, traditionally not so closely associated with the right inside the union, spoke too of 'the obligation which devolves upon us to do our part as citizens in this crisis', and called at Keswick for 'hearty co-operation with the Government'.[49] Under private ownership, he lamented, the MFGB had pursued 'a sort of educational propaganda and agitation for the purpose of creating an atmosphere in which anything like cordial relations between the workers and the owners would be impossible'.[50] Now, however, thanks to the Coal Commission, nationalisation was about to 'bring this class struggle to an end'. The result would be that:

> the position of the miner will be immediately and vitally changed. He will cease to be the hireling of a profit-making syndicate. He will become a public servant, charged with public responsibilities and duties, and in honour bound to have scrupulous regard to the fact that the nation depends upon the efficiency and energy of his labour for its welfare and progress.[51]

---

47   *Western Mail*, 21 July 1919; *South Wales News*, 21 July 1919.

48   Ibid.

49   MFGB Annual Conference 1919, p. 75.

50   The thrust of Hartshorn's article corresponded to his evidence at the Sankey Commission. Output would increase under nationalisation 'if all the miners' leaders, instead of preaching class war and class hatred and class antagonism, as they have been doing for the last quarter of a century, if they could only turn on to developing a social conscience and getting the miners to realise that they were working not for profits but for the community'. CIC 1919, vol. i, p. 366.

51   *South Wales News*, 21 June 1919.

All this would take place only after nationalisation, but, he argued, it should dictate that here and now the miners see the output crisis 'not as a colliery owners' question' (for they were about to be eliminated from the industry), 'but as a question which is of vital importance to the nation and to Labour'. The inference was that the class struggle in the pits must come to an end before nationalisation took place. The miner must show 'a sense of community', and 'a sound moral and social consciousness' before he could 'play his vitally important part in the industrial system manfully and honestly for the sake of the commonweal'.[52] The challenge of nationalisation would be whether the miner was fit to assume this responsibility; the output crisis was the miner's chance to prove that he was.

Hartshorn and Brace were resoundingly defeated at Keswick. However, the central idea encapsulated in Hartshorn's article was not as marginalised inside the MFGB as the vote suggests. Far from it, for the case for nationalisation as presented by the labour representatives at the Sankey Commission was based upon such a premise.[53] According to them, nationalisation was not a selfish or sectional demand, Smillie growling that any such accusation was 'a black and damnable lie'.[54] Rather, they were opposed to private ownership on grounds of national economic efficiency. The chaos of 1,500 competing colliery companies meant that the present system was 'extremely inefficient ... and costly and wasteful, with bad social results'.[55] State imposed rationalisation would lead to better management of resources, an increase in the use of machinery, and a co-ordinated system of distribution. All this would take place within a scheme where:

> pit will be compared with pit, district with district, system with system, manager with manager, cost with cost, in such a way that every part will be tested, and neither inefficient men nor unsuitable systems will be retained.[56]

The main benefits of nationalisation, namely cheaper coal and increased output, would be in 'the interest of the consumer', which was synonymous with the national interest.

52   Ibid.
53   Sidney Webb and Sir Leo Chiozza Money both submitted their own schemes, but these differed from that of the MFGB in organisational form rather than theoretical content. See CIC 1919, vol. ii, pp. 501–44.
54   CIC 1919, vol. ii, p. 322; *South Wales News*, 23 June 1919.
55   Sidney Webb, CIC 1919, vol. ii, p. 493.
56   William Straker, CIC 1919, vol. ii, p. 945.

The MFGB did not deny that the miners would also benefit from national-isation, in particular in the shape of improved safety.[57] However, the hallmark of its scheme was that class and national interests were not mutually exclus-ive, but rather should be one and the same according to 'moral and economic laws'.[58]

Although Straker claimed to view the labour unrest as 'the present hope of the world',[59] the MFGB's proposals to the Commission owed little or nothing to the idea of class struggle. Instead its tone was heavily moral, grounded in Fabianism and the ethical socialism of the ILP according to which private com-petition was not simply inefficient, but 'an evil-producing thing'. Opponents of nationalisation rested their case upon the essentially 'primitive idea' that 'life is an antagonism'. The MFGB, by contrast, held the mark of a civilised outlook to be one where:

> that which draws men together in co-operative activities makes for pro-gress and human welfare; that which keeps men in a hostile attitude one to another wars against welfare and progress.[60]

In the abstract, there was nothing here that revolutionaries could disagree with; out of the antagonism of the classes arises the possibility of classless-ness. However, in the hands of the labour representatives on the Sankey Com-mission, the dialectic vanished, the class struggle stage was removed, and the imperative became co-operation and consensus between the classes. As Smil-lie said at Keswick:

> What we want to do is reconstruction in the interests of the nation ... We want higher ideals introduced into life, not only for our own people ... We want higher ideals for the so-called upper classes than the mere idea that the be-all and end-all of life is to make fortunes and get honours. It ought to be the be-all and end-all of life to co-operate together for increasing human happiness, and not to become merely rich and outdo your fellows. We are quite prepared to enter into any combination which has that as its ultimate result.[61]

---

57   Sidney Webb put much stress on this; CIC 1919, vol. ii, pp. 478–9.
58   William Straker, CIC 1919, vol. ii, p. 945.
59   William Straker, CIC 1919, vol. i, p. 324.
60   William Straker, CIC 1919, vol. ii, pp. 944–5.
61   MFGB 1919 Annual Conference, p. 18.

The Federation's scheme for nationalisation in 1919 has attracted much interest from those wishing to trace the evolution of concepts of workers' control within the British working class.[62] Workers' control had been evolving as an element of syndicalist and revolutionary thought within the British working class over the previous two decades. Its lineage can be traced back to De Leonist ideas imported from the USA by James Connolly, to the syndicalism of Tom Mann's ISEL, to the National Shop Stewards' and Workers' Committee Movement, and to the Unofficial Reform Committee in South Wales. It grew out of, and was developed and refined as a concept in the context of, the class struggle. The Great Unrest of 1910–14 produced *The Miners' Next Step*, the wartime struggles in engineering produced the revolutionary shop stewards movement whose ideas were formalised by J.T. Murphy in *The Workers' Committee*, and the unrest in the pits in 1919 produced the SWSS's *Industrial Democracy for Miners*. Common to all these was the notion that control over the workplace was a staging post in the class struggle, a yardstick by which growing class consciousness, confidence and organisation could be measured, and a bastion from which a final assault on capitalism could be mounted. For Connolly, it meant 'to build up an industrial republic inside the shell of the political state'; for Ablett, it was an act in 'erecting the new society within the lap of the old'.[63]

However, in its scheme for nationalisation as presented by the MFGB at the Sankey Commission, the notion of workers' control was stripped of all revolutionary content. Hartshorn pointed to this as precisely the reason why it should be accepted, saying he was certain 'that unless the demand for state ownership is conceded at the present time, Syndicalism, or if you like, Bolshevism, will take the place of the demand being put forward by the miners', adding, 'now none of us want that'.[64] The guild socialism of Straker and Hodges was the theoretical bridge by which the MFGB took workers' control from the realm of revolution to that of reformism, where, rather than representing a high-water mark in the class struggle, workers' control becomes a kind of beatific state, in which even the colliery manager, of whom, Hodges said, 'we speak as a friend',[65] will be 'purged from the dross of carrying on for profit' and will experience 'a

62   Coates and Topham (eds.) 1970; Coates (ed.) 1974; Pribicevic 1959. The account here owes much to the excellent article 'Mines for the Nation or Mines for the Miners? Alternative Perspectives on Industrial Democracy 1919–21' by M.G. Woodhouse in *Llafur*, vol. 2, no. 3, Summer 1978.

63   Connolly, cited in Coates and Topham (eds.) 1970, p. 13; Egan 1986, p. 23.

64   CIC 1919, vol. i, p. 363.

65   Hodges 1920, pp. 125–8.

sort of psychological exaltation'.[66] The primary aim of joint control – in which Woodhouse points out the miners' representatives would be more a part of the management than of the rank and file – is to inculcate into the miner a sense of 'social industrial responsibility' without which, according to Money, 'you will never get the greatest benefit from your coal'.[67]

Joint control would allow the miner 'to feel the responsibility which would rest upon him as a citizen, and direct his energies for the common good'.[68] Somewhere, over the rainbow of state ownership, lay the end of industrial unrest in the pits. Under the MFGB's proposals there is a metamorphosis of the miner from militant troublemaker to mindful public servant. The demand for nationalisation, whilst arising out of the class struggle in the pits, was lifted out of its murky and slightly sordid depths, and turned into a 'supra-class' issue. This had, as Woodhouse points out:

> profound implications for the actual strategy adopted by the MFGB for achieving nationalisation. If consensus was the hallmark of its approach, this ruled out conflict and the exploitation of the industrial power of the Federation and its allies in the Triple Alliance.[69]

Little wonder that, in the context of a government-proclaimed national emergency, the MFGB leadership was highly susceptible to the appeal to co-operate in increasing output. A perceived Liberal openness to nationalisation also helped strengthen the hand of those inside the MFGB who believed persuasion might bear more fruit than coercion. The *Daily Herald* commented on the opening day of the MFGB Annual Conference on 15 July:

> It is considered that the miners now have an opportunity to improve their tactical position by accepting the Government's offer, and bringing about a substantial increase in the production of coal. If they are able to increase production so as to obviate necessity for any increase in the price of coal, it was suggested that the Government might not be uninfluenced by this fact when they come to take their decision with regard to nationalisation, and that it would remove a solid ground for opposition on the part of large sections of the community.[70]

---

66   Sidney Webb, CIC 1919, vol. ii, p. 488.
67   Sir Leo Chiozza Money, CIC 1919, vol. ii, p. 542.
68   William Straker, CIC 1919, vol. i, p. 324.
69   Woodhouse 1978, p. 98.
70   *Daily Herald*, 17 July 1919.

The apparently indeterminate nature of the attitude of government ministers to the broad principle of nationalisation continued to be an important factor in the strategic calculations of the MFGB during the crisis over nationalisation, keeping alive much longer than it deserved the hope that the majority Sankey Report might be adopted. To some, the mirage of Lloyd George's lingering radicalism also remained seductive. At Keswick, for instance, several delegates expressed the belief that, on the issues of intervention in Russia, conscription, and the use of the military in industrial disputes, Lloyd George was 'in sympathy with our principles'.[71] In apportioning blame for government militarism at home and abroad, Smillie said:

> It is not Lloyd George. The Government has placed Winston Churchill in his present position to deal with strikes, and it is a question of a struggle between Churchill and Lloyd George ...[72]

Influential figures in the labour movement believed Lloyd George's radical credentials extended as far as mines nationalisation. After discussions with him during May, Henderson told Webb that Lloyd George was 'extremely anxious for the [Sankey] report to be in our direction', and advised Webb to use his influence to moderate the tone of the MFGB's proposals as far as possible, so as to maximise their acceptability.[73]

The debate at Keswick on piece rate adjustments under the seven-hour day showed that Smillie did not tend to a view of the government as an implacable foe.[74] He warned delegates against seeing the owners, who were willing to make bigger concessions on this issue than the government, as amicable in any way. Their ulterior motive was to discredit nationalisation, when in fact, Smillie suggested, 'the Government are our best people'.[75] The MFGB leadership clearly considered that the government could be persuaded that nationalisation was in the best interests of British state capitalism, a fact that weighed heavily against a gloves-off response during the output crisis.

Their hopes were reinforced by the political instability of the Coalition and the widespread anticipation that an election was in the offing. Given Lloyd George's proven ability to shift political tack either left or right when the situation demanded, there was a good deal of speculation about which way person-

---

71   MFGB 1919 Annual Conference, p. 146.

72   Ibid.

73   Henderson to S. Webb May 17th 1919, Passfield Papers, (ii 4G, 6g, 62a).

74 . The piece rate question will be dealt with in detail in the next chapter.

75   MFGB Annual Conference 1919, p. 43.

alities and parties would line up. Various permutations of a Lloyd George-led centre party were put forward. The apparent Liberal revival signalled by recent by-elections led some to speculate that he would seek a re-accommodation. Arthur Murray, a Coalition Liberal MP, wrote in late July:

> Lloyd George already sees that the Coalition cannot go on for ever and he will have to make up his mind in which direction he is going to swing. I have always felt that he would turn to the Left and endeavour to rush the Liberal situation in the country.[76]

Others saw the possibility of Lloyd George moving further left, and joining forces with Labour, or at least its more moderate elements. This notion, although not credible in retrospect, held considerable currency in 1919. This was in no small part due to a sensational article in the *Evening News* in mid-April, which had Lloyd George's fingerprints all over it. Under the headline 'A General Election This Year', the correspondent revealed that he had it on good authority that Lloyd George intended to leave the Coalition, and he predicted what the basis of his appeal to the country would be:

> I am tired of these reactionary elements, these vested prejudices. They trim and prune my social programme until the tree of promise looks unlikely to yield fruit in due season. I stick by my programme. I ask you to endorse it. I will choose the men who will carry it out and that quickly.

The correspondent continued:

> He will make a bid for Labour. He will say that the principle of nationalisation is accepted by him so far as mines and railroads, and possibly shipping, are concerned. Labour [he will say] must join forces with him in order to make a success of nationalisation against the enemies of it.[77]

The article was written as a warning to Conservatives not to try and force Lloyd George's hand at the peace talks. To be effective, which it seems to have been, such a threat had to carry credibility with the Tories. Certainly there were plenty in Labour circles willing to believe this kind of radical political

---

76   Cited in Wrigley 1990, p. 201.
77   Cited in Wrigley 1990, p. 178.

realignment was a possibility. In late June, the *Daily Herald* mused about the imminent collapse of the Coalition:

> It is coming. It is very near. Mr. Lloyd George cannot keep his ramshackle Government together any longer. It is probable that, deciding to 'chuck' his reactionary friends, he will come round suddenly as the Friend of Labour, and make a 'dash for freedom'.

Nationalisation of the mines, the article concluded, would be the cement in this new alliance.[78] In reality nothing was further from Lloyd George's mind in the summer. There is no evidence to suggest that he had privately considered an election at all, and furthermore, during his stay at Criccieth in July, he made tentative long-term plans for a Centre Party that would be more anti-Labour than anything else, a permanent extension of the Coalition without, perhaps, the Tory right.[79]

However, in the period of manoeuvre over nationalisation, an approach to Labour was considered possible, and an election likely. On 3 June, Henderson instructed the Executive Committee's Organisational Sub-Committee to begin urgent preparations for an election which he expected 'either at the end of the present year or early in 1920'. It was decided to launch a 'National Campaign' in July and August, focussing on Labour's peace aims, and 'the general items contained in our New Social Order'.[80] Given the majority report of the Sankey Commission, and the publicity it had received, the nationalisation of the mines was the obvious hook on which to hang such a campaign. The Labour Party EC sent its congratulations to the MFGB, and recommended a joint venture of the two organisations, in conjunction with the Parliamentary Committee of the TUC, in 'a public campaign in support of the nationalisation of the mines'.[81]

With direct action threatening to dominate the forthcoming Labour Party Conference, Henderson wrote an article in the *Daily Herald* laying out the leadership's alternative:

> We are a minority in Parliament, and our weakness in the House of Commons is the measure of our failure to win the confidence of the masses whose interests we claim to have at heart. It is from this point of view a

---

78   *Daily Herald*, 28 June 1919.
79   Wrigley 1990, pp. 202–4. Wilson 1970, pp. 205–11.
80   Labour Party NEC Minutes, Box 4; EC Organisational Sub-Committee, 3 June 1919.
81   Labour Party NEC Minutes, Box 4; EC, 26 June 1919, & Joint meeting of EC and PC of TUC, 9 July 1919.

matter of enormous significance that the conference is confronted with a very real working class achievement in the majority recommendation of the Coal Commission in favour of nationalisation ... and recognition of the right of the workers to a share in the industry. In my judgement this will speedily produce a change in the political outlook wholly in favour of our Party. They are calculated to hasten the dissolution of the unnatural alliance of parties that masquerades at present as a Coalition Government. They provide Labour with a first class issue upon which to base its electoral campaign.[82]

The Sankey Commission was, for Henderson, an asset of unparalleled value for the Labour Party. The victory in the local elections in Durham earlier in the year had been an indication of the Commission's electoral potential. Now direct action threatened to rob the Labour Party of it just as an election was in the offing. If, on the other hand, the Labour Party, MFGB and TUC were to collaborate in a campaign for nationalisation:

[N]ot only shall we achieve a great triumph for the principle of public ownership and democratic control of industry, but we shall begin the propaganda which will convert public opinion to the support of the Labour Party in the coming election, which will not, in my opinion, be long delayed.[83]

There was, therefore, tremendous pressure being applied to the MFGB by the Labour Party not to waste this golden opportunity by resorting to a strike over nationalisation. At one point during the Labour Party conference, in the course of a debate on the merits of direct action in general, Henderson even threatened to split the party 'if all affiliated organisations were in favour of direct action'. Significantly, this led Robert Smillie to withdraw the motion in question.[84] The resolution on direct action passed by conference applied only to the issue of government intervention in Russia. On domestic questions such as the nationalisation of the mines, the Labour Party's policy remained solely electoral.

Taken together, these factors – the nature of the MFGB's nationalisation proposals, the possibility that the government might be persuaded to accept the Sankey Report, the belief that an election was imminent in which a favour-

---

82   *Daily Herald*, 26 June 1919.

83   Ibid.

84   Ibid.

able political realignment might occur in Westminster, and pressure from the Labour Party – produced a strong current of opinion within the MFGB, especially amongst the leadership, that winning public opinion behind the miners was of paramount and overriding importance. At the Keswick Conference, George Spencer told delegates that the Sankey Commission had brought them to 'within measurable distance' of that for which the union and the mining communities had been fighting for so many years:

> One feels, if we are going to realise the aim and object which we have been out for for such a long time, we have got to make sure that we have … the strong support of the general community of this country. We cannot afford, as powerful as we are, to run counter to public opinion at this time when public opinion can be easily turned against us.[85]

His argument hinged around the fact that, as he saw it, nationalisation depended on 'the complexion of the House of Commons'.[86] G.H. Warne, from Northumberland, who felt that Bonar Law's offer was 'an insult', nonetheless inclined to Spencer's view of the importance of public opinion:

> We, as a Miners Federation, along with other organisations who are prepared to give us assistance, have got to educate the people of the country, and put into their hands the case we have for the mines being nationalised. If we don't, then I am afraid that the finances at the disposal of these other people, through the medium of the Press and other ways, are going to so prejudice our case in the public mind that we are not going to get nationalisation.[87]

Warne called upon the MFGB to take the earliest opportunity to launch 'a campaign in this country that will educate the people up to nationalisation of the mines'. Here was early mention of what would become the Mines for the Nation Campaign, and all it implied for direct action. But that was still in the future. For now, the class struggle in the pits continued around the output crisis.

The question of falling output had come before the Sankey Commission, with various explanations put forward by both sides. There was some degree of consensus that a fall in output had been inevitable after the war: 'the cream

---

85   MFGB 1919 Annual Conference Report, p. 69. Spencer was just one of many delegates who spoke about the importance of public opinion at Keswick.

86   MFGB Annual Conference 1919, p. 43.

87   MFGB Annual Conference 1919, p. 83.

of the colliers' had been recruited into the army, many had been killed, and of those who had returned, many were disabled and unfit for heavy labour, or had simply 'not yet got back in their stride'. Those who had remained at home to work were tired out after the long years of pressure to maximise production. Furthermore, the return of the soldiers meant that the mines were overcrowded, with up to four times the usual number of men in one working place.[88]

However, beyond this the two camps divided; the owners accused miners of deliberately restricting the output, by operating what was called in Scotland a 'darg', or in South Wales a 'stint'. There was a long tradition of output restriction amongst miners in some areas; by this means coal prices had been kept up in the era of the sliding scale, when prices and wages had been linked.[89] It was also used in the opening up of new seams, during the trial period when the price list was negotiated, so that the piece rate could be maximised, allowing the miner to earn a living wage. This was still a constant source of conflict in 1919. At Ebbw Vale Collieries, for example, there was a month-long strike when the management cut piece rates on a seam where they accused the miners of operating a stint.[90] Output restriction, or 'ca' canny', was advocated by the URC as a method of struggle, and was an important element in its strategy, outlined in *The Miners' Next Step*, of profit reduction and encroaching control.[91]

Although it seems likely that in some areas output restriction was still practiced, the miners' representatives on the Commission argued, reasonably enough, that this was not nearly as widespread since the sliding scale had been scrapped.[92] In any case, the MFGB vigorously denied such a charge, and demanded a full enquiry into falling output.[93] No enquiry took place, but the available evidence strongly suggests that it was the owners and managers who were responsible for falling output. To what extent this was a deliberate and co-ordinated sabotage is unclear, but it was, according to Peter Lee of the Durham Miners' Association:

> Public property at all the mines in the country that the management, dir-
> ectly or indirectly, is trying to stop the output of coal. For one purpose

88   CIC 1919, vol. i, p. 306; vol. ii, pp. 1162, 1167. MFGB Annual Conference 1919, pp. 70, 86.

89   Campbell and Reid 1978, p. 64.

90   CIC 1919, vol. ii, pp. 1100–1. The manager claimed that proof lay in the fact that output started off very low at the beginning of the week, but picked up considerably at the end.

91   TMNS, p. 28.

92   CIC 1919, vol. ii, p. 952.

93   MFGB Annual Conference 1919, p. 67.

only – that they may make it bad for the miners to advocate nationalisation and win through.[94]

Many delegates at Keswick echoed the charge, including Smillie, who said 'there is a strong feeling amongst the miners that output has been lessened ... by a deliberate movement on the part of the mine owners'. The Coal Controller Evan Jones told the Cabinet that the owners were likely to 'do all in their power' to make the period of control unworkable:

> It was all to their advantage to do this, as the purchase of the mines by the state after the three year probationary period depended on the success of this scheme, and such purchase they wish to hinder in every possible way.[95]

Some believed that the owners' behaviour could be explained by apathy. Hedges talked of 'a growing indifference' on the part of colliery companies; there wasn't 'the incentive or the interest on the part of management to get the output there used to be'.[96] Richard Redmayne conceded that the owners 'might not proceed with the same ardour in the management of their concerns'.[97] Miners, however, believed that the owners were guilty of the planned sabotage of production. Lawson, of Durham, felt that impediment of production was 'deliberately done to make nationalisation impossible'.[98] Peter Lee reported that he had been told by owners that they would not undertake investment with nationalisation on the horizon.[99] There were complaints of 'rotten roads, shortage of rail, and bad tackle' from Nottinghamshire.[100] In Yorkshire, some branches complained of 'bad roads, and cannot get along the main roads'. Thirty-five complained of, amongst other things, 'abnormal places, hard coal, faults, bad haulage, old machinery, too many men, bad organisation and distribution of tubs'.[101]

The SWMF circulated a questionnaire to the lodges, the result of which S.O. Davies said 'proves to us absolutely that almost everything conceivable that

---

94   MFGB Annual Conference 1919, p. 73.

95   WC (589), 8 July 1919; CAB 80-1-180.

96   CIC 1919, vol. ii, p. 1186. MFGB Annual Conference 1919, p. 76.

97   CIC 1919, vol. ii, p. 1189.

98   MFGB Annual Conference 1919, p. 76.

99   MFGB Annual Conference 1919, p. 72.

100   H. Hicken, Checkweigher at Williamthorpe Colliery, *Nottinghamshire Evening News*, 21 July 1919.

101   MFGB Annual Conference 1919, p. 87.

could operate detrimentally to output has been carried out in South Wales by the owners'.[102] It was reported that:

> Matters are being aggravated by the employers, who are quite willing for the men to strike on very small questions, questions which during the war would have been settled without any loss of time. The object is quite clear. The employers are out to decrease the output ...[103]

W. Hogg, a delegate from Northumberland, told Conference of 'managers absolutely refusing to cooperate with the workmen on the Output Committees, when it was proposed to discuss ways and means to increase the output'.[104] From Lancashire too came reports of management stonewalling the L&CMF.[105] In Yorkshire, entire shifts were being sent home on spurious grounds; the slightest fall of dirt, arriving five minutes late due to a delayed train, or not having lamps ready well in advance of winding were some examples where this occurred.[106] The Branch Secretary at Brierley Colliery in Yorkshire wrote to Smith that the manager had told him that he did not care about falling output. 'It does not matter to him if the pit stops or not. He knows there is a scarcity of tubs, and does not care a d – [sic] if the pit works or not'.[107]

Shortage of the tubs into which the colliers loaded their coal was by far the most common complaint. For any manager who so desired, this was an easy and effective way of restricting output, for without the means to continually clear his working place, the collier could not work. A YMA survey of branches revealed that in 87 percent of the pits there was a shortage of tubs. At Mitchell Main, they had asked for 1,200, but had been issued with only 200. Ken Bacon, Branch Secretary at Briggs' Pit, said:

> The men are not working five or six hours a shift owing to wanting for empty tubs. A man told me on Sunday they filled three tubs in two days when they ought to fill forty, and the whole sequence is shortage of tubs.[108]

---

102   MFGB Annual Conference 1919, p. 77.
103   Ibid.
104   MFGB Annual Conference 1919, p. 86.
105   MFGB Annual Conference 1919, p. 78.
106   Statement issued by H. Smith after YMA Council meeting of 12 July 1919, in *Sheffield Daily Telegraph*, 14 July 1919; CIC 1919, vol. ii, p. 1198.
107   CIC 1919, vol. ii, p. 1197.
108   CIC 1919, vol. ii, pp. 1195–6.

The Brierley Branch Secretary wrote that 'we have men at our colliery continually coming back short of tubs. It is a very common thing for them to come out at half-shift time'. From Micklefield Main, the Branch wrote regarding the allegations of the owners that the miners were deliberately curtailing production: 'We thought the boot was on the other leg and that it was the owners who were restricting the output by not getting the men plenty of tubs'.[109] In Scotland and South Wales, many pits were only working three or four days per week for the same reason.[110] From Lancashire, Greenall reported 'men being down in the mines day after day and week after week, hours there waiting for tubs: plenty of coal to fill, but cannot get the tubs'.[111]

The sheer volume of such reports strongly suggests that there was a systematic attempt by many owners to reduce output. Redmayne may inadvertently have backed this up when he denied any special tub shortage at the Sankey Commission, saying that during the war the Coal Controller had been constantly pestered for tubs, but now the number of requests from managers was 'practically nil'.[112] Managers may simply have been choosing not to report tub shortages with a decision on nationalisation in the offing. There were even shortages of tubs and rails at some collieries belonging to companies which made these in their own iron and steel plants.[113]

Miners were incensed that, as Herbert Smith said, 'it has gone out to the press that the miner is a bad man, is robbing poor people, not supplying the coal'.[114] The NMA Council issued a statement that:

> It is abominable – in view of the supposed shortage of coal, and all that is being said about the reduction of output – that materials are being studiously kept back, and many waggons withheld.[115]

Miners reckoned that with properly equipped mines they could increase the output by as much as 30–50 percent. This was first and foremost a matter of wages, not public image, though undoubtedly this mattered to them too. For miners who were paid by the piece, long breaks in production involved a drastic reduction in income. At Barmborough Main, for example, an average

---

109   CIC 1919, vol. ii, pp. 1197, 1199.

110   *Daily Herald*, 12 July 1919.

111   MFGB Annual Conference 1919, p. 78.

112   CIC 1919, vol. ii, p. 1186.

113   CIC 1919, vol. ii, pp. 1101, 1187. The examples cited were both in South Wales.

114   CIC 1919, vol. ii, p. 1198.

115   *Mansfield Reporter*, 11 July 1919.

of 248 working shifts a week had been lost since 9 January.[116] In Lancashire, it was estimated that 90 percent of the colliers were on the minimum wage, and there were complaints that 'the miner is no better off than before the war, with food prices doubled and income tax to pay'.[117] Branches here were sending in demands for another 30 percent wage rise, and in the Bolton area there was an unofficial one-day strike against high food prices.[118] At the Annual Conference, there were strong protests against 'profiteers' who had taken advantage of the Sankey Award by increasing the prices of 'practically all uncontrolled goods' in mining areas, and a resolution was passed calling for a reduction in the price of food and clothes.[119]

Where output restriction was at its worst, and miners were losing large amounts of money, angry protest often gave way to industrial action. In order to maintain their wage levels, increasing numbers of miners found that they were forced to engage in a struggle with management over output. At Ashington in Northumberland, where it was 'quite impossible to take the output away from the mine', the men struck for the first time in nine years. This forced management to increase the supply of tubs, and output immediately went up by 20 percent.[120] In Lancashire, there were walkouts in protest at being forced to stand idle underground.[121] In Nottinghamshire, South Normanton and Rufford, colliers were often working only one or two days per week during June and July. Miners in the coalfield reckoned that with sufficient 'trams' they could increase output by up to 50 percent, and here too there were strikes over the issue.[122]

At Keswick, S.O. Davies reacted angrily to the accusation that the miners had anything to do with falling output:

> Many of the men present in this Conference … although direct representatives of the miners have done infinitely more to keep up output in some parts of South Wales than the paid hirelings of the companies.[123]

---

116   CIC 1919, vol. ii, p. 1198.
117   CIC 1919, vol. ii, p. 1186. *Wigan Observer*, 26 July 1919. Letter from a Lancashire miner. Income tax, which had been payable on incomes over £160 before the war, was now payable at £130, with a devalued currency.
118   See Chapter 8.
119   MFGB Annual Conference 1919, p. 121.
120   CIC 1919, vol. ii, p. 982; MFGB Annual Conference 1919, p. 76.
121   CIC 1919, vol. ii, p. 1186. See for example the letter from the checkweigher at Pilkington Colliery.
122   *Mansfield Reporter*, 11 July 1919; CIC 1919, vol. ii, p. 1217.
123   MFGB Annual Conference 1919, p. 77.

He gave the example of a strike at one particular colliery which had been successful in forcing the company, which made tubs and rails in its own iron and steel plant, to bring these into the pits. At Nine-Mile Point, men struck because managers 'were developing the worst seams ... and leaving the best seams until the market conditions became better and the control was lifted'.[124] At several collieries there were 'safety strikes', where poor management or development was creating life-threatening situations.[125] Rhondda District SWMF threatened strikes unless the owners employ rippers and borers to do the blasting, ripping, and clearing top, rather than have the colliers do these tasks themselves, thereby 'allowing [them] to increase the output and at the same time finding employment for the discharged soldiers'.[126]

John Potts gave delegates at Keswick an example of the situation in Yorkshire:

> I was at a colliery a few weeks ago where twenty two men stopped, because it took two or three men to push an empty tub. I asked for mechanical appliances to be adopted, but the manager refused, absolutely. I threatened to stop that colliery. I went down the mine, and I never saw a mine in all my life better adapted for mechanical appliances than in that mine. The manager said that he was not going to spend any money or do anything in that, as it was impossible, until the question of nationalisation was settled ... Ultimately he gave way, and he is now agreeing, through pressure, to put in mechanical contrivances, but it was only through pressure and nothing else.[127]

Nationalisation was not an abstract question for the miners, an added extra to the concessions that had been won in the first stage of the Sankey Commission. Nor was it about the national interest, the testimony of the miners' representatives at Sankey notwithstanding. On the contrary, it was posed extremely sharply as a question of the miners' very livelihood. So long as the issue remained unresolved, the ability of the miners to earn a decent living was under threat. Not only were they less concerned with the output crisis as a matter of public opinion than their leaders, they were more open to the idea of responding to the attack on nationalisation by means of direct action. Furthermore, the issue of

---

124   CIC 1919, vol. ii, p. 980.
125   Ibid. For example, at Risca there was a strike because ropes were cutting into rail and timber.
126   *Rhondda Leader*, 2 August 1919.
127   MFGB Annual Conference 1919, p. 87.

control was raised in concrete form, one which was a world away the official MFGB scheme. To maintain acceptable wage levels in these summer months, many miners had no choice but to engage in a struggle with the management over control of production. To the miner in the pit waiting for a tub into which to load his coal, the interconnected issues of output and nationalisation presented themselves in a very different way than to the well-paid official in union headquarters.

There were, therefore, powerful conflicting forces at work within the Federation when its Annual Conference opened on 15 July at Keswick. As we have seen, the EC shared many of the sentiments voiced by Brace and Hartshorn about the output crisis, and appear to have been unsettled by Bonar Law's offer. The EC admitted that it had not taken a position on the offer, Smillie himself studiously avoided making any reference to it in his introduction to the session, and *The Times*'s labour correspondent reported that 'there are indications ... that the Executive of the MFGB are not indisposed to accept the offer'.[128] Only when the hostility of the majority of the delegates became clear did the EC come down on the side of rejection.[129] This spelled defeat for Brace and Hartshorn's attempt to take direct action out of the nationalisation equation once and for all. Themselves aside, only Watts Morgan MP voted for accepting the offer. Nonetheless, about fifty percent of the speakers in the discussion had been equivocal about acceptance at best, and the final position involved a pledge to the government of 'the whole-hearted support of the leaders of the miners' to increase the output if nationalisation was accepted in principle.[130]

At this point it was already apparent to all but the hopelessly naïve that the government was looking for a way out of nationalisation. Yet Smillie told delegates that he was working on the assumption that the government still intended to carry out the recommendations of the majority report, and went so far as to say that in his opinion 'there was no question of the Government's intention to nationalise the mines'.[131]

> It is not often that we can be in full agreement with the Government, but we would pledge our word on this occasion that we are in hearty sympathy with them in their desire to carry out the recommendations of the Royal Commission, and that we will give them all the assistance in our

---

128   *The Times*, 16 July 1919.
129   MFGB Annual Conference 1919, pp. 65–90.
130   MFGB Annual Conference 1919, p. 67.
131   *The Times*, 15 July 1919.

power to overcome any unlawful assembly of capitalists that is breaking out in rebellion against the Government in the House of Commons.[132]

The idea that Lloyd George might hitch himself to the wagon of direct action in order to overcome Tory opposition to nationalisation of the mines is of course absurd, but the alternative was for the EC to recognise that nationalisation was slipping away, and take industrial action, and that it was not prepared to do. At the MFGB conference, stalemate had been reached between those who staked everything upon public opinion, and the direct actionists. Who would come out on top would be decided in the course of the industrial unrest which had begun even before the delegates left Keswick.

---

132   MFGB Annual Conference 1919. p. 18.

CHAPTER 8

A Second Wave of Unrest

There is no fever more infectious than the strike, and no class of men more susceptible than the miners. The country is evidently beginning to enter another cycle of labour ferment.[1]

∵

The EC had managed to defend its 'wait and see' policy and keep the MFGB invested in the Sankey process. However, question marks were raised over how long it would be able to resist pressures for action in the form of widespread disgruntlement in the collieries over the practical implications of the Sankey Award which came into effect on Wednesday 16 July. On the same day, the month long Yorkshire mines strike began. These grievances, in combination with the output crisis and the 6s increase, were responsible for a growing dissatisfaction with Sankey, and let loose a new wave of unrest which touched every major coalfield and threatened to engulf the Federation. In the week following Keswick, approximately 200,000 miners went on strike over issues arising from the Sankey Commission in Scotland, the North-East, Lancashire, Nottinghamshire, Derbyshire, the Midlands, South Wales and Kent. When the Yorkshire strike is included, the movement at its peak involved around 400,000 miners, or about half of the Federation's total membership. Although outside Yorkshire the strikes were brief affairs, they were a highly important component of the general crisis in the days immediately before the government's final decision on nationalisation. The Yorkshire strike can only be appreciated in the context of, and in comparison with, these unofficial strikes. To facilitate comparative study the unofficial strikes are dealt with separately in this chapter, but they played a supporting role to the Yorkshire strike as the nationalisation drama reached its climax.

The practical implementation of the Sankey Award's reduction in hours left many grades profoundly dissatisfied. In the first place, miners in many districts had obviously interpreted the Award to mean a universal reduction

1 *The Times*, 22 July 1919.

of one hour for all underground workers. 'How can a misconception arise', Robson asked from Durham. 'The Award is not one hour less, but to substitute the word "seven" for "eight" in the Act'.[2] The misconception arose because in several districts where the miners had been able to secure shorter working shifts than the eight hours specified in the 1908 Act, they had assumed that these local differentials would continue. In Kent, where hewers only worked seven hours and forty minutes, they had expected a reduction to six hours forty; in Yorkshire, miners at double shift pits who worked a seven and a half hour day fully intended to see this come down to six and a half. Even in some districts where a full eight hours were worked Monday to Friday, the issue was complicated by a customary short Saturday shift.[3]

An even bigger problem, which affected all areas, was the question of how far piece rates should be increased to guarantee the continuation of the same earnings with a shorter working day. Having reached agreement on this in principle, the government and MFGB had decided, because of local variations in geological conditions and price lists, that the details should be worked out between miners and owners at a district level. Negotiations had been proceeding in all areas, with varying degrees of success, when the Coal Controller, on the instructions of the Cabinet, announced that there would be a 10 percent ceiling on piece rate adjustments in all coalfields.[4] This figure corresponded to the Interim Report's estimate that national average output of coal would fall by 10 percent with the shorter working day.[5] The ceiling was later increased to 12 percent, with the proviso that the national average adjustment did not exceed 10 percent, after the MFGB had pointed out that in the North-East, no increase would be required.[6]

The ceiling was described by James Winstone as a 'bombshell'.[7] Latham of the Midlands Federation declared that 'it has got to be a bigger man than myself to come into my district and tell us to accept 12%'.[8] The miners insisted that the basis for calculating the adjustment should be the loss of hewing time, rather than the projected average fall in national output. The most common figure put forward was 14.3 percent, or the one-seventh increase necessary to bridge the difference between the eight and the seven hour shifts, but even this

2   MFGB Annual Conference 15 July 1919, p. 40.
3   MFGB Annual Conference 15 July 1919, pp. 29–44 and 17 July 1919, pp. 94–110.
4   MFGB Annual Conference 15 July 1919, pp. 27–8.
5   CIC 1919, vol. i, 4(a) Interim Report, p. x.
6   MFGB Minutes of a Meeting with Bonar Law, 5 July 1919.
7   MFGB Annual Conference 15 July 1919, p. 36.
8   MFGB Annual Conference 15 July 1919, p. 29. The owners had agreed to 15 percent.

was seen as the absolute minimum that the affected members would accept. For those miners who had to walk long distances to their workplace from the pit bottom (three miles was not uncommon), the reduction of one hour represented a much greater proportion of their hewing time than one-seventh. Frank Hall and George Spencer calculated on the basis of returns from pits in Derbyshire and Nottinghamshire that the average increase needed was at least 16 percent and that for some men it was as much as 19 percent. Similar claims were made by delegates from Yorkshire, Lancashire and South Wales.[9] Suddenly, the Sankey 'Award' appeared to mean a pay cut.

Surface workers were also very unhappy with the Sankey Award under the terms of which their hours were only reduced to 46.5 per week, widening the gap between themselves and the colliers. Smillie reacted angrily at Keswick to threatened strike action by surface workers, saying that colliers had shown self-sacrifice during the war when they had accepted flat-rate increases which benefited the lower paid workers. In any case, he pointed out, he had warned earlier in the year that:

> We might find it was impossible to fix as short hours for the surface worker as underground worker, because we could not prove the same justification for the above ground man as the below ground man. Our claim all along has been for shorter hours for the miners because of the working conditions underground.[10]

This was, in fact, not true. As some delegates pointed out, the original claim in the Charter was for a six-hour day for all workers in and about the mines.[11] It was only in the hands of the EC that the basis of the claim had been altered. In any case, surface workers believed they were just as deserving of a shorter working day as the colliers were. Bailey from the YMA said that:

> Our surfacemen, especially screen-men, who are manipulators of coal, are working under equally as bad conditions as the bottom man. We have boys thirteen years of age who are working on the screens. In some instances the conditions in some of the screening plants are deplorable. There is dust flying about, and the lads suffer more than the men underground.[12]

---

9   MFGB Annual Conference 17 July 1919, p. 97.
10   MFGB Annual Conference 18 July 1919, p. 130.
11   MFGB Annual Conference 18 July 1919, p. 131.
12   MFGB Annual Conference 18 July 1919, p. 131.

The EC had held out the hope of further concessions for the surface workers over the previous months, but now with the Award's implementation it was clear that these had not materialised, and it looked as if anger might spill over into action. The South Wales delegation received telegrams on the last day of the Keswick Conference that surface workers were out in Monmouthshire. Barker said that 'The surfacemen have now, at the last moment, kicked over the traces, and there is a serious difficulty'. Ten thousand were on strike, 'because their relative position to the underground men has been worsened by the Sankey Award'.[13]

This was an alarming state of affairs for the EC. To have Yorkshire, the second biggest district affiliated to the Federation, out on strike was bad enough. Officials from most districts were warning that they could not keep their members at work much longer for anything less than a 14.3 percent rate adjustment. For Smillie, the possibility that the Federation might be 'forced into a national fight when there is a penny or twopence one way or the other in dispute' was a matter of tremendous concern.[14] Again and again he interrupted delegates at Keswick to warn that 'Probably the whole existence of this Federation may depend upon our being able to keep together on a question of this kind'.[15] During the piece rate debate, he said:

> I would like to say to the Conference that if ever there was a time we required to hold together it is now. I don't want this great Federation to run the risk of wrecking itself on the rocks of what is comparatively a small thing, when this Federation has far bigger national questions than this question.[16]

Smillie's priority was to preserve the organisational integrity and cohesion of the Federation in face of the centrifugal forces which were pulling at its seams; this meant that local and sectional grievances should be subordinated and held in check in the interests of national priorities. He was equally eager to keep the prospect of progress through the Sankey Commission alive; this meant that all districts must strictly observe the Sankey Award. To strike against the terms of the Award, which had been accepted by a ballot vote of the members was both 'a violation of the principles of this Federation' and 'a breach of faith with the

---

13   MFGB Annual Conference 18 July 1919, p. 128.

14   MFGB Annual Conference 17 July 1919, p. 100.

15   MFGB Annual Conference 17 July 1919, p. 109.

16   MFGB Annual Conference 15 July 1919, pp. 43–4.

Government'.[17] It would undermine the credibility of the MFGB as a reliable participant in collective bargaining, and damage its public image. As Smillie argued:

> We cannot face the public in asking the Government and threatening to force the Government to carry out the Sankey Award so far as it gives us any benefits, and in the same breath support men who refuse to carry out the Sankey Award because it does not suit them. The thing would be too ridiculous. We cannot honestly face the country or the men if we do business like that.[18]

Officials from several districts emphasised similar points, and Hartshorn, with the EC's backing, urged the Conference to pass a declaration calling upon 'all surface workers to loyally observe the Award of the Interim Report of Mr. Justice Sankey in accordance with the decision of the ballot vote of the members'.[19] Hartshorn also proposed the EC's resolution for piece rate adjustments, which called for a maximum increase of 14.3 percent. He clarified the EC's position:

> If there is a reduction of one hour per day, then it is intended that we shall not either accept anything less or ask anything more than 14.3%. If the reduction is less than an hour per day, that it shall be in the same proportion. If the reduction is one hour per day on each day, it will be 14.3%. If there is an hour per day on five days and no reduction on the Saturday, there will be five-sixths on that.[20]

Both resolutions were carried, but opposition in the coalfields was far greater than it had been on the conference floor. The surface workers were to receive no support whatsoever from the Federation, nor were piece workers who wished to hold on to local customs like the short Saturday shift. Greenall predicted that 'when the men get to know in Lancashire, they would be out as well with the Yorkshire people', while Barker reckoned they were about to witness 'internecine warfare in the Federation'.[21]

Anger at the inequities of the Sankey Award was so acute that in some areas outside Yorkshire, unofficial strikes were already underway before the

17   MFGB Annual Conference 18 July 1919, p. 133.
18   MFGB Annual Conference 18 July 1919, p. 130.
19   MFGB Annual Conference 18 July 1919, p. 133. It was carried nearly unanimously.
20   MFGB Annual Conference 17 July 1919. p. 94. Carried 601 to 177.
21   MFGB Annual Conference 17 July 1919. p. 104, 18 July 1919, p. 133.

issues were debated at Keswick. Over 1,000 Kent miners at Snowdown and Tilmanstone near Dover, and at Chester near Canterbury had been out since the beginning of the week, fearful that they were 'going to lose every custom and every little privilege we had previously' under the new working arrangements being demanded by the owners, notably the short Saturday shift.[22] In South Wales, practicalities around the seven-hour day had caused 'an epidemic of strikes' in the Afan Valley, parts of the Rhondda, as well as amongst the surface workers of the western valleys of Monmouthshire who were demanding the seven-hour day for themselves; at Blackwood, 2,000 miners were the first of many in Britain to strike in protest at the 6s increase in the price of coal.[23]

On Thursday 17 July, the strike spread northwards to Staffordshire, Derbyshire, the North-East, and Scotland. In south Staffordshire, the strike was once again started by surface workers who wanted their hours reduced in proportion to those of the colliers. The issue of the piece rate adjustment was further complicated by the local continuation of the butty system in south Staffs, and the contractors' unwillingness to pay their drawers for the customary short day on Saturday unless 14.3 percent was conceded. During the following week, the action was spread by roving mass pickets until all 15,000 Cannock Chase miners were out. The result would be, as in Nottinghamshire and Derbyshire earlier in the year, the butty system being scrapped, albeit temporarily.[24] In north-west Staffs, the strike started at Audley where workers demanded 14.3 percent, a further cut in hours for surface workers, and the removal of the 6s increase in the price of coal. 'Mobs of lawless pit lads' pulled out 5,000 miners at nine pits.[25] In Scotland, four pits struck in Dumbartonshire to retain local customs under the seven-hour day, and 2,000 struck at Bowhill in Fife for similar reasons, the action spreading over the following days.[26] Even in the usually calm North-East coalfield where hewers already had a seven-hour day, grades like shifters who had worked under the terms of the Eight Hour Act, and surface workers once again, caused strikes of 15,000 in Northumberland, at Ashington, and in South Durham.[27]

---

22   MFGB Annual Conference 17 July 1919, p. 132.

23   *Western Mail*, 17 July 1919; *South Wales Echo*, 18 July 1919.

24   *Staffordshire Chronicle*, 19 July 1919; *Staffordshire Sentinel*, 24, 25, 28 July 1919; *Cannock Chase Courier*, 26 July, 2 August 1919.

25   *Manchester Guardian*, 24 July 1919; *Staffordshire Sentinel*, 23 July 1919; *Daily Herald*, 24 July 1919.

26   *Daily Herald*, 18, 22 July 1919.

27   *Daily Herald*, 18, 26 July 1919; *The Times*, 22 July 1919.

The miners' unions in Nottinghamshire, Derbyshire and Lancashire, after consultation with the membership via the branches, put in for a 14.3 percent increase in piece rates. Behind the Lancashire delegation's vote against the Executive at Keswick lay profound unrest in the coalfield over the increasing cost of living, with many branches already demanding another 30 percent rise in wages. In the Burnley area there had been a recent one day unofficial strike in protest at high food prices, probably the work of the local Miners' Reform Committee.[28] Whilst the Derbyshire delegation were persuaded by Smillie to vote with the MFGB EC at Keswick, it did so against the specific instructions of the DMA Council to vote for a national strike if no settlement of 14.3 percent had been reached.[29] Restlessness in these counties and Notts was sharpened by their proximity to the Yorkshire coalfield, where, unlike their own officials, the YMA leadership had sanctioned a fight.

The strike in Nottinghamshire and Derbyshire was sparked by the York-shire strike, when Yorkshire owners despatched coal trucks for filling in these adjacent counties. As these arrived in the coal yards, from Cresswell in North-East Derbyshire, to Ripley further south, and across to Mansfield in Notting-hamshire, spontaneous walkouts took place.[30] Once again Mansfield rapidly became the organisational focus of the movement, its marketplace the venue for large daily gatherings of miners from both counties. From here on Sunday 20 July, the familiar figures of Owen Wilcox, Price, Ford and Bromley called upon the 3,000 miners present to spread the strike at pithead meetings the next morning. By the time mass meetings assembled the following evening, 20,000 were already on strike.[31] At Mansfield, Walter Owen said, 'If Yorkshire were out for justice, Notts and Derbyshire were coming out too', whilst Wilcox declared 'that what was fair for Yorkshire was also fair for Notts and Derbyshire, and for that reason they were justified in joining hands'.[32] At Chesterfield, Frank Lee the DMA agent appealed for a return to work, but was met with cries of 'When the Yorkshire miners go back'.[33] At Sutton, the mass meeting on the Lammas recreation ground was more cautious, voting to send a delegation to the NMA headquarters at Basford, but the following evening a reconvened meeting voted

28   L&CMF Monthly Conference 28 June 1919, p. 10; *Yorkshire Post*, 23 July 1919.
29   DMA Special Council Meeting 12 July 1919.
30   *Derbyshire Times*, 26 July 1919; *Nottingham Evening News*, 22, 25, 26 July 1919.
31   *Mansfield Reporter*, 25 July 1919; *Mansfield and North Notts Advertiser*, 25 July 1919; *Notting-ham Evening News*, 21 July 1919.
32   *Mansfield and North Notts Advertiser*, 25 July 1919; *Nottingham Evening News*, 22 July 1919.
33   *Mansfield Reporter*, 25 July 1919.

for 'a sympathetic strike'. Varley reported that the delegation had said as York-shire, Derbyshire, and the big pits around Mansfield were out, 'they felt they were sort of blacklegging'.[34]

The pattern of the strike had much in common with the ones in this area earlier in the year. The centres of militancy were Mansfield in Nottingham-shire and Chesterfield in Derbyshire, with weak points in the Leen Valley and Bolsover. Sutton and Clay Cross lay somewhere in the middle, joining the unof-ficial movement but remaining relatively moderate. Officials declined invita-tions to attend the Mansfield Market Place meetings, no doubt recalling the futility of their earlier forays into unofficial territory, preferring instead to con-centrate their troubleshooting amongst the more moderate miners in villages like Heanor, Shipley, Radford, Newstead and Hucknall.[35] When they did ven-ture into the larger centres, they faced a hostile reception, none more so than the one at the YMCA Hall in Mansfield on Wednesday 23 July, when appeals for patience were howled down with shouts of 'We've had too much!', and Var-ley Spencer and Bunfield were booed off the stage. After the meeting, Bunfield commented that 'it was useless to place any resolution before the men while they were in their present temper'.[36] Even where the officials were able to speak, their resolutions for a return to work were heavily defeated.[37] Once again, the unofficial meetings were, apart from organisational centres, forums for an out-pouring of criticism of the 'antiquated' rules of the miners' associations, and the 'reactionary' nature of their leaders.[38]

At the same time as insisting that 'their leaders had got to be more progress-ive', the unofficial group sought to extend the action from its base in Mansfield, sending 'strike agitators ... all around the districts'.[39] These would address pit-head meetings in a targeted area then organise mass meetings where those pithead votes could be confirmed by all the miners of a particular area. In this way the strike area increased daily until by Thursday 24 July 60,000 miners had joined the movement across the two counties.[40] Mass meetings once more dis-played a transfer of loyalty from official to unofficial leaderships:

34   *Nottingham Evening News*, 23 July 1919; *Mansfield Reporter*, 25 July 1919.
35   *Nottingham Evening News*, 22, 23 July 1919.
36   *Mansfield and North Notts Advertiser*, 25 July 1919; *Sheffield Daily Telegraph*, 24 July 1919.
37   *Nottingham Evening News*, 21 July 1919; *Derbyshire Times*, 26 July 1919.
38   *Nottinghamshire Free Press*, 25 July 1919; *Nottingham Evening News*, 24 July 1919.
39   W. Owen speaking at a mass meeting in Mansfield's YMCA Hall, *Nottinghamshire Guard-ian*, 24 July 1919; *The Times*, 24 July 1919.
40   *Nottinghamshire Free Press*, 25 July 1919; *Derbyshire Times*, 26 July 1919.

The strikers have flouted their leaders and have got out of hand. There was evidence of that at Sutton on Tuesday evening and again at Mansfield on Wednesday evening, when officials holding prominent positions in the local and county mining world were either howled at or listened to in chilly silence. They were given fair play after urgent appeals had been made for it, but it was obvious that they were merely tolerated. On the other hand when the unofficial 'bosses' got on their legs they were acclaimed as heroes, and their extravaganzas aroused the greatest enthusiasm.[41]

Although the Lancashire delegation had registered the dissatisfaction of its members by voting with Yorkshire at Keswick, the L&CMF officials, 'a thoroughly sensible, shrewd, straight-forward set of businessmen', had no intention of following the example of their YMA counterparts in defiance of MFGB policy.[42] As in Nottinghamshire and Derbyshire, the price paid for unity with the Federation was profound division within their own organisation.

The Lancashire coalfield did not fit into the mono-industrial mould of, for example, South Wales or Durham. Most miners lived in areas where they made up less than 30 percent of the working population, often having to travel large distances from home to pit; hence there was less of a sense of 'community' to help sustain trade union organisation. Often it was easier to simply switch to another industry than fight for better conditions, another reason that Lancashire did not have a particular reputation for militancy at this time. 1919 marked something of a turning point in the attitude of the Lancashire miners, particularly those in the Wigan district, who were to gain an albeit short-lived reputation for militancy in the 1920s.[43]

The seriousness of the discontent amongst the membership over the coal price and output crisis became apparent on Sunday 20 July, when at a large 'Hands off Russia' protest in Wigan, calls for action over the government policy on coal came to dominate proceedings.[44] The next morning at Wigan's London and North Western Railway station, about one thousand miners waiting for trains to carry them to work in the pits which lay to the south-east of the town held an impromptu meeting and decided to strike.[45] During the day, pavement chalkings appeared around the town, advertising an unofficial meeting to

---

41   *Mansfield and North Notts Advertiser*, 25 July 1919.

42   *Lancashire Daily Post*, 25 July 1919.

43   Stephen Catterall, in McIlroy et al. 2009, pp. 190–2, 202.

44   *Wigan Examiner*, 26 July 1919.

45   *Wigan Observer and District Advertiser*, 26 July 1919.

'compel the leaders to do their duty'.[46] This meeting voted to spread the strike across the coalfield, as 'they felt instinctively they should be out with the Yorkshiremen'.[47]

Wigan became 'the storm centre in the Lancashire coalfield' due to the presence of an active Miners Reform Committee (MRC). The MRC had built up a base here in the latter stages of the war, holding weekly meetings in Wigan Market Place, and had recently been assisted by a John Maclean speaking tour.[48] Their audience was suddenly greatly enlarged by the Yorkshire strike and the unwillingness of the L&CMF officials to lead similar action. The *Yorkshire Post* reported that:

> Thanks to the action of a number of irreconcilables, who style themselves the Miners' Reform Committee, the influence and authority of these officials have been completely flouted, and their efforts to secure peace spurned.[49]

Wigan Market Place was to Lancashire what Mansfield Market Place was to Notts. Here, a 'daily collection of hotheads' gathered to discuss the progress of the strike and decide on how to extend the action.[50] The *Lancashire Daily Post* reported that 'the leaders of these men, [who] now call themselves the Miners' Reform Committee, revealed their existence in startling fashion', with their ability to involve thousands of miners in picketing designed 'to cause a general stoppage'.[51] Between two and three thousand men were organised into flying columns hundreds strong each morning, and allocated a district for picketing.[52] One observer described the scene:

> The irreconcilables refrained from violence. They began to get busy early yesterday morning, when they assembled in the Market Place. Their leaders impressed upon them the wisdom of being as orderly as possible. Then they fell in like a regiment of soldiers – many of them indeed were exservice men – and marched with martial tread to the Park Lane and Long

---

46   *Wigan Examiner*, 26 July 1919.

47   L&CMF Monthly Conference, 26 July 1919. This explanation of the strikers' motives was offered by P. Newhall.

48   *Yorkshire Post*, 26 July 1919. MacDougall 1927, p. 774.

49   *Yorkshire Post*, 26 July 1919.

50   Ibid.

51   *Lancashire Daily Post*, 25 July 1919.

52   *Leigh, Tyldesley and Atherton Journal*, 25 July 1919; *Wigan Examiner*, 25, 26 July 1919.

Lane collieries of the Garswood Iron and Coal Co., where they persuaded the management to withdraw the men from the pits.[53]

Although extra police were drafted into Lancashire during the trouble, their numbers were insufficient to guarantee protection of plant and equipment, and managers invariably agreed to close the pits.[54] The strike was spread from Wigan to Burnley, where all pits, employing about 4,000 men, were closed.[55] On Tuesday, 8,000 Leigh miners joined the strike wave, which on Wednesday rolled on to Bolton and Ashton-in-Makerfield until between 30,000 and 40,000 Lancashire miners were on strike.[56]

If anything, the antagonism between union bureaucracy and rank and file activists was even greater here than in Nottinghamshire and Derbyshire. L& CMF officials denounced the flying pickets as 'the mob', and 'extremists incapable of being satisfied by any concession'.[57] Encouraged by press attacks on 'the Wigan Dictators' who were causing 'terrorism in Lancashire', one 'prominent miners' leader' went as far as calling on 'the authorities to round up MRC members for the country's good'.[58] In the eyes of the officials, the strikes were subversive and disloyal. Clear majorities in favour of action at mass meetings were overruled by district agents. This heavy-handed approach enraged the MRC and its supporters, who labelled the officials 'the enemy within', and argued for the formation of strike committees 'with full powers to carry on the strike to the end, without the aid of the men's union, whose methods they ought to dish'.[59] At one meeting, after a heated exchange with the officials, a resolution was passed to cease contributing to the union funds.[60]

The most immediate and most general cause of such anger in all three counties was the leaderships' refusal to sanction solidarity action with the Yorkshire miners and fight for their own demands relating to the introduction of the seven-hour day. It was on this basis that the rank and file groups initially spread the action. However, it quickly became apparent that this issue was a

---

53   *Lancashire Daily Post*, 25 July 1919.

54   *Leigh Chronicle*, 1 August 1919; *Manchester Evening News*, 25 July 1919.

55   *Manchester Guardian*, 23 July 1919.

56   *Daily Herald*, 25 July 1919; *Burnley News*, 26 July 1919; *Manchester Guardian*, 29 July 1919; *Bolton Evening News*, 25, 26 July 1919; *Leigh Chronicle*, 1 August 1919.

57   L&CMF Monthly Conference 26 July 1919; *Manchester Guardian*, 24 July 1919.

58   *Manchester Guardian*, 26 July 1919; *Yorkshire Post*, 26 July 1919; *Leigh, Tyldeslev and Atherton Journal*, 25 July 1919.

59   L&CMF Monthly Conference 26 July 1919; *Bolton Evening News*, 22 July 1919.

60   *Bolton Evening News*, 22 July 1919.

lightning rod for a wider discontent. Bonar Law alluded to this in the House of Commons when he told MPs:

> The information we get today is to this effect. Some of them were opposed to the increase of 6s per ton; some were out because food prices were too high; some because we still have troops in Russia; and some because they wanted the abolition of conscription.[61]

The *Daily Herald*, commenting on the situation in Nottinghamshire and Derbyshire, also noted the mixture of political and economic demands which had been thrown into the ring. 'Some are out over piece rates, some over 6s; other pits have political demands on the subjects of conscription, conscientious objectors, and Russia'.[62] The *Nottingham Evening News* listed the causes of the strike as the 6s increase, the non-introduction of nationalisation, piece work rates under the seven-hour day, high food prices, intervention in Russia, conscription, and alleged slackening by mine managers.[63] The resolution to strike passed by a meeting of Woollaton miners in Nottinghamshire:

> was passed unanimously, condemning the raising of the price of coal by 6s a ton (contrary to the findings of the Sankey Commission), supporting the Miners Federation in demanding 14.3% on piece work rates; demanding withdrawal of the troops from Russia, the total abolition of conscription, and any other indirect kind of compulsory military service.[64]

In all three counties therefore, the strikes displayed a tendency to generalise, encompassing wider economic and political demands. The most persistent issue was the government's use of the output crisis to increase the price of coal and discredit the miners' case for nationalisation. At all of the strike meetings, the focus of discussion passed quickly from Yorkshire and the piece-rate question, to that of output and the 6s increase, with bitter denunciations of owners and government. The grievances expressed by many delegates at Keswick were amplified by, and found an outlet in, the strike wave. In Nottinghamshire and Derbyshire, strikers in Sutton, Chesterfield and Mansfield blamed falling output on 'rotten roads, shortage of rails, and bad tackle' and on 'lack of initiative

---

61   118 H.C.Deb.5s, c.1160.

62   *Daily Herald*, 23 July 1919.

63   *Nottingham Evening News*, 23 July 1919.

64   Ibid.

and money in the development of the mines'.[65] In Lancashire, Burnley miners walked out complaining of a lack of tubs, and in Leigh, strike leaders claimed that 70 percent of face workers were being forced on to the minimum wage for the same reason. The *Manchester Guardian* reported that the shortage of tubs, timber and rails in Lancs was worse than in any other coalfield:

> Coal is lying in heaps underground which cannot be brought to the surface because there are so few tubs. Rails and sleepers are stacked at the pithead, but are not laid down where they could be used to provide a greater output. The miners, of course, attribute these things to a sort of 'ca-canniness' on the part of the employers and the government more as an attempt to prejudice the promise of nationalisation.[66]

At a Leigh strike meeting, the district agent Seth Blackledge endeavoured to secure a return to work on the basis that the MFGB was dealing with the piece-rate question. He was interrupted by a miner in the audience who said that 'the 6s per ton had never been mentioned from the platform, and three parts of the men present were not only opposed to 12.5 percent but also to the advance of 6s per ton. (Hear, hear, and applause)'. The majority of the meeting agreed, and Blackledge was voted down.[67]

Quickly the strike movement began to extend beyond its original aims. Almost automatically it seems the fight became not only about the implications of the first Sankey Report, but for the implementation of the second Report on nationalisation. The frequency with which strike meetings in all areas threw up demands relating to the ownership and control of the industry suggests that the political influence of the reform committees and unofficial groups was strengthening once more. The resolution passed by the mass meeting at Mansfield Market Place on 21 July was typical of many: 'That we don't go to work until the 14.3 % is conceded, the whole of the Sankey Award, including nationalisation, is granted and the 6s per ton taken off'.[68]

The strike wave was the expression of widespread doubts amongst the rank and file about the MFGB strategy, and fears that enmeshment in a timetable which was being dictated by the government was causing the atrophy of the

---

65   *Mansfield Reporter*, 25 July 1919; *Nottingham Evening News*, 21 July 1919; *Nottinghamshire Free Press*, 25 July 1919.

66   *Leeds Mercury*, 22 July 1919; *Burnley News*, 23 July 1919; *Leigh Chronicle and District Advert-iser*, 25 July 1919; *Manchester Guardian*, 15 July 1919.

67   *Leigh, Tyldesley and Atherton Journal*, 25 July 1919.

68   *Mansfield and North Notts Advertiser*, 25 July 1919.

Charter. Already, the hours and wages deal encapsulated in the first Sankey
Report had fallen well short of the original demands, and was being further
eroded by the steadily rising cost of living, the government's refusal to sanction adequate piece-rate increases, and by the chronically low level of output
which was everywhere blamed on the management. This last factor marked a
bridging point between disgruntlement with the first Sankey Report, and distrust at the government's intentions regarding the second. Output restriction
and the increased price of coal were understood to be a pincer movement by
owners and government against nationalisation, to which the MFGB's response
appeared dilatory and insufficient at best, and 'reactionary' at worst.[69]

On top of all this, the controversy surrounding the British military support
for the white armies of Koltchak and Denikin tended to confirm fears that
nationalisation was being entrusted to a government which was, despite its
reconstruction rhetoric, pro-capitalist and anti-working class. Partly for this
reason, 'Hands off Russia' meetings held during the strike often turned to
the issue of output and nationalisation, whilst strike meetings usually carried
resolutions for non-intervention in Soviet Russia.

There was considerable anxiety amongst the 'advanced' miners that their
idea of a postwar social and industrial order, fixed so firmly in their sights at
the beginning of the year was in danger of vanishing before their eyes. The
Daily Herald, which had a large circulation in the strike-affected areas, argued
that the general crisis of the summer amounted to a crossroads in the postwar
struggle between capital and labour.[70] Encouraged by the apparently widening
fault lines in the Coalition on the one hand, and the resurgent militancy inside
sections of the working class on the other, it mounted a vigorous campaign for
the urgent use of direct action by the labour movement. Linking the issues
of war in Russia, and nationalisation of the mines at home, it warned that
the prospects of reconstruction along socialist lines would otherwise diminish,
and be replaced by an oppressive alliance of unfettered private capital and
militaristic state.[71]

The strikes displayed a general anxiety about an impending betrayal. I. East-
ham, Chairman of the Burnley Trades and Labour Council, warned Bonar Law
that:

69   L&CMF Monthly Council Meeting, 26 July 1919.
70   *Daily Herald*, 24 July 1919. All copies of the previous day's paper were sold out in Mansfield
      by 10am.
71   See, for example, all issues of the *Daily Herald* between 24–29 July 1919.

the same spirit that was imbued in the minds of the workers and the workers' sons when they pushed the Germans across the Rhine would be just as operative at home as across the water when the occasion required it.[72]

Those who had fought in the war often voiced their bitterness. At Leigh, for example, ex-Lieutenant Goulding said:

Last Saturday they were celebrating a capitalistic peace as a result of a capitalistic war ... There was one law for the rich and one law for the poor. This was the free country which he and five million other mugs went to fight for in 1914. (Applause).[73]

At Mansfield, Thompson condemned the 6s increase 'which hit the poor man who had been to the front', and said:

The peace celebrations were taking place at the same time as one of the greatest industrial crises in the country. During the war the miners had shown as much sacrifice as anybody, and had been promised that something good should be done for them. He was glad to see that throughout the country the ball was already rolling.[74]

Encouraged by the strike wave's tendency to generalise from economic to political and vice versa, the outbreaks of militancy elsewhere in the working class, the campaign of the *Daily Herald*, and the Triple Alliance Conference on direct action over Russia, rank and file leaders clearly felt that they were in the van of a major labour offensive, which might escalate into the long awaited national strike movement. At Mansfield Market Place, where there was 'a good deal of revolutionary talk', and at Wigan Market Place where 'speeches were made by revolutionaries', they heralded the beginning of mass direct action. At Mansfield, in a speech characterised by 'unparliamentary language', Bromley said that:

if the workers are going to be further burdened with the 6s a ton, the miners were going to kick against it, and when they kicked, the capitalist

---

72   *Burnley News*, 26 July 1919.

73   *Leigh Chronicle and District Advertiser*, 25 July 1919.

74   *Nottinghamshire Free Press*, 25 July 1919.

system would be tottering. To cease to be wage slaves they must own and control that which they produced, and when that was achieved they would have attained the brotherhood of man.[75]

Councillor Alec Norris provides another example:

They had now reached a revolt of the workers of Britain, which meant to destroy the capitalist system, and to set up an institution wherein they would all work for the common good.[76]

Officials fought to sever the link between the rank and file and the unofficial groups. Striking miners were told that Yorkshire was not their concern, that they could not be expected to 'crucify themselves on behalf of other sections of the community' over the 6s increase until 'the people as a whole developed a political consciousness', and that the Triple Alliance was going to take up this issue, and nationalisation, as well as British military intervention in Russia.[77] The central message of the district bureaucracies was that continued unofficial action meant 'an end to this Federation', and was 'stabbing Smillie and Hodges in the back'.[78] Varley told the Mansfield miners: 'It came to this ... they had to decide whether they belonged to the Federation or not. If Notts could manage things better, they could secede from the Federation and go on their own'.[79]

It seems that at this point in the campaign for the Charter, many were indeed questioning whether continued membership of the Federation was desirable if it undermined their capacity to fight over immediate issues like the local implications of the introduction of the seven-hour day. Some complained of 'the unwieldiness of our organisation, through which we were unable to deal quickly', and there were occasional calls at mass meetings for smaller, regionally based breakaway unions. At Mansfield Market Place, strike leaders floated the idea of an amalgamation of Nottinghamshire, Derbyshire, Leicestershire and Yorkshire. As we shall see, in Yorkshire the end of the strike there would also occasion widespread demands for a breakaway from the Federation.[80]

75   *Mansfield and North Notts Advertiser*, 25 July 1919.

76   *Mansfield Reporter*, 25 July 1919.

77   These arguments were put forward by Fred Varley, *Nottingham Evening News*, 22 July 1919;
      Seth Blackledge, *Leigh Chronicle and District Advertiser*, 1 August 1919; Price, *Nottingham-
      shire Guardian*, 25 July 1919.

78   L&CMF Monthly Meeting 26 July 1919.

79   *Nottinghamshire Guardian*, 24 July 1919.

80   L&CMF Monthly Meeting 26 July 1919; *Nottinghamshire Free Press*, 25 July 1919.

There was much more at stake in the strike wave than a percentage point here or there. In essence, it was a crucial test of the legitimacy of the MFGB leadership, on a national and district level, at a time when the forward momentum of the campaign for the Charter had stalled. Were it to fail, rank and file pretenders were waiting in the wings, advocating full use of the miners' industrial power, and eager to don the mantle of leadership. Fortunately for Smillie, the government recognised the importance of maintaining the authority of union executive power, and on 25 July, conceded the Keswick demands on piece rates almost in full. This deal cut the ground away from under the feet of the rank and file groups, whose enthusiasm had blinded them to the limitations of the strike wave. Following mass meetings in all areas over the weekend of 26–27 July, the miners drifted back to work, and the rank and file rebellion subsided as suddenly as it had appeared.

In Nottinghamshire, the unofficial group was quick to recognise that the piece rate deal had restored the authority of the officials, and its members advocated the return to work, consoling themselves with the thought that 'a ballot of the Triple Alliance was to be taken in the next three or four days', and that 'the next time they stopped they would stop with the railwaymen and fight for the national ownership of all the means of life'.[81]

In Lancashire a hard core of a few thousand young militants refused to accept the inevitable, and made a desperate attempt to keep the strike going by marching from pit to pit, threatening to destroy winding gear, or throwing tubs down the pit shafts where they found men working. At Diggle's Pit, Westleigh, when banksmen refused to haul the men up, 'the gang raided the haulage house and stopped the haulage rope'. As soon as underground workers heard of trouble on the surface, they scrambled for the cages, afraid of being stuck below, and so relatively small numbers of miners were able to cause great disruption.[82]

By the time the L&CMF Council met on Saturday 26 July, 'the whole coalfield was a seething mass of discontent', with miners who wanted to work threatening to deal with the pickets by physical force.[83] At mass meetings on the Sunday, 'It was stated that resentment is so great that many collieries may strike rather than work with the mob leaders'.[84] Delegates associated with the unofficial action, like P. Newhall of Garswood Hall, were given a roasting at the Council meeting. Chris Holding, reporting back to his Plank Lane branch, said:

---

81   *Nottinghamshire Guardian*, 25 July 1919; *Mansfield and North Notts Advertiser*, 25 July 1919.
82   *Bolton Evening News*, 24, 25 July 1919; *Manchester Guardian*, 26 July 1919; Manchester Evening News, 25 July 1919.
83   L&CMF Monthly Meeting 26 July 1919; *Manchester Guardian*, 26 July 1919.
84   *Daily Herald*, 28 July 1919.

He could tell them frankly that in the Conference on Saturday at Bolton they had that matter up hot. One man, indeed, who was present, got it pretty warm, and he deserved all they said of him.[85]

Once the MFGB had secured the piece rate settlement, support for the rank and file leaders fell away dramatically, and where they tried out of desperation to use coercive tactics to sustain the strike's momentum they faced a severe backlash from the members. Newhall's public ostracism by the L&CMF was a measure of this reaction, and allowed his employers to sack him shortly after the strike. Whilst prior to the deal a majority of the strikers had proved relatively impermeable to accusations that rank and file leaders were union saboteurs, now they were more receptive. Once again, the bureaucracy was in the ascendant; 'when they came up against the Government they must be a united army behind their leaders and not go into the fight sectionally'.[86] The officials hammered home this lesson in the aftermath of the strikes. As Spencer told the Broxtowe miners:

If there is to be any fighting, if your blood boils for a fight, let the fight be a sensible, well organised, well thought out, well drilled, and a determined fight not by sections striking here and sections standing there which is doing no good.[87]

Though short-lived, the strike movement had shown that a direct action offensive was still possible in the summer of 1919. What requires explanation is the obvious contradiction between the general political demands expressed by the strikers, and the fact that the return to work was occasioned by a deal solely concerning piece rates. Baffled commentators attempted to simply explain the strike away as 'an unaccountable attack of midsummer madness'.[88] The reality, of course, was far more complex, and the variety of factors which allowed the MFGB leadership to reassert its hegemony will become clear in the light of the Yorkshire strike.

---

85   L&CMF Monthly Meeting 26 July 1919; *Leigh Chronicle*, 1 August 1919.

86   G. Spencer, *Nottinghamshire Guardian*, 24 July 1919.

87   *Nottingham Evening News*, 28 July 1919.

88   Ibid.

CHAPTER 9

Yorkshire

Yorkshire had its coat off. (Cheers) Since the Government had interfered with them, they were out to teach the Government a lesson it would remember.[1]

The Yorkshire miners were on strike on a question which should have been a national one. If the employing classes were to fight them by sections, as in the Yorkshire dispute, it would not be long before the whole country was defeated.[2]

∴

1   Political and Industrial Traditions of the Yorkshire Miners' Association

The importance of the Yorkshire Miners' Association to the MFGB is shown by the prominent roles its leaders played in the national Federation. Herbert Smith was Vice-President in 1919, later to rise to the Presidency during the trials of the 1920s. In an earlier period, Ben Pickard had been a key figure in attempts to establish a national union, and he was President of both the YHA and the MFGB until his death in 1904.[3] The political and industrial characteristics of the YMA and its members found expression in these two leaders. Ben Pickard was a Lib-Lab; the 'iron man of Barnsley',[4] he could be both pragmatic and pugilistic, his moderation belying an ability to display considerable industrial aggression. In a period of relative economic stability, growth and prosperity, this blend brought results and Pickard attracted an uncommon devotion and loyalty from his members. This was an acute problem for the ILP pioneers in the coalfield, one of whom complained that the miners 'have only one political

---

1   Frickley Branch delegate to mass meeting on 28 July 1919 in the early stages of the strike, *Mexborough and Swinton Times*, 2 August 1919.

2   W. Carter of Brodsworth branch as defeat approaches *Doncaster Gazette*, 15 August 1919.

3   Baylies 1993, p. x.

4   Howell 1983, p. 18.

belief – and that is a belief in Ben Pickard'.[5] Pickard himself saw such loyalty
as a pre-requisite for effective trade unionism, and often interpreted dissent
as betrayal, a tendency graphically illustrated in his withering assault on Pete
Curran and the ILP in the Barnsley by-election of 1897.[6]

Herbert Smith was at the forefront of a new generation of YMA leaders whose
political affiliation was to the ILP. He came to prominence following the deaths
of the triumvirate of long time Lib-Lab officials Frith, Cowey and Pickard, all of
which occurred between December 1903 and February 1904. Their passing was
attributed by YMA members to the strain of the long-running Denaby-Cadeby
dispute, Yorkshire's test of the Taff Vale judgement.[7] Disputes like this, and
the one at Hemsworth in 1906, were long and bitter struggles involving severe
hardship for the mining communities, including mass evictions from company
houses.[8] These signalled the break-up of the institutionalised industrial relations that had typified Lib-Labism, and the beginning of a new era of intensely
harsh industrial conflict that ultimately spelled its defeat.[9] Smith's election as
President in 1906 was the first clear sign of an emerging trend that had become
well-established by our period. In 1919, most of the key figures in the YMA leadership were ILPers, and were veterans of the era of Lib-Lab domination. Edward
Hough, Vice-President, had been active in Pete Curran's campaign; Alf Smith,
Agent, had been the first ILP member on the Normanton Urban District Council in 1906; John Potts, Treasurer, had converted from Liberalism to the ILP
during the Hemsworth dispute during which he had been the Branch Secretary.[10]

Whilst the ILP's rise to prominence within the YMA signalled the changes
that had occurred since Pickard's day, there were strong elements of continuity
as well. This could be measured by the continuing influence within the union
of non-socialists who had been prominent during the Lib-Lab era. Samuel
Roebuck, Junior General Secretary and a member of the MFGB EC in 1919 had
been a delegate for Darfield Main Branch in 1890.[11] John Hoskin, Financial Secretary, was already Treasurer by 1904, and had built his career on his bookkeep-

5   Howell 1983, p. 19; Baylies 1993, chapter 9.
6   Rubinstein 1978, pp. 102–34. Baylies 1993, p. 245.
7   Baylies 1993, chapter 11; Howell 1983, p. 22.
8   Baylies 1993, chapter 12.
9   Howell 1983, pp. 20–1.
10   For E. Hough, see Bellamy and Saville 1972–84, vol. iii, p. 117; for A. Smith, see Bellamy and
     Saville 1972–84, vol. iii, pp. 175–6; for J. Potts, see Bellamy and Saville 1972–84, vol. ii, p. 311,
     and Baylies 1993, chapter 12.
11   Bellamy and Saville 1972–84, vol. iv, p. 152.

ing skills, taking no active part in the political life of the YMA, thus comfort-
ably making the transition from one regime to the other.[12] Fred Hall and John
Wadsworth, Agent and General Secretary respectively, had donned the mantle
of Lib-Lab leaders following the deaths of Pickard and company. Hall had been
MP for Normanton since 1906, as had Wadsworth for Hallamshire. In 1910, following the MFGB's affiliation the previous year, both had been obliged to sign
the Labour Party constitution and sever links with their Liberal Associations,
but for neither of them did this involve an ideological conversion.[13]

Continuity between the two periods did not end there. Smith, like Pickard,
developed a leadership style which embraced both industrial toughness – he
was all grit and gristle – and political moderation. Like Pickard, his trade
unionism did not flow from his ideological beliefs; rather, his brand of socialism
was a pragmatic political expression of his industrial activity as a miners' leader,
and subordinate to it. Thus he had joined in the attacks on ILP newcomer John
Potts in 1906, when the latter had criticised the candidacies of Lib-Labs Hall
and Wadsworth. Smith saw this as damaging to the cohesion of the YMA, which
was in principle more important than party politics. In the 1920s, he was to
be renowned for his intolerance of communist critics within the MFGB; new
targets for old hostilities, the CPers were to Smith what the ILPers had been to
Pickard.

Taylor has explained how the similarity between Pickard the Lib-Lab and
Smith the ILPer was not just a matter of 'an aggressive, often authoritarian
leadership style'.[14] The transfer of political allegiance of the YMA, and Smith,
from Lib-Lab to Labour was in essence a pragmatic adaptation, designed to
maximise industrial effectiveness and political leverage when it became clear
that these were no longer possible in association with the Liberals. This represented not so much the triumph of socialism over Lib-Labism, as a synthesis
of competing ideologies on the basis of a changed industrial reality, with Herbert Smith as its visible expression. This synthesis produced a 'practical, non-
ideological approach to union affairs' best described as labourism.[15] It was as
the embodiment of labourism that Herbert Smith commanded such loyalty
amongst the Yorkshire miners, for they embraced the same ideals and values,
which sprang from shared industrial and community experiences. As Hobsbawm has written, 'Amongst the millions of men in caps he was certainly exceptional: but he was exceptional only as a particularly majestic tree in a large

---

12   Bellamy and Saville 1972–84, vol. iv, p. 98.
13   Gregory 1968, pp. 108–13.
14   Taylor 1992, p. 250 and *passim*.
15   Taylor 1992, pp. 232, 248–9.

forest'.[16] In 1919, during the Yorkshire strike, one observer wrote that 'In stature, appearance, and in speech, he completely typifies the men he leads, and in all circumstances he is absolutely a man of the people'.[17]

Smith was above all concerned to nurture what Taylor has called 'the powerful political culture of the Yorkshire coalfield which stressed the supreme importance of solidarity and loyalty'.[18] Like Pickard before him, maintaining organisational cohesion was at the top of Smith's agenda. This, of course, is always and everywhere an important consideration for any union bureaucracy, but in Yorkshire it had always been given a special emphasis due to the problems of organising such a large workforce with such diverse conditions and pressures.

At the war's end there were about 200,000 mineworkers in the county, producing over 15 percent of Britain's total output, a figure exceeding that of the Scottish coalfields. Whilst the YMA sought to organise all miners into one trade union, the county was effectively divided into two districts, a fact reflected by the existence of two separate employers' organisations, the West Yorkshire Coalowners' Association (WYCOA), and the South Yorkshire Coalowners' Association (SYCOA). The YMA itself had come about in reaction to the weakness of the South and West Yorkshire miners' associations prior to 1881. Conditions in the south and west of the county could hardly have been more different. The West Yorkshire coalfield was the older of the two, and was already well into its long, slow decline. About one third of the county's miners worked here, in small pits belonging to family concerns, characterised by low productivity and outdated mining methods. The South Yorkshire coalfield by contrast was new and expanding. Centred on the Barnsley seam in the Doncaster area, it had deep, large scale, capital-intensive mines that were amongst the most advanced in the world. The workforce in West Yorkshire lived in old established mining communities, whereas in the south a large proportion were newcomers, attracted into the area from surrounding coalfields and from the countryside by the large scale investment and the wages which were amongst the best for mining in the country. In 1905, Doncaster had still been a largely rural area; between 1901 and 1931, its population more than doubled. The miners here were developing a reputation for being more militant than their brothers in West Yorkshire, and would continue to do so beyond our period of study.[19]

---

16   Hobsbawm 1984, p. 212.

17   *Western Mail*, 23 July 1919.

18   Taylor 1992, p. 253.

19   Taylor, in Campbell et al. 1996, pp. 227–8; Baylies 1993, chapter 1, esp. pp. 16–22; Supple 1987, pp. 18–28; Neville 1974, pp. 24–5.

Size of membership, variations in conditions, and different patterns of work and community experience were potentially disruptive factors for the YMA. Only South Wales was bigger and more diverse. But the development of the SWMF had involved the creation of eighteen separate districts, each with its own union structures and full-time agent, and each possessing a fairly high degree of autonomy. This arrangement allowed the union to absorb the centrifugal tendencies generated by size and diversity; district agents could respond to their members' local grievances, and even call strikes, without necessarily dragging in the rest of the Federation. This was not the case in Yorkshire, where there was only one agent and seven full-time officials for 127,000 members, compared to 24 agents and sub-agents, and 28 full-time officials altogether in South Wales for 160,000 members. Whereas the SWMF had achieved stability and permanence with the aid of an elaborate machine, the YMA had had to rely to a greater extent on moral and ideological means, by stressing notions of loyalty, solidarity, and mutual preservation. Smith embodied these principles, his tough leadership style deriving from 'his profound belief that the union was the servant of its members and that the members owed a reciprocal obligation of loyalty to the union'.[20]

Both rank and file and officials had an acute sense of the importance of unity, of not allowing divisions which could be exploited by the owners; this was important when two sets of coalowners often conspired to set men from one part of the county against the other. The apparatus of the YMA was designed to prevent such divisions; branches that wanted to take industrial action had first to request permission to ballot from the Council, and, if granted, had to request permission again to lodge notices once a ballot majority had been secured. In both cases, the Council could not give the go-ahead until all other branches had been consulted.[21] Significantly, this careful and lengthy procedure was, with few exceptions, observed by the Yorkshire miners during the course of 1919, and there were hardly any instances of the unofficial strikes which so plagued other areas.[22] This was due above all to the fact that, at the end of the consultative process, Yorkshire officials tended to comply with the wishes of the membership if they wanted action. As Baylies says, 'The YMA did not indulge in empty threats'.[23] That this remained true in 1919 marks out the YMA

---

20   Taylor, in Campbell et al. 1996, p. 226.
21   Baylies 1993, p. 170. The majority required for action was two-thirds of at least three-quarters of the branch members.
22   YMA 1919.
23   Baylies 1993, pp. 236–7.

bureaucracy from others within the Federation, at district and national level, and testifies to the confidence it had in the loyalty of its members.

The membership had a considerable voice in the general affairs of the union. In addition to the YMA Council, on which each branch had a delegate, 17 of the 24 members of the Executive Committee were elected from the rank and file on a panel basis.[24] The membership was also consulted about the make-up of delegations to MFGB conferences, in particular as to how many officials should go. The potential here for rank and file control of the union must have been the envy of the South Wales URC members, who often complained about the preponderance of officials in conference delegations and sought to restrict their numbers, but this delegation of authority in the YMA's constitution was never exploited by the membership. Tradition meant that the officials were always elected to go by an overwhelming majority, even during 1919.[25]

The YMA was thus characterised by a powerful consensual ethos, functioning via a type of social contract between leaders and led, under the terms of which mutual loyalty was rewarded with organisational stability and gradual material progress. This caused a severe restriction of the space in which a rank and file movement could grow, something that would remain true through the 1920s, when the Miners' Minority Movement failed to gain more than the most tentative of footholds in the coalfield. When rank and file agitation did occur, it tended to take the form of a sporadic insurgency that reflected the temporary breakdown of channels of communication. This happened during the first half of the decade in which the transition from Lib-Lab to Labour took place, resulting in uncharacteristic criticisms of the leadership and a series of unofficial panel meetings.[26] With the transition completed, traditional channels were re-opened and unofficial activity faded away.

The extent to which patterns of loyalty were re-established after the decline of Lib-Labism was illustrated during the war. In 1916, Smith signed voluntary no-strike agreements with both the WYCOA and the SYCOA which endured virtually unscathed until the end of hostilities. In contrast to the industrial unrest that affected some other coalfields in the war's latter stages, there were only ten collieries affected by strikes between January 1916 and November 1918, and these were relatively minor affairs. Furthermore, these were official strikes – the leaders were flexible enough to break their own agreement when necessary. The absence of unofficial activity, and the relative tranquillity of the coalfield,

24 Neville 1974, p. 872. There were 17 panels, each consisting of a number of branches.
25 YMA Ordinary Council Meetings 12 May 1919, 7 August 1919.
26 Taylor, in Campbell et al. 1996, p. 247.

was reflected in the fact that mining was scarcely mentioned in the report of the Yorkshire division of the 1917 Industrial Unrest commission.[27]

Excessive political radicalism and unorthodox industrial strategies were rejected as incompatible with the pragmatic approach upon which organisational stability and permanence had been built. Traditions of consensus and loyalty should not be read as pusillanimity, however. On the contrary, progress had been paid for in bitterly fought industrial struggles in which the Yorkshire miners had displayed a fighting spirit which, if not unmatched, was certainly unsurpassed. The political and industrial culture of the YMA informed and shaped the course of events in 1919, both within Yorkshire's borders and beyond, and explains an apparent paradox; in a year hallmarked by unofficial insurgency, Yorkshire, a relative oasis of harmony between officials and rank and file, accounted for the majority of strike days lost. Direct action, viewed by officials elsewhere as a radical and even revolutionary challenge to conventional trade unionism, was officially endorsed and put into practice by a YMA bureaucracy not given to political or industrial extremism. The form this took, especially in July and August, decisively affected both the outcome of the struggle for the Miner's Charter, and the fortunes of the direct action movement itself.

2 Political Moderation and Industrial Aggression: The Social Contract in Early 1919

The social contract between officials and the rank and file, characterised by the familiar blend of political moderation and, when necessary, industrial aggression, survived the strains of the war. It was put to the test in the early months of 1919, but in contrast to the other areas under study, was ultimately strengthened by the experience.

Four and a half years of compulsory arbitration had seen the accumulation of a catalogue of grievances that demanded attention as soon as the war ended. There were serious protests from several districts regarding the supply and quality of food and at numerous collieries there was unrest and the threat of strikes over anachronistic and rickety price lists. These sorts of problems were, as was the case in all coalfields, exacerbated by the problems associated with demobilisation. Reports came in to the YMA of ex-soldiers being refused work, or being dismissed for protesting at being placed on work which they

---

27   Commission of Enquiry Into Industrial Unrest, 1917: No. 7 Division: Report of the Commissioners for the Yorkshire and East Midlands Area, PRO CAB 24/23 cd.8664, *passim*.

were physically incapable of doing through disability.[28] Here once again, the ingredients were present for chronic industrial unrest at the war's end. As we have seen, in other areas this took the form of a rash of unofficial strikes, as an impatient rank and file clashed with cautious union bureaucracies. In Yorkshire, however, official trade union procedures proved remarkably resilient, and industrial relations remained well ordered. Pit level militancy over price lists, for example, was conducted through official channels, with all branches following careful procedures for balloting. In the context of the general wage militancy in mining in 1919, it is significant that this lengthy process was observed to the letter. In no instance did it break down and lead to wildcat strikes.[29]

Objectively, the problems presented by the return of soldier-miners were no less acute in Yorkshire than elsewhere; the proportion of the workforce that had enlisted or been called up was similar to other coalfields.[30] The difference lay in the manner in which the YMA sought to reincorporate the returning men into the industry. It moved quickly to establish a scheme for dealing with the pressures of demobilisation, and by 18 December 1918, District Demobilisation Committees, consisting of five owners' representatives and three workers representatives, had been established in South and West Yorkshire.[31] The smooth functioning of the scheme was in part due to the co-operation of the coalowners' Associations, but more important was the co-operation of the YMA members, who agreed that all men who had become miners during the war should be laid off in rotation to make way for men who had joined the colours.[32] In South Wales, Nottinghamshire, and Fife and Lanarkshire this had proved unacceptable, and rank and file militants were able to agitate effectively against any scheme which involved one miner making another redundant. In Yorkshire, however, just as there had been no effective opposition to the war or even to the YMA's no-strike agreement, now there was no audible demand for the post-1914 miners to have the right to work in the pits. Whether or not the reform committees campaigned over any of these issues is uncertain, but if they did, then the impact they had was insufficient to leave any trace in the sources.

Of the four areas under study then, Yorkshire was unique in that the demobilisation crisis was resolved without the development of a breach between rank and file and officials. Consequently there was less pressure for immedi-

---

28   YMA Ordinary Council Meeting, 7 January 1919; *Barnsley Chronicle*, 11 January 1919; *Barnsley Independent*, 4 January 1919.

29   YMA 1919.

30   Committee into coal production at the end of the war.

31   Neville 1974, pp. 524–5.

32   MFGB Special Conference, 14–16 January 1919, pp. 36–7.

ate national action to secure the Miners' Charter. This had important national ramifications; the large Yorkshire delegations consistently provided the national leadership with a bedrock of support in its strategy of progress through negotiation and compromise. The Yorkshire delegation voted with the MFGB EC at all the crucial moments during the first quarter of the year.[33]

There was strike action in this period in Yorkshire – in January – but again, unlike elsewhere, unity between leaders and led was the order of the day as officials promptly responded to pressure from below. The dispute in question was over the implementation of the November 1918 national agreement for a 49-hour week for surface workers.[34] The deal was introduced without a hitch in West Yorkshire, but in South Yorkshire the owners refused to agree to the provision for a 20-minute 'deadstop' when working would cease to allow 'snap' to be taken.[35] Smith said that 'if industries could not be worked with a twenty minute stoppage for meals, then something was seriously wrong', and abruptly informed the owners that were it not conceded by 7 January, then the YMA surface workers would down tools at 2.30 pm and implement it directly.[36]

At Denaby Main and Cadeby Main, surface workers carried out Smith's threat, and stopped work to take their food. On Friday 10 January, the management sent them home, along with the underground workers. The following day, the men at Bullcroft and Brodsworth followed suit and were also locked out. During the next week, the action spread branch by branch though the Doncaster and then the Barnsley areas, until by the Saturday something approaching 30,000 South Yorkshire miners had been locked out at 26 collieries.[37]

Although they had favoured further negotiation, the YMA officials did not attempt to stop the branch level militancy, and on 18 January recommended that 'direct action' be taken to force a settlement, and 150,000 miners struck work.[38] At a meeting shortly before the strike, the Coal Controller had told the miners he was powerless to instruct the SYCOA to concede the deadstop.

---

33   MFGB Special Conference 14–16 January 1919, p. 99; MFGB Special Conference 12–13 February 1919, p. 37; MFGB Special Conference 26–27 February 1919, p. 53.
34   Neville 1974, pp. 526–7.
35   *Barnsley Chronicle*, 25 January 1919. 'Snap' was a colloquialism for a light meal.
36   *Sheffield Weekly Independent*, 25 January 1919; Joint Committee Meeting of SYCOA and YMA 23 December 1918.
37   *Mexborough and Swinton Times*, 18 and 25 January 1919; *Barnsley Independent*, 18, 25 January 1919.
38   YMA Ordinary Council Meeting 18 January 1919.

Twenty-four hours into the strike, he wired the YMA reversing his position.[39] The miners returned to work victorious, and over the following weeks the YMA secured further concessions for surface workers in negotiations.[40]

The deadstop strike was a vindication of the YMA's social contract; although only a small minority of the membership (YMA surface workers in South Yorkshire) were directly affected, the union responded with a vigorous display of solidarity. In a manner reminiscent of the South Wales strike of 1915, the government, in the shape of the Coal Controller, had 'capitulated with humiliating promptitude'[41] in the face of the industrial power of the Yorkshire miners. Prior to the dispute, 18,000 surface workers had remained outside the union. By the end of February, an estimated 97 percent of surface workers had joined the YMA.[42]

In other districts, unofficial action in January raised the curtain on a year of rebellion. In Yorkshire, by contrast, the January strike announced an uncommon unity between leadership and led. A surface workers' demand for a 20-minute deadstop might not seem particularly pressing when stood next to the grand themes of the Miners' Charter, but by refusing to allow bread and butter concerns of the membership to be eclipsed by national priorities, the YMA officials maintained its allegiance, and closed down any space in which unofficial pretenders could operate.

Nonetheless, the general atmosphere of antipathy towards officials did not leave Yorkshire entirely unscathed; there was an active Miners Reform Committee of indeterminate size, in South Yorkshire at least, which was represented at the London Workers' Committee Conference in June.[43] Its intervention in the deadstop strike illuminates the difficult situation it faced in Yorkshire in 1919. On 24 and 25 January, there were mass meetings in Doncaster Market Place, which seem likely to have been called by branches in the Doncaster area where the MRC had some measure of influence. In line with the reform committees in Fife and Lanarkshire, MRC members sought to extend the strike into all-out action for the six-hour day. Geoff Whittles said that:

> In his opinion there were too many officials in connection with their various organisations. What was needed was rank and file settlements, also that when there was trouble the entire machinery should be put

39   *Barnsley Independent*, 25 January 1919.
40   Neville 1974, p. 530.
41   *Mexborough and Swinton Times*, 25 January 1919.
42   *Yorkshire Post*, 23 January 1919; *Doncaster Chronicle*, 28 February 1919.
43   *Workers' Dreadnought*, 21 June 1919.

in motion, and not one or two pits to come out while the others were working.[44]

In coalfields where officials dragged their heels over local grievances, this sort of agitation met with considerable success, and gave the Reform Committees opportunities to push the idea of immediate direct action over the Miners' Charter. However, in the context of the deadstop strike, it was utterly futile; the entire membership (except for the pump-men and winding-engine men) had been called out on strike, and it was consulted at branch meetings before the settlement was finally accepted.[45] Far from the dispute driving a wedge between rank and file and officials, it cemented an already amicable relationship, and reinforced the social contract.

Similarly, no success was forthcoming from MRC attempts to secure rule changes along the lines advocated by *The Miners' Next Step*. One, from Bullcroft branch, sought to add to the YMA's list of objectives:

> To continually agitate in favour of increasing the minimum wage and shortening the hours of work until we have extracted the whole of the employers' profits.[46]

Another sought to turn the Executive Committee into a rank and file body, where officials would have only an advisory role. Neither of these met with much support from the rest of the YMA delegates, however.[47] Claims made by a South Yorkshire delegate to the London Workers' Committee Conference regarding the strength of the MRC are therefore surprising:

> A miner from South Yorkshire reported that the Workers' Committee movement is making great progress there. In some collieries the workers had as many as eighty members. His branch was prepared to endorse any drastic action.[48]

Either these numbers are exaggerated or they underline the basic point; even with such a respectable membership, the unity of officials and members in

---

44   *Workers' Dreadnought*, 24, 25 January 1919.

45   *Workers' Dreadnought*, 31 January 1919.

46   YMA Ordinary Council Meetings 26 May 1919, 31 May 1919. The resolutions were defeated by 129 to 10, and by 2,040 to 173, respectively.

47   Ibid.

48   *Workers' Dreadnought*, 21 June 1919.

the YMA rendered the MRC ineffective. The only partial exception to this rule came in late March, when there was a brief unofficial strike of up to 30,000 miners in the Sheffield area, in protest at the MFGB Conference decision to recommend acceptance of the Sankey Report. Leaflets were distributed in the run up to the ballot, calling for the rejection and 'D ... [sic] the consequences'.[49] However, the strike quickly fizzled out without having any discernible impact elsewhere in the coalfield. Baylies's general point about examples of unofficial and anti-official activity is germane to the Sheffield strike: 'Such incidents are important by their relative rarity in revealing the essential continuity of the broader internal solidarity which characterised the YMA'.[50]

The Yorkshire MRC simply could not find an audience in 1919; it was smothered by the pragmatic, consensual approach of the miners and their leaders. Its difficulty was visible, ironically, in the expulsion of J. Walton MP by the YMA Council in May 1919 for having issued a leaflet calling for a 'no' vote in the MFGB ballot for a national strike.[51] The *Workers' Dreadnought* was ecstatic, seeing it as evidence that 'the workers are beginning, at last, to rise up against the bureaucrats who have sold them'.[52] In fact it simply demonstrated that Walton had transgressed the codes of unity and solidarity upon which the YMA had been built; anyone who threatened these was a pariah.

The Yorkshire miners were not unaffected by the general shift to the left which took place in the working class. Many political resolutions were sent by branches to the YMA Council during the year: for the abolition of conscription and the withdrawal of troops from Russia; against 'the imperialistic document miscalled the Peace Terms as drafted by the aristocratic dictators of the Big Four'; for the boycotting of the peace celebrations and many others.[53] However, just as industrial militancy was channelled through the YMA's official structures, so too was political radicalisation. The miner's rejection of unofficial industrial activity set the parameters for this radicalisation, which took place within the received Labourite tradition of the YMA rather than against it. There is little trace of the revolutionary politics we have seen in other areas. The Yorkshire miners were content to express political demands within their union, and entrust their resolution to the officials.

---

49   *Sheffield Independent*, 28, 29 March 1919; *Sheffield Daily Telegraph*, 29 March 1919; *Yorkshire Telegraph and Star*, 31 July 1919.

50   Baylies 1993, p. xi.

51   YMA Ordinary Council Meeting 12 May 1919.

52   *Workers' Dreadnought*, 31 May 1919.

53   YMA Ordinary Council Meeting 7 January 1919; YMA Adjourned Council Meeting 17 May 1919; Ordinary Council Meeting 26 May 1919.

3   The 'Fresh-Air Strike'

Historians have ignored the Yorkshire strike of 1919, preferring, perhaps under-
standably, to concentrate on the more incendiary events on the Clyde in January. The YMA strike was a peaceful affair; there were no mass pickets, no violent
clashes with the state. Yet, in a sense, what did *not* happen in the 'uneventful'
Yorkshire strike is as important as what *did* on the Clyde six months before.

One is struck first of all by its timing. Running from mid-July to mid-August,
it took place at the height of the national crisis; it was during these four
weeks that speculation about the government's possible downfall was loudest,
that the riots in Luton and elsewhere took place, that the police struck and
chaos ensued on Merseyside, that the strikes of bakers and North-East Railway
workers happened, and that the Triple Alliance decided to ballot on direct
action. This helps to account for the gravity with which the decision of the
Yorkshire miners to strike was received, for it seemed that the infectious fever
of industrial militancy had spread to one of the MFGB's biggest affiliates.

Many believed they were witnessing the beginning of the long awaited showdown between the miners and the state. YMA officials thought that the unrest
that was coursing through the coalfields in Nottinghamshire, Derbyshire, the
Midlands, Lancashire and South Wales would force the leaders there to call
all their members out.[54] Furthermore, although the MFGB EC had carefully
avoided direct action at Keswick, outside the conference hall some leaders,
Hodges in particular, were talking up the prospects of a fight with the government over nationalisation.[55] From Westminster, parliamentary correspondents
reported that MPs from all parties 'now appear to have made up their minds
that we are to have a big coal strike'.[56] Another wrote:

> The view of the lobby last night was that the Yorkshire dispute over time
> rates will now be kept going to embarrass the country, so that at any
> moment it can be merged into a national dispute. Coalowners in the
> House and the business community generally, now expect the miners
> to declare a national strike and hold that it must come soon. A number
> of blast furnaces have been blown out already in the expectation of the
> cutting off of coal supplies.[57]

---

54   *South Wales News*, 26 July 1919. Interview with an unnamed YMA official.
55   *Daily Herald*, 14 July 1919; *The Times*, 14 July 1919.
56   *Manchester Evening News*, 22 July 1919.
57   *Sheffield Daily Telegraph*, 18 July 1919.

The strikes in other areas, and the fighting rhetoric of some leaders, meant that the Keswick conference had not dispelled the popular perception that the decisive battle was about to be joined. Certainly this view held currency in government circles. Lloyd George called for an urgent report from the Cabinet Committee on Industrial Unrest. Auckland Geddes was fearful that in the prevailing atmosphere, and in light of the unofficial strike on the North Eastern Railway, the Yorkshire strike 'might result in a complete strike of all the miners, and possibly this would be followed by the Triple Alliance coming out'.[58] Lloyd George told the Cabinet on 21 July:

> The present situation was practical, and not theoretical, Bolshevism, and must be dealt with a firm hand … The whole of the future of the country might be at stake, and if the Government were beaten and the miners won, it would result in Soviet Government. A similar situation might result to that of the first days of the Revolution in Russia, and although Parliament might remain, the real Parliament would be at the headquarters of the Miners' Federation in Russell Square.[59]

Insofar as the anticipated conflict over nationalisation of the mines lay at the heart of the pervading atmosphere of crisis and instability, the Yorkshire strike took on a significance that transcended its regional status. Indeed the Yorkshire strike would prove to be the decisive industrial conflict in mining in 1919, but not in the way so many imagined at its outset. For although it was fought over an issue entirely unconnected with nationalisation, by the time it ended it had wrought a profound impact on the question of ownership of the industry.

Following the MFGB's decision that disputes arising from the introduction of the shorter working day should be dealt with by the districts, the YMA had entered into negotiations with the coalowners' associations. Initially the YMA sought a piece rate adjustment of 16.6 percent to compensate those who had a long walk to work underground, but subsequently it decided to accept 14.3 percent in order to show, as the Silverwood branch delegate explained, 'that they were not out for the pound of flesh'.[60] Both sets of owners agreed to the

58   CAB 27–59/IUC, 19 July 1919.

59   WC (596A), 21 July 1919; CAB 23-15-136/7.

60   *Barnsley Chronicle*, 9 August 1919. According to Smith, 6,000 miners had to walk three miles
      to and from working places, leaving them only five hours effective hewing time under the
      seven-hour day. According to his calculations, these men needed 25 percent to avoid a loss
      of earnings; *Rotherham Express*, 12 July 1919.

figure, so at this stage the piece rate adjustment issue appeared to have been resolved to the YMA's satisfaction.[61]

The alteration of hours proved more contentious. Under the Eight Hours Act, the union had successfully established a seven and a half hour day at most double-shift pits, and a short Saturday shift of six and a half hours, and it wanted these factored in to the working of the Seven Hour Act. As a branch official explained,

> What the men had asked for was two hours a day less in the bowels of the earth and they had been granted one on paper, but only half an hour in fact, according to the owners' interpretation.[62]

In South Yorkshire, progress was made in negotiations, as the SYCOA agreed to a six-hour Saturday, and to a 46-hour week inclusive of meal times for surface workers, rather than the 46 and a half hours exclusive that was stipulated in the Sankey Award.[63] In West Yorkshire, however, the owners were less willing to make concessions, and the situation was complicated further by a separate dispute over winding-enginemen's pay and hours which had remained unresolved since February.[64] On 12 July, the YMA Council voted to strike if no settlement had been reached within three days, but the YMA officials were not committed to precipitate action, rather seeing the strike vote at this stage as a lever with which to extract further concessions. Potts explained a few days later, at the MFGB Conference, that at this stage they had been confident that a peaceful resolution of the hours question had been within reach.[65]

Between these two dates the Coal Controller made his second, and decisive, intervention. On 14 July, he instructed the owners again that the 12.5 percent ceiling was in no circumstances to be breached. The SYCOA intimated that

61   *Daily Herald*, 5 August 1919, *Barnsley Chronicle*, 26 July 1919; *Yorkshire Telegraph and Star*, 26 July 1919. The complicated issues raised by the introduction of the seven-hour day, the basis upon which piece rate adjustments were calculated, and the course of negotiations leading up to the strike are best explained in the interviews and statements made by Smith and Roebuck contained in these issues. These should be read in conjunction with the debates at MFGB Annual Conference. See also Cole 1923, pp. 105–7.

62   *Rotherham Express*, 12 July 1919.

63   MFGB Annual Conference 17 July 1919, p. 101; MFGB Annual Conference 18 July 1919, p. 132.

64   Neville 1974, p. 552.

65   MFGB Annual Conference 17 July 1919, p. 88.

as far as it was concerned, it was still amenable to 14.3 percent should Jones change his mind, but the WYCOA withdrew its offer and talks broke down. The piece rate adjustment had now once again become the major issue in dispute, although the related demands concerning hours remained very much on the table.[66] This was the situation when the YMA Council met on 15 July, the day on which notices were set to expire. Smith urged the delegates to confine the strike to West Yorkshire, presumably hoping that there was the prospect of progress in South Yorkshire as long as the reasonable attitude of the SYCOA continued, but he was outvoted by a narrow margin as the Council voted for a county-wide stoppage.[67]

The Council meeting was then adjourned until the following day, so members could be consulted at branch meetings. In South Yorkshire, there was apparently 'some indignation ... at the recommendation of Herbert Smith', and 'in the spirit of unionism' the branches here voted for solidarity action with their West Yorkshire colleagues, resulting in a much larger majority for a total stoppage at the reconvened Council meeting.[68] Solidarity aside, the Coal Controller's intervention had made a strike policy against one set of employers redundant. As the WYCOA's Secretary pointed out in a letter to Smith, the dispute was now 'entirely between your Association and the Coal Controller'.[69] When, at 10 pm on Wednesday 16 July, all miners on the afternoon shift struck work, they were walking into a conflict not with the owners primarily, but with the government.

The strike was already being set in motion when the MFGB Annual Conference opened at Keswick on 15 July, and the YMA delegation, minus Herbert Smith, went seeking the support of the national union. It was dismayed to discover that the EC's position, whilst at variance with the 12.5 percent figure, was in essence the same as that of the Coal Controller and the owners when it came to the interpretation of the hour's reduction under the Sankey Award. Alf Smith, recently elected Agent in place of Fred Hall MP, protested that 'we were all of opinion that the Sankey Award did mean an hour's reduction'. To Smillie's swift rebuttal that '[t]hat is not a Federation demand', the Yorkshiremen countered frankly that, in which case, 'we are trying to get better conditions than the Sankey Award'. Smillie claimed that Yorkshire, by defying a national

---

66   *Barnsley Chronicle*, 15, 16 July 1919; *Leeds Mercury*, 15, 16 July 1919; *Doncaster Chronicle*, 15, 16 July 1919.

67   YMA Special Council Meeting 15 July 1919. The vote was 1,270 to 1,046.

68   *Leeds Mercury*, 17 July 1919; YMA Adjourned Council Meeting 16 July 1919. *Yorkshire Post*, 5 August 1919. The vote was 1,881 to 448 in favour of county-wide strike.

69   YMA Special Council Meeting 15 July 1919.

ballot result, was threatening the existence of the MFGB, and instructed YMA delegates to vote with the EC on the issues of piece rates and surface workers. They stood their ground.[70]

Under the MFGB's interpretation, the only way in which the YMA could realise the 14.3 percent was by giving up its local customs and working a longer Saturday shift than had been the case prior to the Sankey Award. Furthermore, it would mean losing the favourable terms that had been secured for the surface workers in South Yorkshire. Along with Lancashire and Kent, Yorkshire voted against the MFGB's resolution on piece rates, and it refused to participate in the declaration to surface workers, which was directed as much against its own members as those on strike in Monmouthshire.[71]

During one heated exchange, Alf Smith protested that 'Yorkshire are fighting not only the owners but the Federation'.[72] This was no exaggeration. During the course of the Conference, a gulf had opened up between the YMA and the MFGB EC, one that widened and deepened as the strike wore on. Herbert Smith, as Vice-President of the MFGB, was placed in a difficult situation, but, naturally, his first loyalty was to the YMA, and he had no qualms about continuing the strike. On behalf of the YMA he sent a strongly worded missive to the National Executive, warning it 'not to interfere with the Yorkshire position at all' as enough of a 'disservice' had been done to its cause at Keswick, and pointing out that 'we know our position better locally than the Miners' Federation does'.[73] In public, Smith was no less dismissive of the Federation, saying that the Keswick resolution on piece rates 'was the first time he had heard of trade unionists voting for a maximum. It was a new school of thought entirely to him'. His bottom line, that 'the Miners' Federation have nothing whatever to do with the position in Yorkshire', was shared by the YMA membership, and remained unchanged until the end of the strike.[74]

Smith's unambiguous position testified to the parochial solidarity character-istic of the YMA. Whilst other district leaderships might criticise the MFGB EC at Keswick, and even in the case of Lancashire and Kent vote against it, only the Yorkshire officials were prepared to side unequivocally with their members in practice. Once again, the YMA's social contract had come into effect, overriding Smith's doubts about the advisability of the strike. As he said:

---

70   MFGB Annual Conference 17 July 1919 pp. 96–102.

71   MFGB Annual Conference 18 July 1919, pp. 130–2.

72   MFGB Annual Conference 17 July 1919, p. 96.

73   YMA 24 July 1919.

74   *Barnsley Chronicle*, 19 July 1919; *Leeds Mercury*, 23 July 1919.

Whatever might be my own opinion as to how this dispute should have been manoeuvred, the men have decided the line of policy in a constitutional manner, and I shall follow out that line to the best of my ability.[75]

Ben Turner would comment in the *Daily Herald* at the end of the strike:

Mr. Smith urged the Yorkshiremen to work. The rank and file (Smith is always willing to consult, and be guided from, the rank and file) insisted upon a complete stoppage. It was done democratically in every sense of the word. It was a tribute to their loyalty to each other.[76]

Of the two hats on offer, Smith had no hesitation in choosing his Yorkshire cloth cap. For the duration of the strike, he effectively left his post as Vice-President of the MFGB, considering it to be inconsistent with the wishes and interests of his members.

With support from the Federation completely out of the question, the YMA looked to its own resources to bring more pressure to bear on the government. On 19 July, in line with the decision of the YMA Council earlier in the week (but again, according to Geddes, against Smith's advice), the pump-men and winding-enginemen were pulled out, thus opening up the danger of the pits flooding, and taking the total number on strike to 200,000.[77] In doing so, the Yorkshire miners were drawing on the experience of the deadstop strike, when the Coal Controller had caved in within 24 hours.[78] This time, however, the government was determined to stand firm. Lloyd George told ministers: 'He was rather inclined to agree that the mine owners were right and that a fight had got to come'.[79] From Westminster it was reported that:

The feeling is actually one of relief that an end will be put to a state of things which has existed for some time. It seems to be recognised that a trial of strength between authority and the extreme section of Labour must come sooner or later, and the general tendency is to regard the

---

75   *Leeds Mercury*, 21 July 1919.
76   *Daily Herald*, 21 August 1919.
77   YMA Special Council Meeting 16 July 1919; *Sheffield Independent*, 19 July 1919.
        CAB 27–59; UC 3, 23 July 1919, 'Telephonic Report by Sir Eric Geddes'.
78   See the speech of the Manvers Main delegate to his Branch Meeting on 16 January, *Mexborough and Swinton Times*, 19 July 1919.
79   WC (596A), 21 July 1919; CAB 23-15-136/7.

present time as the best. The suspense, it is argued, has got to be put an end to.[80]

The Board of Trade issued instructions to divert ships carrying coal to home ports, to restrict exports, and for railway companies and local authorities to stockpile as much coal as possible. The Industrial Unrest Committee, which met six times during the strike, feared solidarity action by the railwaymen, and considered putting into effect contingency plans to establish a volunteer transport corps. Horne pushed hard for such a move, but it was rejected by the Committee on the advice of George Roberts who felt that, with a Triple Alliance ballot on direct action impending, such provocative action would undermine 'the more moderate leaders [who] were endeavouring to persuade the men to vote against'.[81] However, the Committee did decide to send 2,500 naval ratings to the coalfield to pump the mines. They, and any miners who were prepared to strike-break, would be protected by police and troops. Horne was concerned that there were insufficient military personnel in Britain:

> The number of troops available for protection appeared to him to be altogether inadequate. The docks at Liverpool, Cardiff and Bristol alone would require large military forces. He thought it would prove necessary to bring to this country at once the whole or the greater part of the four Divisions on the Rhine who were stated to be available for this purpose.[82]

Two divisions were brought back to Britain and stationed at Clipstone Camp, from where battalions were sent to Leeds, Pontefract, and Dewsbury.[83]

Lloyd George emphasised to the Cabinet that 'if the Government chose the present moment they must be certain that they were on firm ground and had pubic opinion behind them'.[84] The government presented the decision of the YMA to pull out the pump-men and winding-enginemen as an act of vandalism. It developed the theme of a national crisis in which the miners were now not only selfish and lazy, but had become dangerous and destructive, a threat to 'the community'.

Lloyd George announced the decision to send in the navy in the House of Commons on Monday 21 July. He emphasised the threat that the strike

80   *Manchester Evening News*, 22 July 1919.

81   CAB 27–59; IUC Minutes, 19 July and 6 August 1919.

82   CAB 27–59; IUC Minutes, 19 July and 25 July 1919.

83   CAB 27–59; UC 8, 22 July 1919, 'Telephonic Report from Sir Eric Geddes'; Wrigley 1990, p. 199.

84   WC (596A), 21 July 1919; CAB 23-15-136/7.

posed to the economic health of the nation, exaggerated the risks of flooding, and underlined Yorkshire's importance to domestic industry. In a situation in which production was 'the need of the hour', the withdrawal of the pump-men and engine-men was an act of 'sabotage'.[85] The usual analogies with the war were drawn, with a 'very high up member of the Government' saying:

> This is precisely the same thing as done by the Germans in the case of the mines of northern France. Not only did they prevent the French from getting coal, but by destroying the mines as far as possible, they endeavoured to prevent an economic recovery.[86]

George Roberts, Food Controller and ex-Labour MP, went as far as suggesting that German agents might be behind the strike as 'Germany has planned for just this position'.[87] The press was quick to seize upon this theme:

> It is incredible that the Yorkshire miners, in cold blood, should do so deliberate act of terrorisation – what the Germans did as a savage act of warfare in the coal mines of France.[88]

The *Leeds Mercury* was apocalyptic in tone:

> Language is inadequate to convey the gravity of the present situation. Not in the blackest days of the war were we faced with such an appalling prospect. Our very existence as a community is threatened.[89]

If the strike was to continue for long, 'the black figure of famine will stalk the land, accompanied by the fell sisters, pestilence and crime'.[90] The enemy was within, and the Yorkshire miners were cast in the role of a menacing Prussian bully.

Government and press played up notions of a classless community, based upon the shared values of the sanctity of property and of legality, from which the miners had excluded themselves. Their behaviour, *The Times* suggested,

---

85   118 H.C.Deb.5s, c.916–18, 21 July 1919.
86   *Leeds Mercury*, 22 July 1919.
87   *Sheffield Daily Telegraph*, 24 July 1919.
88   *Leeds Mercury*, 22 July 1919.
89   Ibid.
90   Ibid.

'shows a recklessness and callousness which is literally appalling'.[91] Whilst the Yorkshiremen were portrayed, to use a phrase coined by George Roberts, as 'the forces of disorder', the government was keen to present itself as the saviour and protector of the 'community', to which it appealed for support.[92] Eric Geddes was sent by the Cabinet to co-ordinate the government's rescue mission. His very public brief was neither to mediate nor to strike-break, but rather 'to safeguard the vital interests of the general public'.[93] Upon arriving at his headquarters in Leeds, Geddes issued the following statement to the press:

> The Government is determined that all the resources of the State, whether of citizens or armed forces of the Crown, shall be used to prevent ruin to the community. It is not to take part in fighting the strike, but to save the life of the nation as far as we can.[94]

Portrayed as being above the sordid infighting between the YMA and the MFGB which had caused the strike, Geddes was dubbed 'Minister for the Man-in-the-Street'.[95]

Propaganda notwithstanding, the government was pursuing a high-risk strategy by sending troops into the midst of the Yorkshire strike. Ever since the army had been deployed on Clydeside at the beginning of the year, the possibility that the military might again be used in industrial disputes had been a concern of the trade union movement. Churchill's Spring circular to Army Commanders, which enquired as to the suitability of troops for use in strikes, had sent shock waves through the labour movement, and the renewal of the military service acts had compounded fears of a militaristic state at home and abroad. The YMA Council was particularly concerned, as the episode resurrected memories of the 'Featherstone Massacre' during the strike of 1893, when troops had fired upon a crowd of miners, leaving two dead.[96] Geddes reported that the owners were afraid that the intervention of the armed forces 'would result in the wrecking of the pit machinery and of the villages in the neighbourhood of the pits'. He himself felt that the possibility of rioting was

91 *The Times*, 22 July 1919.
92 *Sheffield Daily Telegraph*, 24 July 1919.
93 Ibid.
94 Ibid.
95 *Leeds Mercury*, 23 July 1919.
96 YMA Ordinary Council Meeting 10 February 1919; see Baylies 1993, pp. 119–25 for an account of the 'Featherstone Massacre'.

increased by the shortage of food in the area due to a strike of the Co-operative societies in the South Yorkshire coalfield.[97]

The Triple Alliance Conference met to discuss the use of direct action to combat the threat of militarism on the very day that 3,000 troops armed with machine guns appeared on the streets of Leeds to protect the 1,500 naval ratings that were being deployed in the Yorkshire pits. Although Geddes stressed that this was 'by no means an offensive measure', a confrontation seemed all too possible.[98] Hartshorn said: 'Nobody can tell what will happen. But if there is a collision between the men of the Fleet, and the miners in Yorkshire, it may be the first spark of civil war'.[99] Nottinghamshire miners on strike at Wollaton Colliery agreed; they passed a resolution condemning the use of the troops, 'which we consider is heading for civil war'.[100] It was also possible, according to Captain Munro of the Admiralty, that with leave long overdue, some naval ratings might be susceptible to fraternisation by miners 'of an unsettled mind and revolutionary ideas. It was possible that naval ratings might also absorb some of these ideas themselves'.[101]

In the event, neither violence nor fraternisation occurred.[102] The way was eased for the government by Herbert Smith, who instructed his members to '[l]et them [the naval ratings] severely alone. Do not interfere with them in any way'.[103] Fearing a repeat of 'the 1893 business at Featherstone', he repeated the instruction again and again throughout the strike. With one eye on the convulsions elsewhere in Britain, he told a Grimethorpe mass meeting: 'If there is any rioting to be done, let me do it and don't put yourself in that position', while at Sheffield he warned: 'If we cannot win a strike without rioting, we had better not strike'.[104] The most extreme reaction to the arrival of troops was 'mild excitement' in Doncaster as a couple of charabancs full of sailors passed through the town.[105] Overall, the troops were received with 'an indifferent air'.[106]

---

97   CAB 27–59; UC 6, 10.30am, 22 July 1919, 'The Coal Strike, Telephonic Report by Sir Eric Geddes'; Ibid, UC 7, 11.50am, 22 July 1919.

98   *Leeds Mercury*, 24 July 1919.

99   *Daily Herald*, 22 July 1919.

100   *Nottingham Evening News*, 23 July 1919.

101   CAB 27–59, lUC Minutes of Meeting, 6 August 1919.

102   Cole 1923, p. 108 refers to 'isolated attempts to interfere', but I have been unable to find confirmation in any of the sources.

103   *Barnsley Chronicle*, 23 July 1919.

104   Ibid; *Sheffield Daily Telegraph*, 23 July 1919.

105   *Doncaster Chronicle*, 25 July 1919.

106   *Leeds Mercury*, 22 July 1919.

The outstanding feature of the strike was its passivity. In Sheffield it was reported that:

> The attitude of the miners to the whole subject is curious. They appear to be really apathetic and indifferent ... They do not, at the moment, seem to worry about anything. They are not holding meetings and not discussing the strike very much.[107]

In the mining villages of West Yorkshire:

> The men on strike were enjoying the calm of a delightful summer evening. Some smoked the pipe of peace serenely at their own doorways; others sauntered along the quiet lanes; and at the street corners little groups passed on the gossip of the day.[108]

In and around Wakefield there was an 'almost total absence of violent talk or violent action',[109] whilst the South Yorkshire towns and villages 'took on a holiday air'.[110] In the Mexborough district there was 'little incident and no public discussion', and the *Doncaster Gazette* reported that:

> The strike, so far as this district is concerned, will go down on record as one of the quietest known. The police express grateful astonishment at the extreme good order which is prevailing.[111]

With only about 250 strike-breakers in the whole county, and given that the miners had decided to follow Smith's advice to leave the navy alone, there was scarcely a disruption to this tranquil scene. There was occasional picketing at the few pits which were not entirely solid, as well as the usual allegations of intimidation, but the most violent incident of the whole strike occurred at New Monkton Colliery, where a demonstration of women, youths, and children broke some of the lamp room windows, frustrated at their inability to stop pit deputies from going in to work. Perhaps this section of the mining community felt less bound to observe Smith's instructions than the YMA members'

---

107   *Sheffield Daily Telegraph*, 23 July 1919.
108   *Leeds Mercury*, 23 July 1919.
109   *Wakefield Advertiser and Gazette*, 12 August 1919.
110   *Yorkshire Post*, 17 July 1919.
111   *Mexborough and Swinton Times*, 26 July 1919; *Doncaster Gazette*, 8 August 1919. Geddes wired the Cabinet with similar news, CAB 27/59. UC 15. 24 July 1919.

themselves.[112] In any case, apart from this incident, and a marked increase in petty theft of food as the strike approached its end, the police had nothing to report.[113] There appears to have been only one 'clash' between miners and state, when a few strikers out walking the lanes in the Barnsley area were told to go home by a police constable. Smith told the press:

> We are not going to tolerate this sort of thing. We are citizens, and so long as we conduct ourselves right, we shall expect proper courtesy from the police or anyone else.[114]

The YMA Council resolved to take the matter up with the West Riding County Council. When set against earlier fears of violence and upheaval, that this incident merited the special attention of the YMA speaks volumes.

YMA officials worked hard to maintain peace and passivity. Herbert Smith packed in up to twenty mass meetings around the county on weekends, at which he repeated his appeal for calm. YMA Council delegates told their members 'to conduct themselves simply as if they were on a general holiday',[115] instructions which were followed to the letter. 'Sport, walking, club frequenting and gardening are practical attractions preferred to the search for the naval men'.[116] Many families took advantage of the strike by taking a seaside holiday, and the 'younger element', the source of unrest in so many other coalfields, 'played an immense amount of cricket'.[117] Once the initial vote to strike had been taken, the active involvement of the rank and file in the strike was over, until one month later they were again called upon to vote on the YMA Council's recommendation to return to work. Between times they filled their days with leisurely outdoor pursuits, leaving the running of the dispute to the officials. The *Rotherham Express* said miners would remember the summertime dispute as 'the Fresh Air Strike'.[118]

The MFGB EC secured the national deal on piece rate increases on the tenth day of the Yorkshire strike. This paved the way for agreements in the region

112   Pickets were mounted at Dinnington Main, Rossington Main, Wolley and Darnton, see *Doncaster Gazette*, 8 August 1919; *Leeds Mercury*, 22 July 1919; For New Monkton's 'ugly incident', see the *Sheffield Independent*, 31 July 1919.
113   *Doncaster Gazette*, 8 August 1919.
114   *Sheffield Daily Telegraph*, 25 July 1919.
115   *Mexborough and Swinton Times*, 9 August 1919.
116   *Sheffield Daily Telegraph*, 23 July 1919.
117   *Sheffield Daily Telegraph*, 24 July 1919; *Rotherham Express*, 26 August 1919.
118   *Rotherham Express*, 9 August 1919.

of 14.2 percent to be quickly reached in most coalfields, and deepened Yorkshire's isolation within the Federation. The YMA representatives on the national executive were conspicuously absent from negotiations with the government. Herbert Smith found out the details of the agreement in the press, and his opinion was that it didn't affect the YMA at all.[119] The formula of an 11.1 percent increase for the loss of 47 minutes only gave Yorkshire a maximum increase of 12.2 percent, a figure which the YMA Council regarded as utterly unacceptable, and after referring the question to the branches, it was decided to fight on regardless.[120]

In the wake of the national agreement, the press attacks on the YMA increased in intensity, especially on Herbert Smith, who was accused of keeping the strike going 'to emphasise his power and dignity as the head of the Yorkshire Miners' Association', and as a snub to the MFGB.[121] Personal attacks on Smith were ineffective, however. *The Daily Herald* wrote:

> The most remarkable thing in the whole strike is the loyalty of the 200,000 miners to Herbert Smith. All the men are convinced that their President is voicing their real and legitimate claims, and their confidence in him is unshaken. They will never forgive the attacks upon him, for they know that at this moment his son lies ill as the result of his services with the Army during the war. By making unjustified attacks on Herbert Smith, the press is stiffening the miners' resistance.[122]

Smith must have been under enormous pressure. In between EC meetings, occasional negotiations and an almost constant tour of the mining communities, he was tending to his son Ernest who had been wounded at Ypres in March 1918. He died shortly after the strike.[123]

Nothing, it seemed, could subvert the unity of the YMA. Under siege, it closed ranks. Smith said:

> Whatever outsiders may think, Yorkshire will continue this fight because of its justice ... The YMA and its members will decide its policy whatever may be the feeling in other counties.[124]

---

119   *Leeds Mercury*, 28 July 1919.
120   YMA Special Council Meeting 2 August 1919; *Doncaster Gazette*, 8 August 1919.
121   *Wakefield Advertiser and Gazette*, 5 August 1919.
122   *Daily Herald*, 4 August 1919; *Daily Herald*, 21 August 1919.
123   *Barnsley Independent*, 13 September 1919.
124   *Leeds Mercury*, 21 July 1919.

Tom Smith of Nunnery branch told a Sheffield mass meeting:

> They were told that the strikes were caused by a few unfeeling men called
> extremists or Bolsheviks. He didn't know whether he was an extremist or
> Bolshevik, but he did know that when there was a fight to be waged for
> justice, he was in it if the miners of Yorkshire were in it.[125]

In fact, Herbert Smith's support for the strike came not from personal conviction, but rather because 'the members were loyal to the union, and he did not
wish to do anything to damage that loyalty'.[126] He told Manvers Main miners:
'When you decide on a line of policy, I will either do as you say, or get out of
the movement'.[127] At Grimethorpe, he said: 'You know our policy, which you
decided. You had a perfect right to decide that. I am not complaining, and
you will decide what is the next step. I am no Kaiser'.[128] At no point did Smith
attempt to use his personal authority to undermine or curtail the action. In fact
he reiterated again and again that 'this is your strike, not Herbert Smith's, and
I want to advise you neither one way nor the other'.[129] Even towards the end of
the strike, when he was concerned about the hardship affecting mining communities, he refused to force a return to work.

> This is your strike, not mine, but we are all in it. Any man does wrong if he
> leads his men on when they are not being properly housed and fed. Your
> benefits are my benefits, your downfall my downfall. I am appointed by
> you and must take my instructions from you and not dictate in return.[130]

It is questionable whether Smith would in ordinary circumstances have been
so compliant and tolerant of a strike with which he clearly had disagreements,
but the unrest in other coalfields had shown what officials could expect when
they refused to accommodate the militancy of their members. With Smith and
his colleagues unwilling to antagonise the Yorkshiremen, the strike remained
solid, and settled down into a stand-off between the YMA and everybody else
involved, punctuated only by a few abortive sorties into negotiations. The
miners remained quiet but determined whilst depleted coal stocks caused

---

125 *Sheffield Daily Telegraph*, 11 August 1919.
126 *Barnsley Chronicle*, 9 August 1919.
127 *Mexborough and Swinton Times*, 19 July 1919.
128 *Barnsley Chronicle*, 9 August 1919.
129 Ibid.
130 *Mexborough and Swinton Times* 19 August 1919.

widespread factory closures. Already by the end of July, 89,830 people were on the unemployed register for Yorkshire and the East Midlands area, compared to 38,132 before the strike began.[131] Eric Geddes issued instructions to local authorities in all the major towns in Yorkshire, Lancashire and the East Midlands to 'take extreme measures' to preserve coal stocks, which led to restrictions on electricity, gas and water supplies.[132] *The Daily Herald* said:

> The aspect of the West Riding is an impressive tribute to the power of Labour. Everywhere one passes silent collieries and idle sidings. Tram services are curtailed, towns are in darkness, and factories and works are closed or closing.[133]

Despite its impact on the regional economy, the strike had clearly reached an impasse. With the safety of the pits guaranteed by the navy, the government was under little pressure to seek an accommodation which would jeopardise the national piece rate formula.[134] Within the YMA neither rank and file nor officials were willing to be the first to break the consensus by which the strike had been launched and sustained. However, as it became increasingly clear that Yorkshire was not going to be able on its own to break the government's resolve, there were moves to spread the action. Whilst the Oulton and Rothwell miners called for the MFGB to call a national strike, in Doncaster and Bentley meetings passed resolutions calling for the YMA to send 'propagandists' to other coalfields, to explain Yorkshire's case, and to call for joint action.[135] This was the only occasion on which Smith directly opposed a demand raised by the rank and file during the strike. 'We are already looked upon as villains. If we go into other counties it will be stated we are trying to cause unrest'.[136] Whilst he was prepared to defy MFGB policy and support his members, he refused to go against the national executive and spread the strike to other areas.

131   *Barnsley Chronicle* 16 August 1919. Towards the end of the strike, 27,000 were reported to have been thrown out of work in the steel, iron and engineering industries of Sheffield and Leeds, 25,000 in the textile industry in Bradford, and 5,000 in the steel trade of Rotherham, *Sheffield Daily Telegraph* 9, 12 August 1919.

132   *Sheffield Daily Telegraph* 24 July 1919.

133   *Daily Herald* 2 August 1919.

134   By the end of the strike 500 naval ratings had been used at two dozen pits in West Yorkshire, and one dozen in South Yorkshire, *Barnsley Chronicle* 16 August 1919.

135   *Doncaster Gazette* 8, 15 August 1919; *Sheffield Daily Telegraph* 12 August 1919.

136   *Mexborough and Swinton Times* 9 August 1919.

Had Smith given his approval for delegations to be sent to other areas, it is unlikely that they would have met with active solidarity on the basis of the YMA's demands. By this stage, the national formula had already led to settlements everywhere apart from Lancashire, and whilst surface workers might have been attracted to the possibility of fighting the Sankey Award, the chances of being able to pull in the colliers was slim. The only possibility of extending the strike lay in the way in which the rank and file unrest in the neighbouring coalfields had raised demands relating to nationalisation. By now, the first Sankey Report was a dead issue everywhere except Yorkshire, but the questions of output, the 6s, and the government's interminable delay in pronouncing on nationalisation were not. The MFGB's 'wait and see' policy became less tenable with every day that passed. An alliance between the YMA and the rank and file groups in other districts against the MFGB's policy with regards to *both* the First and Second Reports was, in theory at least, a possibility. Some of the ingredients which had led to political generalisation in strikes in surrounding areas were present in Yorkshire; there was anger over the 6s price rise (the YMA Council had passed a resolution of protest),[137] the output crisis was being acutely felt before the strike (most of the evidence presented to the Sankey Commission about management ca' canny came from here), and furthermore there was an established Miners' Reform Committee. Nonetheless, there was not a single instance of nationalisation being raised as a strike demand in Yorkshire, and therefore no possibility of escalating the strike beyond its borders.

There was no unofficial dynamic to the Yorkshire strike, and no political radicalism. By prioritising his members over the MFGB, he ensured their loyalty. There were no unofficial meetings, no leaflets, no resolutions to the YMA Council, no heckling of officials at strike meetings.[138] In the Sheffield area, where the MRC had led a brief strike in March, it was reported that there was 'very little evidence of the extremists' influence'.[139] Perhaps the MRC's situation is captured in this cameo appearance at a strike meeting at Bentley where 'various questions were asked by members, and one man, who attempted to introduce politics, was shouted down'.[140] At the end of the strike, the *Mexborough Times* observed:

---

137   YMA Ordinary Council Meeting 12 July 1919.
138   The only reference to possible organised activity by the MRC is contained in a paragraph in a strike report in the *Doncaster Chronicle*, 1 August 1919, which mentioned 'inflammatory speeches' by ILP propagandists in the Doncaster area. (Occasionally, the press used 'ILP' and 'MRC' interchangeably).
139   *Sheffield Weekly Independent*, 26 July 1919.
140   *Doncaster Chronicle*, 8 August 1919.

The conduct of the miners as a whole has been exemplary. They have obeyed their leaders' injunction to be peaceable and law abiding to the letter, and they have submitted themselves loyally and steadily in all things to the advice of their leaders.[141]

Without escalation, defeat was inevitable. Neither officials nor rank and file were willing to acknowledge this, so the strike continued until financial hardship became serious enough to make talk about surrender possible. As the strike approached the one month mark, there was genuine distress in mining communities. Some risked danger to supplement their meagre strike pay by mining outcrop coal which ran in thin seams near the surface in the Barnsley district.[142] In Worksop, miners began to apply to the workhouse for out-relief, whilst in the Doncaster area miners demanded, amidst reports of infant deaths through malnutrition, that the Board of Guardians begin an emergency scheme for feeding children.[143] The West Riding Education Committee heard reports from its members of widespread hunger in mining areas, including examples of homes where children had no food whatsoever.[144] Whilst strike pay was insufficient to sustain individual families, it nonetheless represented a huge drain on the YMA's resources. By the end of the strike it had paid out £370,000, which meant the liquidation of a fund which had taken a quarter of a century to accumulate.[145]

On Friday 8 August, following the SYCOA's reaffirmation of its willingness to pay 14.3 percent, the Coal Controller agreed to a meeting with the two owners' associations and the YMA. Frank Hodges was also present, at the invitation of Edgar Jones. The meeting proved fruitless for the YMA, as Jones continued to insist on 12.2 percent as laid down in the national agreement with the MFGB.[146]

On Tuesday 12 August at the YMA Council meeting, delegates voted by 120 to 36 to recommend the branches to call off the strike.[147] Following branch meetings on the Wednesday, the Council reconvened on Thursday 14 August to find that their recommendation had been accepted. On a card vote, an immediate resumption of work was approved by 2,213 to 528.[148] The miners

---

141 *Mexborough and Swinton Times*, 16 August 1919.
142 *Yorkshire Telegraph and Star*, 9 August 1919.
143 *Sheffield Daily Telegraph*, 12 August 1919; *Doncaster Chronicle*, 15 August 1919. Such a scheme was put in place in the last few days of the strike.
144 *Sheffield Daily Telegraph*, 9 August 1919.
145 *Sheffield Daily Telegraph*, 13 August 1919.
146 *Leeds Mercury*, 12 August 1919; *Barnsley Chronicle*, 12 August 1919.
147 *Sheffield Weekly Independent*, 13 August 1919; YMA Special Council Meeting 12 August 1919.
148 YMA Adjourned Council Meeting 14 August 1919.

returned to work empty-handed, forced to accept a 12.2 percent piece rate adjustment, no universal hour's reduction, and the 46.5 hour week for surface workers exclusive of mealtimes as laid down in the Sankey Award.[149]

The *Doncaster Gazette* commented on the abandonment of the strike:

> The sudden change of front speaks much for the loyalty of the men to their local officials and to the Council at Barnsley ... [H]ad the leaders urged a 'carry on' policy, there is but little doubt the men would have loyally responded, tightened their belts, and remained out.[150]

However, the consensus that had sustained the strike showed signs of cracking under the pressure of defeat. Whilst most of the branch meetings held across the county voted in favour of calling off the action, there was some opposition. At collieries around Doncaster, Barnsley, Pontefract, and Featherstone, there were majorities for fighting on.[151] Miners were demoralised and angry at what had been a 'ghastly failure'.[152] Local newspapers reported that 'at all the meetings there was keen disappointment at the suddenness and completeness of the debacle'.[153] At Silverwood, the branch delegate 'admitted a mistake had been made somewhere, and the miners were humiliated in this matter'.[154] In the Sheffield area, the opinion was that:

> It has been a great blunder. The men feel it has been a mistake all through. A great majority of the men are astonished and disgusted at the complete somersault that has been effected by the officials at Barnsley.[155]

At some places, one miner explained, the branch meetings were very small because 'We've been badly sold ... and our men are so disgusted with the whole business that they would not bother about voting or anything else'.[156]

---

149   Ibid; *Leeds Mercury*, 14 August 1919.

150   *Doncaster Gazette*, 15 August 1919.

151   *Rotherham Express*, 26 August 1919; *Mexborough and Swinton Times*, 16 August 1919; *Doncaster Gazette*, 15 August 1919; *Leeds Mercury*, 14 August 1919; *Sheffield Daily Telegraph*, 14 August 1919. The branches that voted to continue the strike were Barnburgh, Goldthorpe, Askern, Brodsworth, Bullcroft, Pontefract, and one of the two Featherstone branches.

152   *Doncaster Chronicle*, 15 August 1919.

153   *Mexborough and Swinton Times*, 16 August 1919.

154   *Rotherham Express*, 26 August 1919.

155   *Sheffield Weekly News*, 16 August 1919.

156   Ibid. At Askern, near Doncaster, only 200 out of 1,400 voted, and at Bullcroft Main, only 130 out of 2,000.

If the YMA officials did not escape the fallout of defeat, neither was the YMA's tradition of unity left entirely unscathed. Sectional interests and pressures began to make themselves felt. A survey of branch meetings found that day workers, whilst they had stood by the surface workers for a month, voted solidly for a return to work.[157] Defeat, and the prolongation of the strike over winding-enginemen's pay, also strained the solidarity between miners from different parts of the coalfield. One South Yorkshire miner said:

> I blame the West Yorkshire men. They must have known they could not get the support of the Federation, and without that we could not possibly win. Lots of our men did not want to come out, but they wished to be loyal to the officials, and this is the result.[158]

Although defeat tugged at the bonds that held the YMA together, they withstood the pressure, and the social contract between officials and members remained intact. Most branches protested bitterly at the return to work, but bowed to the inevitable, and passed votes of confidence in Smith and his colleagues.[159]

The bulk of the recriminations for the defeat were directed at the MFGB. In the last two weeks of the strike, the intensity of feeling against the Federation leadership grew, and with the defeat it reached a crescendo. The *Mexborough Times* reported on 9 August that 'At every mass meeting this week, miners in the audience have said they have been let down by Smillie'.[160] In the same part of the coalfield at the end of the strike:

> The general feeling among the miners ... is that they could have won the strike if the Federation had helped them, either covertly or openly, directly or indirectly.[161]

G. Probert, Hickleton Main's delegate, told his members: 'Without a doubt, the attitude of the Federation had operated powerfully against the Yorkshire miners, and had taken away any chance they had of winning the fight'. At this meeting, there was 'a good deal of angry discussion, directed mainly at the Federation, and some scores of voices called for a breakaway from the Federa-

---

157   *Rotherham Express*, 26 August 1919.

158   *Sheffield Weekly News*, 16 August 1919.

159   *Doncaster Chronicle*, 22 August 1919; *Leeds Mercury*, 16 August 1919.

160   *Mexborough and Swinton Times*, 9 August 1919.

161   *Mexborough and Swinton Times*, 16 August 1919.

tion'.[162] In the Doncaster district, there was 'strong feeling against the MFGB for leaving the Yorkshire miners in the cold', and at Bullcroft, 'a suggestion at the mass meeting at the end of the strike to break away from the Federation was warmly received among the men'.[163] This was a general response: 'The present rank and file feeling in Yorkshire is that the YMA should cast off from the Federation and fight its own battles in future'.[164] Smith had a job on his hands to maintain the union's affiliation to the national body.

The Yorkshire strike was a watershed in the campaign for the Miners' Charter. If a strike for nationalisation was ever to happen it was in these summer months, when almost one in two miners were involved in action in defiance of the policy of the Federation Executive. However, the pressures for escalation and political generalisation inherent in the situation, and exhibited in the rank and file strikes, left Yorkshire totally unaffected. The strike remained stubbornly on the terrain of the First Sankey Report, and showed no interest in the more political ground of the Second. The defeat left the YMA financially weakened, physically drained, and with an acrimonious relationship with the Federation. The priority of the YMA officials in the coming weeks and months was to repair the damage that the strike had caused – to replenish depleted funds and patch up relations with the MFGB. Remarkably, at the YMA Council Meetings on 19 August and 3 September, Lloyd George's rejection of nationalisation was not even discussed.[165] Insofar as a national strike over the question of ownership remained a theoretical possibility, the YMA was out of the game.

The strike marked a divide between two epochs of industrial conflict in the mines. It began amidst renewed predictions of a national strike, possibly involving the Triple Alliance, of gigantic social upheaval, clashes with troops, and civil war. It ended in defeat, with 'direct action' dead in the water. The strikes in neighbouring coalfields in the early stages of the Yorkshire dispute had demonstrated the same disdain for authority, ebullience and volatility which had characterised the unrest of the first quarter of 1919. However, in the hands of the YMA, direct action was utterly stripped of its image of insurgency. The miners obeyed their leaders' appeal for calm, stayed clear of the armed forces, and had little or no involvement in the running of the strike. The passivity of the miners, the government's careful preparations through the IUC, and

162 *Mexborough and Swinton Times*, 16 August 1919.
163 *Doncaster Chronicle*, 1, 15 August 1919.
164 *Mexborough and Swinton Times*, 16 August 1919.
165 YMA Council Meetings 19 August, 3 September 1919.

the hardship and distress which a long strike entailed for the mining com-munities, are characteristic of the coming confrontations of 1921 and 1926.[166]

The Yorkshire strike was a landmark victory for the government. It stopped the rot which had begun with the 1915 South Wales strike, when Lloyd George arrived in Cardiff, recognised the impotence of the government's position, and promptly surrendered. At the time, he told Runciman: 'We shall have to come to grips with them some other time'.[167] Between July 1915 and August 1919, the government remained on the defensive, unable to resist the power of the miners. Now, finally, things had changed. At the end of the strike, the Edlington branch delegate told his members: 'We have lost a trench, but not the battle'.[168] In fact, the failure of the strike had set the seal on the fate of nationalisation; in doing so, it marked not only the end of the road for the Miners' Charter, but the beginning of the road that led, eventually, to Black Friday.

---

166   Ralph Desmerais, in his study of the Industrial Unrest Committee, sees the Yorkshire strike
        as 'important insofar as it provided a case study for the Government's preparations for the
        larger strikes to follow'; Desmerais 1975, p. 9.
167   Runciman Papers, WR 303, 31 August 1919.
168   *Doncaster Gazette*, 15 August 1919.

CHAPTER 10

The Demise of Direct Action and the Reconsolidation of Labour

On the Clyde, in South Wales, in Lancashire, in various places where you have had an experiment in the working of direct action, has it brought the victory which you expected it would? The Government did not fall, and the great civil authorities were not on their knees in response to the clamour that there was in the streets. Come back to the fountain head, the real source of all civilised government, representative institutions.[1]

.. .

In July, Lloyd George expressed grave unease about the potential social and political implications of the growing unrest, singling out the strikes in mining as the major threat to his government. When barely one month later, on 18 August, he stood up in the House of Commons to announce that nationalisation had been rejected, he could be reasonably confident that direct action would not be the miners' response. This concluding chapter looks at how the change in fortunes of direct action came about, and makes some suggestions about the political consequences of its demise. In pursuit of an explanation as to how the government came out so firmly on top in the battle over nationalisation, I shall reassess the impact of the summer strike wave, and re-evaluate the politics and strategies of both official and unofficial leaderships within the Federation. Particular reference will be made to the Yorkshire strike, both as an alternative paradigm of 'direct action' in 1919, and as a strategically crucial defeat for the entire MFGB in which the Federation Executive bore primary responsibility. The chapter, and the book, end by setting the denouement of the campaign for the Miners' Charter in its wider context: the decline of direct action within the labour movement and the consequent strengthening of the electoral politics of the Labour Party. The case will be put for a fundamental reconsideration of this period in labour historiography and for recognition of the critical place occupied by the year 1919.

---

1   J.R. Clynes speaking on 23 July at the Triple Alliance Conference, MFGB 1919, pp. 41–2.

At the Keswick Annual Conference, the MFGB EC had made it clear that, whatever its internal differences over using direct action as a last resort, it remained committed to working for a peaceful outcome to the nationalisation question, and that it would brook no action which might upset the apple cart. Its policy came under serious attack from two sources within the Federation during July and August. Firstly from the Yorkshire miners who, by taking action against the terms of the Interim Report, disrupted the ongoing Sankey process, and secondly from the rank and file groups in other areas who sought to exploit this development. Agitation for strike action amongst the grades most disaffected by the terms of the Interim Report opened up a space within which they could encourage generalisation and escalation. In several areas they found a constituency of support for extending the strikes beyond limited sectional demands. Each day of the strike wave saw the call for nationalisation become louder, and it seemed that at the eleventh hour the long awaited showdown between miners and the government might occur. During the crisis in February, Smillie's health had suffered. Once again, the pressure took its toll.[2] During talks between the miners' leader and Lloyd George at the height of the unrest, one observer wrote:

> Anybody who saw, as I did, Mr. Robert Smillie as he came from the conference room in 10 Downing Street would have been struck by his air of utter weariness and profound dejection. Sucking at his pipe, which it was too much trouble to light, he stood alone for several minutes, rather a pathetic figure, with nothing of the demagogic passion about him.[3]

Hodges told the *Daily Herald* on 31 July, 'the marvel is that the whole of the miners of the country were not on strike on Monday of this week. Only a few men realised how near we were to general stoppage'.[4] Despite its limitations, the unrest had succeeded in bringing matters to a head within the Federation. With half of the members of the national union on strike, there was never likely to be a better chance for direct action over nationalisation. The choice facing the MFGB EC was clear; carry on what the rank and file groups had started by endorsing the strike wave and extending it to the other coalfields in

---

2   *Mansfield and North Notts Advertiser*, 1 August 1919. Spencer told Nottinghamshire miners that they were 'driving a bayonet into his heart. They had a mighty leader, but he was worn out. He should not have been called upon to go immediately into battle again, but sent away to rest'.

3   *Leeds Mercury*, 25 July 1919.

4   *Daily Herald*, 31 July 1919.

a decisive offensive against the government, or cling to the fading hope that there was still mileage in the Sankey process, and that Lloyd George might somehow pull the rabbit of nationalisation out of the hat of a Tory-dominated government.

For the Federation leadership, a national strike in the summer of 1919 was no more attractive a prospect than it had been in February or March. All the factors which had led the EC to look for a way out of conflict during that crisis were present once more; the influence of rank and file groups in the strikes which were affecting several important coalfields, the general atmosphere of unrest and insurgency outside the mining industry, and the widespread predictions of a clash between the Triple Alliance and the government in which it was once again threatening to use all the forces of the state. Taken together, these were more than enough to resurrect fears that a national strike might easily escape official control. *The Leeds Mercury* reported from a meeting of the MFGB EC that:

> Conversations with the miners' representatives gave the impression that they had no wish to go to the bitter extremity. They fully realise what that would mean, not only for the country, but for themselves.[5]

Leaving aside unease over the possible consequences of a national conflict, there were other reasons why the MFGB leadership chose not to use the strikes as a platform from which to launch a decisive offensive against government and owners. In the first place, class conflict cut against a philosophy which saw nationalisation as an expression of co-operation between miners and state for mutual benefit. Direct action ran counter to a belief that the state could be used for liberal and progressive purposes, and a hope that powerful members of the government might be prepared to use it thus. In fact, during the summer months it had become increasingly plain, to anyone prepared to see it, that the government was not going to nationalise the mines unless forced to do so. If the mauling of the Ways and Communications Bill had not made that clear, then the MAGB's campaign amongst backbenchers, the government's counter-offensive in the shape of the 6s rise and the widespread press reports that Duckham's scheme of trustification was the favoured option, should have done so.

However, the MFGB EC remained blinkered to political developments as it continued to try and pick a way forward by worrying the fault lines within the

---

5 *Leeds Mercury*, 25 July 1919.

Coalition. Its strategy – to rally sufficient levels of public opinion that the Coalition might collapse in the hope that a pro-nationalisation government might emerge from the fallout – was optimistic at best. While it had not formally abandoned direct action, a growing section of the Federation leadership had been moving towards this strategy ever since the first sittings of the Sankey Commission had produced a wave of public sympathy for the miners' cause. The strike wave of the summer bucked this developing trend and threatened to force the leadership to face up to that which it was studiously trying to ignore – that Lloyd George's government was about to renege on its promises.

In seeking to keep alive the possibility of a peaceful outcome, the Federation Executive was not just placing false hopes in Lloyd George; the logic of its position meant that the MFGB had to defend the sanctity of the Sankey Commission and mobilise its resources against the strikes. In attempting to hold the line over Sankey, the Federation leadership turned on sections of its own membership and exposed its internal divisions to the government, surrendering the initiative at the critical moment, and inviting counter-attack.

It could almost be said that the MFGB EC conspired in the defeat of the YMA. Whilst in public it blamed the strike on the 'shilly-shallying and dilly-dallying of government officials', in private it made no attempt to conceal its hostility towards the YMA from the government.[6] Robert Horne informed Lloyd George that 'Hodges told Tom Jones that he was not going to fight for a fraction', and that Brace and Hartshorn had told him that they and the rest of the Executive were willing to make a deal even though Yorkshire would 'probably complain bitterly that it was being let down'.[7] The government did not deliberately provoke the Yorkshire strike, though many miners believed it had, but it was quick to take advantage of the situation once it realised that the MFGB would sacrifice the YMA for a deal that would undercut the strikes in the other coalfields. Lloyd George told the cabinet on 25 July, hours before the piece rate deal was struck: 'If the Government had the Miners' Federation behind them in fighting the Yorkshire Association it would be a very great advantage'.[8] Once the unofficial strikes had subsided, there was little incentive for the government to seek a quick solution in Yorkshire. In fact, there were positive benefits to be gained from leaving the YMA twisting in the wind. During the first week in August, Lloyd George told the Cabinet:

---

6   Smillie to the NUSMW Annual Conference, cited in *Leeds Mercury*, 14 August 1919.
7   Horne to Lloyd George 25 July 1919, F/27/6/23.
8   WC (599) 25 July 1919; CAB 23/9/63.

He thought it was just possible that the Yorkshire strike might collapse, in which case the Yorkshire miners might be so indignant with Mr. Smillie for his attitude towards them that they would desert him. The government would then be in a more advantageous position. It might be preferable to wait and see whether the Yorkshire strike did not fizzle out, before we risked provoking a general strike, before declaring war against nationalisation.[9]

In the meantime, the government pressed home its advantage by mounting a public relations offensive against both the YMA and the MFGB. A series of letters from Horne to Hodges, published widely in the press, asked what, in light of the national agreement on piece rates, the MFGB leadership was doing to bring order and discipline to its ranks.[10] Given their own part in creating the situation, the complaints of Federation leaders that the strike – and its defeat – was the government's fault sounded shrill.[11] It was their own policy that enabled the government to isolate, strangle and defeat the Yorkshiremen, weakening the Federation into the bargain. 'We have been deceived, betrayed, duped', Hartshorn famously protested on hearing of the government's decision against nationalisation. At the MFGB Conference on 3 September, he offered the following post-mortem:

> The Prime Minister is at the head of a Tory House of Commons. He could have done either of two things. He could have said that nationalisation of the mining industry is absolutely inevitable, and if the present Parliament will not carry it through I am prepared to go to the country, and in so doing he would have had the whole-hearted support of the Labour movement at his back. As a matter of fact he said: I can only go so far as the Tory majority in Parliament will allow, and they will not allow him to go in this direction ... What has happened is simply this, that the Prime Minister and his Government have surrendered to the mass of shareholders.[12]

It might be said in the MFGB EC's defence that the Coalition did eventually fall apart, so its strategy had not been entirely groundless. However, to gamble

---

9   WC (606 A) 5 August 1919; CAB 23/1/158.
10   See, for example, *The Times*, 30 July 1919.
11   *Daily Herald*, 31 July 1919. Hodges complained that the government was seeking 'to place the onus upon the MFGB to secure a resumption of work in a district in which the government is entirely responsible for a strike'.
12   MFGB Special Conference 3 September 1919, p. 16.

on the chance that the government would collapse – and collapse in time – was risky in the extreme, especially when one considers the size of the Tory majority in the House of Commons. By the time the Coalition did disintegrate, the challenge of direct action had evaporated, the miners were in headlong retreat, and the political realignment which followed at Westminster was to the right, not the left.

The limitations of the unofficial movement, visible during the first quarter of the year, resurfaced again in the unrest of the summer. The parochial nature of the unofficial groups once more meant that the strikes remained disconnected, local affairs. The demand for direct action over nationalisation, put forward so often in all areas during the strikes, could not be expressed nationally as there was no national organisation to do so. Again, whilst all coalfields were affected to some degree by the militancy, important areas like Durham, Northumberland and Scotland remained relatively quiet. The most noticeable gap was in South Wales, the spearhead of the unofficial movement. The strikes here were brief, sectional and uncoordinated. There is no evidence that the rank and file movement in South Wales made any attempt to follow the example of its counterparts in Lancashire and the Midlands, and link up with the action underway there. In fact, during these critical weeks the SWSS and its supporters were expending their energy in physical confrontations with followers of C.B. Stanton's British Empire Union. A series of pitched battles took place in the Rhondda, Aberdare and Neath valleys in July and August, as the militants tried to disrupt the anti-Bolshevik propaganda tour of the Russian Social Revolutionary, Luboff.[13] Perhaps the SWSS was sidetracked by its enthusiasm to defend the Russian Revolution; perhaps the anti-state ownership stance of *The Miners' Next Step* and *Industrial Democracy for Miners* saw no significance in strikes which were raising the demand for nationalisation. Whatever the reason, the abstention of the SWSS represented a critical weakness in the unofficial front.

Even in those coalfields where the strikes were most widespread, they were uneven in character. In Nottinghamshire and Derbyshire, Mansfield and Chesterfield respectively provided high profile centres of militancy; the mass meetings here were dramatic events, virtual no-go zones for officials. Meanwhile, the smaller villages of the Leen Valley provided quietly contrasting foci of loyalty. In Lancashire, Wigan miners were more prone to militancy than their counterparts in Atherton and Tyldesley, who did not join in the action.

Within the militant centres themselves there was heterogeneity. Younger miners, for example, were more prone to spontaneous action and militancy than their older colleagues:

---

13   *Western Mail*, 15 to 29 July 1919; *South Wales News*, 16 to 29 July 1919.

The older miners are alright, they have the sense to do what is right; but the young hotheads, who have had no experience, think it is a great lark to go on strike and dislocate business. They do not stop to think of the consequences, they are like a bull in a china shop, and the elderly and more cool-headed men have great difficulty in restraining these young bloods.[14]

Geographical and generational inconsistencies help to account for limitations in industrial militancy and political radicalism. Another determining factor was the space in which direct actionists had to operate. Their ability to influence, generalise and politicise the summer strikes was largely contingent upon official inaction over the implication of the Interim Report. A comparison with the official strike in Yorkshire provides clarification. Here, by giving approval to industrial action to rectify limited grievances, union leaders reconfirmed their legitimacy as custodians of the membership's collective economic and political interests. Consequently, whilst there was more industrial action here than anywhere else, there was hardly a trace of the radicalism displayed elsewhere.

As the strikes of the summer subsided, the position of the militants weakened throughout the Federation, and with the defeat of Yorkshire it became acute. In Lancashire and the Midlands, there was a backlash against the rank and file groups. In Nottinghamshire and Derbyshire, they were booed and heckled at some of the mass meetings which voted to end the strikes. Owen Ford, boasting that he had personally pulled out several collieries in the Mansfield area, was met with a cry of: 'you can cause some misery, can't you?' At Sutton, miners 'shouted down an extremist who was trying to dissuade them from passing a resolution to go to work'. At Lower Hartshay colliery, a resolution was passed 'disapproving of pithead meetings being held in the morning, thus stopping the pits, and declaring that in future stoppages must only take place after proper notice has been given'.[15] At a Special NMA Council meeting, a resolution was passed: 'That we highly appreciate the action of our officials in standing by the resolution passed at the Annual Conference at Keswick'.[16]

In Lancashire, we have seen how the Reform Committee, in taking desperate measures to keep the strike going, provoked an angry reaction from many miners. The L&CMF leadership took advantage of this to re-stamp its authority

14   *Leigh Chronicle and District Advertiser*, 25 July 1919.
15   *Nottinghamshire Guardian*, 25 July 1919; *Mansfield Reporter*, 25 July 1919; *Nottinghamshire Evening News*, 25 July 1919.
16   NMA Special Council Meeting 24 July 1919.

over the membership, and reinforce its view of legitimate union practice. 'Leadership and constitutionalism were essential', a branch official told his members, 'otherwise disorder and chaos would prevail'.[17] Greenall and his colleagues steered the reaction against the 'lawless course' that the reform committee had taken into an offensive against the idea of direct action itself. Prior to the strikes, on 19 July, the L&CMF Council had instructed its Triple Alliance to vote in favour of direct action over conscription, troops in Russia, and so on. Now the Council reversed its decisions, and recommended a 'no' vote in the Triple Alliance ballot, still scheduled to take place.[18]

In Fife and Lanarkshire, the Reform Committees had been struggling to make any headway since the aborted general strike at the beginning of the year. In April, an attempt was made to co-ordinate activities when representatives from Fife, Lanarkshire and the Lothians met in Glasgow to establish the Scottish Mineworker's One Union Movement, with James MacDougall as its Secretary.[19] In May and June, the organisation pushed for action in support of miners imprisoned for picketing offences committed in January, against the proposed increase in Income Tax, and for various changes to their union constitutions. None of these efforts met with any success, and by June MacDougall was complaining that a shortage of speakers on industrial unionism meant the number of meetings was falling.[20] In August, calls were made in the miners' section of *The Worker* for solidarity with the strikers in Yorkshire, but these were half-hearted and met with no response.[21] By September, the One Union Movement was in a slump. The Glasgow shop stewards Campbell and Gallacher toured Fife in September to try and resuscitate it, but their report made depressing reading. They had found militants subdued, inactive and complaining about the apathy of their workmates.[22]

In South Wales too, enthusiasm for direct action was on the wane. The SWSS suffered a number of defeats at the SWMF Annual Conference in June. In a break with union tradition, W.H. Mainwaring, A.J. Cook and Noah Tromans stood against Brace, Winstone and Onions for the posts of President, Vice-President and Treasurer, but they were roundly beaten.[23] So too were

---

17 *Leigh Chronicle*, 1 August 1919. Chris Holding, Plank Lane Branch.
18 L&CMF Council Meeting 9 August 1919.
19 *The Worker*, 24 May 1919.
20 *West Fife Echo*, 30 June 1919. *The Worker*, 31 May and 7 June 1919.
21 *The Worker*, 9 and 16 August 1919.
22 *The Worker*, 20 September 1919.
23 SWMF Annual Conference, 16–21 June 1919. Voting was as follows: William Brace 177, W.H. Mainwaring 61; J. Winstone 227, A.J. Cook 24; A. Onions 173, N. Tromans 71.

resolutions seeking to bar MPs from holding union office (which would have removed Brace, Richards and Onion from their posts), and to dismiss Tom Richards for his conduct in the run up to the strike ballot in February. The discussion over these resolutions became a debate 'between the syndicalists and the Parliamentarians about syndicalist aims and methods'. S.O. Davies made a 'fiery speech in support of syndicalism and direct action and against antiquated constitutional methods' and William Brace summed up the case for the parliamentary road. The rank and file movement in South Wales had been strong enough to force a vote in the highest forum of the SWMF about the ideas contained in *Industrial Democracy for the Miners* but not strong enough to win it. The result was 177 to 76 in favour of the parliamentarians.[24]

Following this, the press made much of a resolution passed by the miners of Cwmtillery on 24 July that stated: 'All persons acting under the red flag of revolution be immediately expelled from representing this lodge'.[25] This was taken, in conjunction with various other developments (the warm reception afforded by Rhondda miners to the Prince of Wales upon his visit, and the collapse of an unofficial strike over victimisation at the Lewis Merthyr colliery), to mean the end of the road for the 'extremists'. The *Echo* noted that a 'marked change is coming over the South Wales miner', and the *Western Mail* reported that in the valleys, 'the common belief is that the day of the Red Flag extremists is passing'.[26]

Press reports gave a one-sided view of the changing mood in the coalfield and the alleged backlash against the militants was doubtless exaggerated.[27] The 'extremists' of South Wales would continue to play an important part in mining politics in the 1920s and beyond. Nonetheless it seems that the press had drawn attention to a real phenomenon – the passing of the syndicalist moment in its South Wales heartland. Its legacies were complex, but the various strands are traceable through the subsequent careers of Noah Ablett, A.J. Cook, Arthur Horner and Nye Bevan. Ablett, heartbroken by the defeats of the 1920s, spent the last years of his life wracked by illness and alcoholism. Cook spurned the Communist Party, choosing to influence events from within the official structures of the SWMF and the MFGB. Horner combined communism and trade union officialdom, whilst Bevan joined the Labour Party and crossed the river to fill his bucket in parliament. All of them were outstanding individuals, and their own biographies fill important pages in the history of the labour

---

24   *Western Mail*, 18 June 1919.
25   *South Wales Echo*, 26 July 1919.
26   *Western Mail*, 26 July 1919.
27   See, for example, *South Wales Echo*, 24 July 1919.

movement. But in an important sense the different roads they took had a common source – a desire to compensate for the failings of the syndicalist movement from which they had all sprung. Significantly, this carried them out of the rank and file in which they had pinned their earliest hopes, and, with the exception of Ablett, into prominent national office. None of them became 'gentlemen' in the way that *The Miners' Next Step* had talked about leaders, but their positions involved them all in making compromises and adaptation in the name of a pragmatism which as younger men they had scorned.

The subsidence of the unofficial strikes and the defeat of Yorkshire were accompanied by other retreats and defeats in the industrial world. By the end of the first week in August, the police strike was in a tailspin. The solidarity strikes on the London tubes and at the Nine Elms depot had not spread; nor had Tom Mann proved capable of launching a second front in Liverpool, where branch meetings of dockers and railwaymen declined to answer his call for action. Isolated, the police strike collapsed, its participants victimised wholesale, and trade unionism rooted out of the force. The strike on the NER, whilst a victory insofar as it led to the abolition of the unpopular eyesight tests, failed to escalate into a national movement for the abolition of the Conciliation Board, as the Vigilance Committees had hoped it would. On top of all this, on 12 August, the Sub-Committee of the Triple Alliance Executives met, lost its nerve, and voted to 'postpone' the ballot on direct action over conscription and military intervention in Russia. Taken together, these developments amounted to a turning of the industrial tide. Hartshorn wrote:

> The lack of far sightedness and discipline on the part of the Yorkshire miners, and the collapse of the farcical strike of the Police Union, have for a time destroyed the driving force of the direct action movement.[28]

The Times agreed, its editorial on 5 August proclaiming the collapse of the police strike and the slow strangulation of the Yorkshire miners as a 'turning point in industrial warfare'.[29] A week later, following the Triple Alliance's u-turn, it reported the demoralisation of trade union militants:

> During the past fortnight, feeling among the organised workers has been settling steadily against the use of the strike weapon for other than industrial aims. The movement of which the police strike was symptomatic,

28   *South Wales News*, 30 August 1919.
29   *The Times*, 5 August 1919.

failed so signally to win support that the extremists were becoming confused, and appeared to be at a loss to know how to retrieve the situation.[30]

Within the MFGB, the balance of forces shifted decisively against the direct
actionists. This both reflected, and encouraged, the shift which took place
within the labour movement as a whole. In the first week of September, the
MFGB and Triple Alliance both met in conference to discuss the way forward,
but the result was a foregone conclusion. The Triple Alliance voted to confirm
the decision of the Sub-Committee, and shelved the idea of taking direct action
to halt the reactionary foreign adventures of Winston Churchill.[31] At the Federation conference, the Executive announced its 'regret' at the government's
decision, but advised the miners against taking industrial action 'at this stage'.
It was, Smillie explained:

> Our duty to consult our fellow trade unionists, and, further, it is our duty to
> the nation itself to give time to put it before the nation before any drastic
> action is taken on this matter.[32]

The resolution which the MFGB put before the TUC Conference on 10 September did not mention direct action over Russia, asking only that the Parliamentary Committee 'immediately interview the Prime Minister on the matter'.[33]

The by now familiar pattern of delay and procrastination continued, over
the issues of Russia, the mines and much else besides. With the nationalisation
crisis behind him, Lloyd George was in no hurry. He postponed meeting the
TUC delegation until 9 October, by which time the railway strike had been
settled. His position on nationalisation was, of course, unchanged. When the
TUC reconvened, a full two months later, it was decided that the time was still
not right for direct action, as the public was not yet fully behind the miners.
The upshot was the Mines for the Nation campaign, which dragged on into the
New Year, arousing scant interest outside mining areas. Even within the labour
circles there was little enthusiasm. The Labour Party's 1920 Annual Report
noted that the campaign of demonstrations 'was conducted under somewhat
difficult circumstances, owing to the impossibility of ... securing the services
of as many of the Party's front rank speakers as they desired'.[34] The farcical

---

30   *The Times*, 13 August 1919.
31   MFGB, Triple Alliance Conference Report 4 September 1919, p. 35.
32   MFGB Conference 3 September 1919, p. 15.
33   TUC Annual Conference 1919, pp. 259–74.
34   Labour Party Annual Report 1920, p. 8.

pretence that direct action was still under serious consideration was finally ended on 11 March 1920, when a Special TUC Conference voted against a general strike, in favour of a continuing propaganda campaign in preparation for the next general election, whenever that might be.[35]

As the idea of direct action ceded ground within the labour movement, parliamentary reformism gained it, a process encouraged by the series of by-elections at Swansea East, Bothwell, Pontefract and Widnes during the summer months. Whilst Bothwell was the only genuine mining constituency, all the others bordered major coalfields, and were dominated by the issue of mines nationalisation. MFGB leaders sought, with considerable success, to turn the contests at Swansea and Bothwell into referendums on the issue. Smillie and Hodges were the main speakers in David Matthews's campaign in Swansea, in which he narrowly failed to get elected. At Bothwell shortly afterwards, John Robertson trounced his opponent, a Liberal coalowner. Robertson's success invited comparison with the 'disastrous' strike underway in Yorkshire:

> It is not often that the defeat of a Government candidate has a steadying influence. The Victory for mines nationalisation at the ballot box has demonstrated the power of political action, and it has thus provided a sound argument for members of the Labour Party who are strongly opposed to direct action.[36]

The Labour Party looked to follow up the Bothwell breakthrough with victories at Pontefract and Widnes in early September. Pontefract was a large and industrially diverse constituency straddling the eastern edge of the South Yorkshire coalfield. Polling took place after Lloyd George's announcement on nationalisation, and, with a sizeable mining vote in the constituency, the Labour Party's election agent confidently stated that the result was a 'foregone conclusion'.[37] In a shock result, the Coalition candidate beat the YMA checkweigher Isaac Burns by a fairly comfortable margin.[38] The YMA blamed the result on the Widnes by-election, where the Labour Party had concentrated its resources to campaign for Arthur Henderson's return to parliament, but according to the *Goole Times*: 'The non-mining section of the Labour Party reply that the

---

35   Arnot 1953, pp. 217–18.

36   *Sheffield Weekly Independent*, 9 August 1919.

37   *Yorkshire Post*, 8 September 1919. Although it was not strictly speaking a mining con-
      stituency, the 8,000 miners constituted the main occupational group, with dockers and
      seafarers numbering 7,000, and agricultural workers 6,000.

38   The voting figures were W. Forrest (Coalition Liberal) 9,920; I. Burns (Labour) 6,445.

Yorkshire strike had more than anything else to do with the result, because it antagonised many workers and their wives'.[39] During the campaign, Isaac Burns was smeared as an irresponsible trouble-maker, 'a direct actionist and revolutionary' who was 'unrepentant' about the strike, or the widespread disruption it had caused to the regional economy.[40]

To Labour Party leaders seeking ammunition against direct action, the Yorkshire strike and the Pontefract result provided a compelling case. Not only had direct action been ineffective, they claimed, it had also shown itself to be electorally damaging. The Widnes contest provided a sharp contrast. Henderson's campaign steered well clear of the direct action controversy, emphasising instead his qualities as a statesman and parliamentarian. In his acceptance speech, he welcomed his election for the effect he hoped it would have on the labour movement:

> I regard the result as important because it strengthens the hands of those working class leaders who desire to restore the confidence of the workers in constitutional machinery and rehabilitate Parliament in their eyes as the legitimate instruments of reform.[41]

Local election results consolidated the process by which representative politics recovered ground in the labour movement. The Labour Party Executive's Report in June 1920, which looked back over the period since the general election, gave special mention to the mining areas of Durham and Monmouthshire, where the Party had gained majorities on the County Councils.[42] In the November 1919 municipal elections, Labour had increased its number of councillors by 1,087, to 1257.[43] In London, the result had been 'phenomenal'; Labour won 572 seats out of a possible 1,362, compared to only 46 in 1912.[44] 'Throughout the country', the Executive Committee's Report stated proudly, 'the local authority where Labour governs is by no means a rare exception, and public bodies where Labour is unrepresented, are becoming a vanishing quantity'.[45] The *Labour Leader* could again speak confidently about the 'March of Labour', while in South Wales, the ILP leader Emrys Hughes was able to take a tougher stance

---

39   *Goole Times*, 26 September 1919.
40   *Goole Times*, 5 September 1919.
41   *The Times*, 13 September 1919.
42   Labour Party Annual Conference 1920, Report of the Executive Committee, p. 35.
43   *Labour Leader*, 6 November 1919.
44   Labour Party Annual Conference 1920, Report of the Executive Committee, p. 37.
45   Ibid.

against direct action. In an article written for the *Merthyr Pioneer* in April 1920, he articulated the political transformation that had occurred in the politics of the South Wales labour movement over the previous months:

> Nationalisation will come when the Labour Party comes into power, or when Labour is strong enough to influence government policy in that direction. It does not require a great deal of imagination to see the way things are going to trend during the next decade. The increase in the Labour Party's vote shouts clearly that the workers are beginning to realise the importance of the political method.[46]

The turn from direct action towards electoralism within the labour movement, well underway by the end of 1919, was a crucial element in the stabilisation of British society in the postwar period. In the sense that men like Clynes, Henderson and Thomas had been its most vocal advocates, the swing of the pendulum towards the Labour Party was their victory. However, when Lloyd George had looked for help to steer British capitalism through stormy waters, he had turned not to these moderates, but to Robert Smillie. At the height of the unrest in March, he had warned his Cabinet colleagues against proposals to arrest the miners' leader. 'Smillie is an extreme man', he told Bonar Law:

> But the fight he put up against the extremists induced them to accept the [Sankey] Commission, and proves that he has some measure of statesmanship in his equipment. If the leaders are under lock and key, the movement will pass into the hands of hot-heads ... of the Noah Ablett type.[47]

Shortly afterwards, he explained to those assembled at the Paris Peace Conference how his government had avoided the extremes of social conflict by making careful concessions to British workers through their trade union leaders:

> As a result, trade unionists such as Smillie ... who might have become formidable, have in the end helped us to avoid a conflict. The English capitalists, thank God, are frightened; this makes them reasonable.[48]

---

46   *Merthyr Pioneer*, 24 April 1920.
47   Letter to Bonar Law on 20 March 1919, cited in Wrigley 1990, p. 164.
48   Mantoux 1964, p. 34.

Lloyd George singled out Smillie with good reason; as leader of the most powerful group of British workers, he was pivotal to the management of unrest and militancy in 1919. Smillie's reputation as a trade union leader was unmatched. Not only had he led the first ever national miners' strike, but he had also refused to condemn the South Wales strike of 1915, and given his backing to the 1917 Lanarkshire strike against profiteering.[49] 'They become gentlemen' and gain 'considerable social prestige', *The Miners' Next Step* had warned about trade union leaders, yet Smillie had defied the stereotype. Refusing offers of government posts during the war and retaining his working class roots and lifestyle in Larkhall, he was no social climber. Militants looking for the tell-tale signs of *embourgeoisement* which foretold betrayal found none.

Perhaps this explains why Smillie was not subject to the unofficial leadership's critique of officialdom in 1919. Following the aborted attempt at a general strike in Scotland, Reform Committee members wrote that they placed their trust in Smillie and confidently expected that he would lead them to victory.[50] As late as July, John MacLean gave his opinion that 'Smillie may yet be the glory of the British revolution'.[51] Smillie was the national figurehead of a direct action 'movement' surrounded by an aura of rebellion and shot through with the language of revolution. And yet not only was he politically opposed to revolution, he was against the exercise of workers' industrial power in the volatile atmosphere of 1919. His energies as a trade union leader in 1919 were directed overwhelmingly towards the avoidance of conflict and upheaval. It was his militant credentials that enabled him to persuade the miners to accept the Sankey Commission, to abide by its interim report, and to wait for the government to pronounce on nationalisation. Apparently, he later confessed that accepting the Sankey Commission had been 'the greatest mistake of his life', though it is doubtful whether he would have done much differently had he the chance over again.[52] When offered a final bite at the cherry of direct action in 1919, at the Triple Alliance Conference in September, he again declined. Although it was figures like Thomas and Sexton who pushed hardest for calling off the ballot on direct action over Russia, Smillie used his influence as Chairman to back them up.[53]

That Smillie was the figurehead of the direct action movement gives lie to the myth, common at the time, that it was a revolutionary movement. As was

---

49   *The Call*, 9 August 1917.

50   *The Worker*, 15 February 1919.

51   *The Worker*, 19 July 1919.

52   *Workers Dreadnought*, 14 August 1920. Interview with Tom Watkins about Smillie.

53   MFGB, Triple Alliance Conference 4 September 1919, p. 33.

suggested at the beginning of this study, one cannot really talk about a national direct action 'movement'; certainly it wasn't something that one could join. In political or philosophical terms, direct action as a concept was equally loose and amorphous. All things to all people, it could be anything from a means of overthrowing parliamentary government, to a means of rescuing it. Direct action was an ideological house with many rooms, and under its expansive roof it could accommodate men as diverse as Will Hay and Bob Smillie, John MacLean and Ramsay MacDonald.[54]

It is not that workers are necessarily the militant 'other' to moderate trade union leaders. Smillie, for example, spent his life far to the left of those he was seeking to represent. However, if the potential for a revolutionary direct action movement in mining did exist in 1919, then it lay with the miners themselves. The unofficial strike wave could reflect the sectionalism and prejudices of the communities from which they sprung, but equally they could exhibit notions of class consciousness, workers' control and internationalism. Ultimately, these elements did not crystallise into a revolutionary consciousness amongst more than a small minority. Industrial struggle never attained a momentum or reached a level sufficient to spontaneously generate a revolutionary consciousness, as described by Rosa Luxemburg in her depiction of the 'mass strike'.[55]

For a genuine revolutionary direct action movement to have emerged in mining, it would have required the establishment of explicitly revolutionary organisations in all the major coalfields, linked up nationally, and prepared to do battle not just with the right of the MFGB leadership, but with the left as well. The reality was, of course, very different. In practice, the unofficial groups remained hamstrung by their parochialism and by a less than watertight theory of the trade union bureaucracy. Above all they were not revolutionary organisations. Individual revolutionaries played an important part in setting them up and developing their programmes, but overall the groups were composed of miners who held a wide range of political beliefs. Ultimately what they had in common was a faith in industrial militancy, but this alone was not enough to carry the day, and as a consequence that faith itself diminished.

---

54   Alex MacDonald wrote in the *Merthyr Pioneer*, 19 July 1919: 'Kerensky is an indirect actionist, Lenin and Trotsky are direct'. He then went on to list as 'indirect actionists' Henderson, Clynes, Thomas, Brace, Hodge, Barnes, Tillet, Thorne, Adamson and Wardle. Direct actionists included Smillie, Williams, Hodges, Cramp, Paul, McManus, Murphy, Kirkwood, MacLean and 'hosts of others who stood by their class when they were most needed'.

55   For a discussion on the dynamics of class consciousness and the role of the trade unions and strikes in this respect, see Kelly 1988, pp. 32–146, and Cliff and Gluckstein 1986.

However, as a popular idea rooted in an impetuous militancy, direct action was profoundly important in 1919. In an often chaotic and spontaneous way, it served as a rallying cry that threatened to carry the large battalions of the labour movement with it. The experience of war had politicised workers, so direct action in 1919 went beyond the 'Non-Political' syndicalism of 1910–14, eroding divisions between politics and economics within the labour movement.[56] As Straker told Triple Alliance delegates:

> Just as the working classes are becoming enlightened, they are recog-
> nising the close relationship between these two questions, and that polit-
> ics have now got into the economic field, where trade union questions
> became political questions, and political questions became trade union
> questions.[57]

At the beginning of the year, all roads seemed to lead away from parliament and towards direct action, and the Miners' Charter lit the way. This posed a threat to coalowners and government, but also to the Labour Party as it struggled to carry out its postwar readjustment. Some even entertained the possibility of a split. The *Daily Herald* pointed out that 'in the ranks of Labour there is a tendency for two parties to arise. Direct Action chants the one; Political Action sings the other'.[58] Just when the forces of direct action were at their greatest, the MFGB gave the Labour Party a foothold in the shape of the Sankey Commission. The wave of public sympathy to which the Commission's hearings gave rise alerted the miners' leaders to an alternative strategy, and gave succour to embattled Party leaders as winning public opinion became the end in itself.

Since its inception, the Miners' Charter had evolved from being the particu-lar programme of one occupational group into the rallying cry of an entire class. It entered 1919 as the powerful symbol of direct action; it departed it as the lame Mines for the Nation campaign, an adjunct to Labour's *New Social Order*, to be achieved through the ballot box. As the Charter atrophied so too did the idea, central to direct action, that social change and socialism itself might best be brought about by the collective industrial power of workers.

In the mythology of the labour movement, direct action died along with the Triple Alliance on 'Black Friday', 15 April 1921, a date welcomed by Ramsay MacDonald as the end of an epoch:

---

56   This was the slogan of Tom Mann's *Industrial Syndicalist*.
57   MFGB, Triple Alliance Conference 23 July 1919, p. 29.
58   *Daily Herald*, 1 August 1919.

We may well look back upon these dark days with gratitude. The events of this strike ought to settle for a long time the influence of those who preached the doctrine that Labour can emancipate itself by industrial means alone.[59]

It took this defeat, and that of 1926, before the syndicalist and revolutionary idea of direct action was fully marginalised within the labour movement, but as we have seen, it was in headlong retreat long before Black Friday. In this regard, the calling off by Triple Alliance leaders of the ballot for direct action over Russia on 12 August 1919 presaged things to come.

We may confidently assume that the power of the Triple Alliance as a force, either in industry or politics, is gone forever. All over the country the workers in the different industries had come to look upon the Triple Alliance as being the vanguard of the revolutionary forces of the now thoroughly awakened workers, but in the space of a few hours their hopes were dashed to the ground.[60]

The demoralisation to which this had given rise would be mitigated to a degree by 'Labour's Fling' in 1920,[61] but a compelling case can be made that the Council of Action was by that stage pushing at an open door.[62] When the real test came eight months later, it was the climb down of 1919 that provided the precedent.

We have, it is worth reminding ourselves, been dealing with only one section of the working class, and the experience of the miners cannot be taken as synonymous with the experience of the class as a whole. Despite the broad spread of industrial unrest in 1919, sectionalism between groups of workers remained and each section had its own particular experiences, its own industrial specificities. Nonetheless, with a million members the MFGB occupied a special place in the labour movement. Its fate was decisive to the process by which the direction of labour politics was decided in the postwar period. Various historians have pondered alternative outcomes. Woodhouse, for example, holds that not only might a national miners' strike in 1919 have won its far-reaching

---

59   *Socialist Review*, July – September 1921, p. 197. MacDonald wrote this in the wake of the miner's defeat.

60   *The Worker*, 23 August 1919.

61   Miliband 1964, chapter 3, (iii).

62   White 1974.

demands, but it also could conceivably have developed 'insurrectionary over-tones'.[63] As Hutt saw it, the government was, until March at least, 'confronted with the alarming prospect of a general strike fraught with revolutionary implications'.[64] Certainly it would have provided the general labour unrest with a much needed focus, and might have succeeded in forging a real national direct action movement out of the frenetic and uncoordinated strikes which were taking place across the industry. Had this happened, then the question of state power, of 'who rules?', would doubtless have arisen. In Saville's view, a successful miners' strike would have transformed the whole history of the 1920s. Nationalisation would have wiped out the most reactionary group of British industrialists and boosted the industrial and political power of labour in society as a whole. The disasters of 1921 and 1926 would not have taken place, and, though socialism would not have been the result ('the political leadership was too strongly labourist'), 'the victory of the miners in 1919 might have helped to make Britain a more civilised society'.[65] Other suggestions might be made here; when the Communist Party came into existence in the 1920s, it was during a period of retreat for the labour movement. Had it been formed on the back of a major industrial and political victory, its size and influence would have been far greater, and revolutionary politics in Britain less marginal.

Such counterfactuals are useful only insofar as they help to illuminate in new ways the actual course taken by history. The miners and the Triple Alliance had declined the opportunity of launching an offensive national strike in the favourable conditions of the postwar boom and the general atmosphere of militancy that accompanied it. As a consequence, when the battle did come two years later, the miners were on the defensive and fighting in hostile terrain. Almost inevitably the radicalism and optimism of the year in which everything had seemed possible came to be buried under this defeat, and the political orthodoxies to which it gave rise. Nationalisation of the mines did eventually come, almost thirty years later. The Sankey Commission provided the MFGB with a political agenda that endured until 1948, but the demand for state ownership was never again twinned with radical ideas of worker's control.

We leave the last word on perspectives to two retired South Wales miners who lived through this period as young men. One looks back to 1919 as a golden age for the miners; the other highlights the themes of betrayal and lost opportunity. Mel Thomas from Maesteg recalled:

---

63   Woodhouse 1969, p. 158.
64   Hutt 1937, p. 18.
65   Saville 1988, pp. 46–7.

They felt that the Sankey Commission had solved all their problems anyhow, that was the general feeling. And even today of course, if you talk to the older man and you ask him a question as to what was the outstanding agreement he ever knew, and Sankey was the best man the miners ever knew, Sankey.[66]

James Griffiths, Ex-President of the SWMF:

The slide down began after 1919; pits were closing down and unemployment was growing, and in some areas there was a drop in wages already. The export trade in coal was never recovered after 1918 in fact. For the miners, 1919–20 was a big turning point, not only economically but politically as well. We were told 'don't strike', the Sankey Commission will look into everything. Lloyd George pledged that what the Sankey Commission decided the Government would accept. So the word 'Sankey' is writ large in mining politics. There was no question of misunderstanding. We stayed at work then because the pledge was absolutely definite that whatever Sankey recommended would be accepted by the Government. Before we knew it, almost, we were fighting for survival.[67]

---

66   Interview with Mel Thomas 17 May 1973, SWML MT/60/6.
67   Leeson 1973, p. 74.

Bibliography

Private Collections

Edgar L. Chappell Collection, Box no. 7, Royal Commission on Industrial Unrest, 1917 (7 folders of minutes of evidence and written submissions), National Library of Wales, Aberystwyth.

S.O. Davies Papers, Glamorgan Record Office, Cardiff.

Tom Jones C.H. Collection, National Library of Wales, Aberystwyth.

David Lloyd George Papers, House of Lord Record Office.

Ramsay MacDonald Papers, PRO 30/69, Public Record Office.

John MacLean Papers, National Library of Scotland, Edinburgh.

Monmouthshire and South Wales Coalowners' Association Records, National Library of Wales, Aberystwyth.

E. Sylvia Pankhurst Papers, Internationaal Instituut voor Sociale Geschiedenis, Amsterdam.

Passfield Papers (includes Beatrice Webb's Diaries, 1/35), British Library of Political and Economic Science.

Runciman Papers, Newcastle University Library.

Steel Maitland Papers, Memorandum: 'Coal Mines Nationalisation Controversy, 1919'. GD 193/80/206–17, Scottish Record Office, Edinburgh.

Trade Union and Party Political

Derbyshire Miners' Association, Minutes of Council and Executive Committee Meetings, National Union of Mineworkers (Derbyshire Area), Chesterfield.

Durham Miners' Association, Minutes of Council Meetings and Conferences, Bound Volumes of Circulars, National Union of Mineworkers (North-Eastern Area), Durham.

Independent Labour Party, Conference Reports, John Rylands Library, University of Manchester.

Labour Party Conference Reports, Working Class Movement Library, Salford.

Labour Party Correspondence Files, Volumes 1–3, Labour Party Archives, National Museum of Labour History, Manchester.

Labour Party National Executive Committee Minutes, Labour Party Archives, National Museum of Labour History, Manchester.

Lanarkshire Miners' Union, Minutes of Council and Executive Committee, NLS/PDL/31. National Library of Scotland, Edinburgh.

Lancashire and Cheshire Miners' Federation Records, National Union of Mineworkers (Lancashire Area), Bolton and Leigh.

Miners' Federation of Great Britain, Minutes of Conferences, Executive Committee Meetings, Financial Statements, Triple Alliance Conference Reports, Minutes of Meetings between the Federation Executive Committee and the Government.

National Union of Scottish Mineworkers, Minutes of Annual Conference, 1919, NLS/PDL/45, National Library of Scotland, Edinburgh.

Nottinghamshire Miners' Association, Minute book, 1919, Miners' Offices, Basford.

South Wales Coalfield Archive, Swansea:

    i. C.1, Minutes of South Wales Miners' Federation Executive Council and Annual Special Conferences.

    ii. D.1, Financial Records.

    iii. E.2, Correspondence – Bound Volume of Circulars.

    iv. K.10, 13, 20, Various District, Area and Combine Records.

    v. Lodge Records, various.

    vi. Personal Collections – Jesse Clark (diary and notebooks); S.O. Davies (letters and newspaper cuttings); Arthur Horner (letters and newspaper cuttings); D.J. Williams (pocket diary), 1919.

    vii. Transcribed interviews with ex-miners, SSRC Coalfield History Project, South Wales Miners Library, Swansea.

TUC Conference Reports, Working Class Movement Library, Salford.

Yorkshire Miners' Association, Minutes of Council Meetings, Executive Committee Meetings and Financial Statements, National Union of Mineworkers (Yorkshire Area), Barnsley.

Cabinet Records

Cabinet Records from Prime Ministers to the Crown, Reel 13, 1915–16 (Microfilmed by Macromedia Ltd. For the Harvest Press, 1974), John Rylands Library, University of Manchester.

CAB 23, Minutes of Cabinet Meetings, 1919, Public Record Office.

CAB 24/23, Reports of the Commission of Enquiry into Industrial Unrest, Cd.8662 x.v.1 to Cd.8696 x.v.149, Public Record Office.

CAB 24/78/387, Report on Revolutionary Organisations in the UK, Cabinet Paper GT 7619.

CAB 27/59 UC Series, Report, Proceedings and Memoranda of War Cabinet Committee on Industrial Unrest, Public Record Office.

Parliamentary Papers

Coal Industry Commission, Minutes of Evidence from the First Stage of the Enquiry and Interim Report, Cmn.359; Minutes of Evidence from the Second Stage of the Enquiry, and Final Reports, Cmd.360; Appendices and Index, Cmd.361.

Departmental Committee on Conditions Prevailing in the Coal Mining Industry due to the War, Part 1, 1st General Report, Cd.7939 xxviii, 1, 1914–16; Part 2, Minutes of Evidence and Index, Cd.8009 xxxviii, 5, 1914–16; 2nd General Report, Cd.8147 xxxviii.321, 1918.

Department of Employment, British Labour of Statistics, Historical Abstract (HMSO, 1971).

General Reports and Statistics of Mines and Quarries in the UK, British Colonies and Foreign Countries, 1919; Part One, Cmd, 925, 1.729; Part Two, 1920, Cmd.1007, 1.933.

Hansard, House of Commons Debates, 5th Series.

Report of the Departmental Committee appointed to consider the position of the Coal Trade after the War, C.9093, xiii.321, 1918.

Report of the Provisional Joint Committee presented to the meeting of the Industrial Conference Central Hall, Westminster, 4 April 1919. Cmd.139 xxiv.1.

Similar Report, with an Appendix on the Causes and Remedies for Labour Unrest, and a Provisional Scheme for Trade Union Representation on the National Industrial Council, 1919 Cmd.xxiv.21.

Newspapers and Periodicals

1   *National*
Colliery Guardian
Daily News
Economist
Financial Times
Labour Gazette
Manchester Guardian
New Statesman
Sunday Pictorial
The Times

2   *Regional and Local*
Aberdare Leader
Alfreton and Belper Journal and East Derbyshire Advertiser
Ashton Reporter

Barnsley Chronicle
Barnsley Independent
Bellshill Speaker and North East Lanarkshire Gazette
Bolton and Bury Catholic Herald
Bolton Evening News
Bolton Journal and Guardian
Burnley Catholic Herald
Burnley Express
Burnley News
Cambrian Daily Leader
Cambuslang Advertiser
Cannock Advertiser
Cannock Chase Courier
Cowdenbeath and Lochgelly Times and Advertiser
Derbyshire Advertiser
Derbyshire Courier
Derbyshire Times
Derbyshire Worker
Derbyshire and Chesterfield Reporter
Doncaster Chronicle
Doncaster Gazette
Doncaster and Thorne Advertiser
Dover and County Chronicle
Dover Express
Dover Standard Kentish Advertiser and Colliery News
Dover Telegraph
Dunfermline Journal and Advertiser for the West of Fife
Dunfermline Press
Durham County Advertiser
Fife Free Press
Fifeshire Advertiser
Glamorgan Free Press
Glasgow Citizen
Glasgow Daily Record
Glasgow Observer
Glasgow Weekly Mail and Record
Goole Times
Goole Weekly Journal
Hamilton Herald and Lanarkshire Weekly News
Heanor Observer

Ilkeston Advertiser
Ilkeston Pioneer
Keighley News and Bingley Chronicle
Kirkcaldy Times
Lanarkshire and Hamilton Herald
Lancashire Daily Post
Leven Advertiser and Wemyss Gazette
Leeds Catholic Herald
Leeds Mercury
Leeds Weekly Citizen
Leigh Chronicle
Leigh, Tyldesley and Atherton Journal
Llais Llafur
Manchester Evening News
The Mail for Kirkcaldy, Central and West Fife
The Mail for Leven, Wemyss and East Fife
Mansfield Chronicle
Mansfield and North Nottinghamshire Advertiser
Mansfield Reporter and Sutton in Ashfield Times
Merthyr Express
Mexborough and Swinton Times
Monmouthshire Evening Post
Monmouthshire Weekly Post
Motherwell Times
Nottingham Evening News
Nottingham Evening Post
Nottingham Guardian
Nottingham Journal and Express
Nottingham and Midland Catholic News
Nottingham, Notts and Bulwell Local News
Nottinghamshire Free Press and Derbyshire Chronicle
Nottinghamshire Guardian
Nottinghamshire Weekly Express
Pontefract and Castlefield Express
Rhondda Leader
Ripley News
Rotherham Advertiser
Rotherham Express
Rothwell Courier and Times
St. Helens Catholic Herald

St. Helens Examiner
St. Helens Newspaper and Advertiser
St. Helens Reporter
Sheffield Daily Telegraph
Sheffield Independent
Sheffield Mercury
Sheffield Weekly Independent
Sheffield Weekly News
Sheffield Weekly Telegraph
South Elmsall and Hemsworth Express
South Wales Daily Post
South Wales Coal Annual
South Wales Echo
South Wales News
Staffordshire Advertiser
Staffordshire Catholic News
Staffordshire Chronicle
Staffordshire Sentinel
Staffordshire Weekly Sentinel
Tyldesley and Atherton Chronicle
Walsall Observer
Walsall Pioneer and District News
Wakefield Advertiser and Gazette
Wakefield Express
West Fife Echo
West Yorkshire Pioneer
Western Mail
Wigan Catholic Herald
Wigan Examiner
Wigan Observer
Yorkshire Catholic Herald
Yorkshire Evening News
Yorkshire Evening Post
Yorkshire Evening Press
Yorkshire Gazette
Yorkshire Herald
Yorkshire News
Yorkshire Post
Yorkshire Telegraph and Star
Yorkshire Weekly Post

3   *Labour and Socialist*

The Call
Daily Herald
Forward
Herald
Industrial Syndicalist
Labour Leader
Merthyr Pioneer
Plebs Magazine
Rhondda Socialist
The Socialist
Socialist Review
Solidarity
The Worker
Workers' Dreadnought

Books, Articles, Chapters in Books

Ablett, Noah 1915, 'Wagery in the South Wales Coalfield', *Plebs Magazine*, 7(8).

Adams, T. 1998, 'Working Class Organisation, Industrial Relations and the Labour Unrest, 1914–21', Unpublished PhD Thesis, University of Leicester.

Allen, Victor 1960, *Trade Unions and the Government*, London: Longmans.

———— 1981, *The Militancy of British Miners*, Shipley: The Moor Press.

Anderson, Kevin 1995, *Lenin, Hegel and Western Marxism*, Chicago, IL: University of Illinois Press.

Armitage, Susan 1969, *The Politics of Decontrol of Industry, Britain and the United States*, London: LSE Research Monographs 4.

Askwith, George 1920, *Industrial Problems and Disputes*, London: John Murray.

Bagwell, Phillip 1963, *The Railwaymen: A History of the National Union of Railwaymen*, London: Allen and Unwin.

———— 1971, 'The Triple Industrial Alliance, 1913–22', in *Essays in Labour History, 1886–1923*, edited by Asa Briggs and John Saville, London: Macmillan.

Baylies, Carolyn 1993, *A History of the Yorkshire Miners, 1881–1918*, London: Routledge.

Bellamy, Joyce and John Saville (eds.) 1972–84, *Dictionary of Labour Biography*, London: Macmillan.

Bevan, Aneurin 1952, *In Place of Fear*, London: Heinemann.

Birchall, Ian and Norah Carlin 1983, 'Eric Hobsbawm and the Working Class', *International Socialism*, 2(21): 88–116.

Black, Carmel 1991, 'The Origins of Unemployment Insurance in Queensland, 1919–23', *Labour History*, 60(May): 34–50.

Blackledge, Paul 2005, 'Brian Manning: Historian of the People and the English Revolution', *Historical Materialism*, 13(3): 219–28.

———— 2006, *Reflections on the Marxist Theory of History*, Manchester: Manchester University Press.

———— 2011, 'Why Workers can Change the World', *Socialist Review*, 364.

———— 2012, 'Eric Hobsbawm (1917–2012)', Socialist *Review*, 374.

———— 2013, 'Left Reformism, the State and the Problem of Socialist Politics Today', *International Socialism*, 2(139): 25–56.

Boyns, Trevor 1987, 'Technical Change and Colliery Explosions in the South Wales Coalfield, c.1870–1914', *Welsh History Review*, 13.

Briggs, Asa and John Saville (eds.) 1971, *Essays in Labour History 1886–1923*, London: Macmillan.

British Socialist Party 1919, *Hands off Russia*.

Bullock, Alan 1960, *The Life and Times of Ernest Bevin, Vol. 1*, London: Heinemann.

Burgess, Keith 1980, *The Challenge of Labour: Shaping British Society, 1850–1930*, London: Croom Helm.

Callinicos, Alex 1995, *Socialists in the Trade Unions*, London: Bookmarks.

———— 2007, 'Leninism in the Twenty-First Century?', in *Lenin Reloaded: Towards a Politics of Truth*, edited by Sebastian Budgen, Stathis Kouvelakis, and Slavoj Žižek, Durham, NC: Duke University Press.

Campbell, Alan 1978, 'Honourable Men and Degraded Slaves: A Comparative Study of Trade Unionism in Two Lanarkshire Mining Communities, c.1830–1874', in *Independent Collier*, edited by Royden Harrison, Hassocks: Harvester Press.

———— 1989, 'From Independent Collier to Militant Miner: Tradition and Change in the Trade Union Consciousness of the Scottish Miners, 1874–1929', *Scottish Labour History Journal*, 24: 8–23.

———— 1994, 'The Communist Party in the Scots Coalfields in the Inter-War Period', Unpublished paper presented at the Opening of the Books Conference, University of Manchester.

———— 1996, 'The Social History of Political Conflict in the Scots Coalfields, 1910–1939', in *Miners, Unions and Politics, 1910–47*, edited by Alan Campbell, Nina Fishman and David Howell, Aldershot: Scolar Press.

———— 2000, *The Scottish Miners 1874–1939: Volume Two: Trade Unions and Politics*, Aldershot: Ashgate.

———— 2009 [2004], 'Scotland', in *Industrial Politics and the 1926 Mining Lockout: The Struggle for Dignity*, edited by John McIlroy, Alan Campbell and Keith Gildart, Cardiff: University of Wales Press.

Campbell, Alan, Nina Fishman and David Howell (eds.) 1996, *Miners, Unions and Politics, 1910–47*, Aldershot: Scolar Press.

Campbell, Alan and Fred Reid 1978, 'The Independent Collier in Scotland', in *Independent Collier*, edited by Royden Harrison, Hassocks: Harvester Press.

Campbell, John 1987, *Nye Bevan and the Mirage of the British Socialism*, London: Weidenfeld and Nicholson.

Campbell, John and William Gallacher 1919, *Direct Action*, Glasgow: Scottish Workers' Committee Pamphlet no. 1.

Carr, Edward Hallet 2001, *What is History?*, London: Palgrave.

Carter, G.R. 1915a, 'The Coal Strike in South Wales', *Economic Journal*, 25: 453–65.

—— 1915b, 'The Sequel of the Welsh Coal Strike and its Significance', *Economic Journal*, 26: 521–31.

Catterall, Stephen 2009 [2004], 'Lancashire', in *Industrial Politics and the 1926 Mining Lockout: The Struggle for Dignity*, edited by John McIlroy, Alan Campbell and Keith Gildart, Cardiff: University of Wales Press.

Challinor, Raymond 1972, *The Lancashire and Cheshire Miners*, Newcastle: Graham.

—— 1977, *The Origins of British Bolshevism*, London: Croom Helm.

Chappell, Richard and Alan Clinton 1974 (eds.), *Trotsky's Writing on Britain, Vol. 1*, London: New Park.

Clegg, Hugh 1985, *A History of British Trade Unions since 1889, Vol. 2*, Oxford: Oxford University Press.

Cliff, Tony 1975, *Lenin: Building the Party*, London: Pluto Press.

Cliff, Tony, and Donny Gluckstein 1986, *Marxism and Trade Unions: The General Strike of 1926*, London: Bookmarks.

Coates, Ken (ed.) 1974, *Democracy in the Mines: Some documents of the controversy on mines nationalisation up to the time of the Sankey Commission*, Nottingham: Spokesman Books.

Coates, Ken and Tony Topham (eds.) 1970, *Workers' Control*, London: Panther.

—— 1991, *The Making of the Labour Movement, Vol. 1: The Emergence of the Labour Movement 1870–1922*, Oxford: Basil Blackwell.

Cole, G.D.H. 1923, *Labour in the Coalmining Industry 1914–1921*, Oxford: Oxford University Press.

—— 1948, *A History of the Labour Party from 1914*, London: Routledge and Kegan Paul.

Cole, Margaret (ed.) 1952, *Beatrice Webb's Diaries, 1912–1924*, London: Longmans.

Coleman, D.C. and Peter Matthias (eds.) 1984, *Enterprise and History*, Cambridge: Cambridge University Press.

Coombes, Bert 1939, *These Poor Hands*, London: Gollancz.

Cowden, Morton H. 1984, *Russian Bolshevism and British Labour, 1917–21*, Boulder, CO: East European Monographs.

Cowling, Maurice 1971, *The Impact of Labour, 1920–1924: The Beginning of Modern British Politics*, Cambridge: Cambridge University Press.

Craig, Frederick (ed.) 1968, *British Parliamentary Election Statistics 1918–68*, Glasgow: Political Reference Publications.

Craik, William 1964, *The Central Labour College, 1909–29*, London: Lawrence and Wishart.

Cronin, James 1979, *Industrial Conflict in Modern Britain*, London: Croom Helm.

————— 1984, *Labour and Society in Britain 1918–79*, London: Batsford.

————— 1988, 'Strikes and Power in Britain, 1870–1920' in *Industrial Conflict: Papers Presented to the Fourth British-Dutch Conference on Labour History*, edited by Lex Heerma van Voss and Herman Diederichs, Amsterdam: IISG.

Cronin, James and Jonathan Schneer (eds.) 1982, *Social Conflict and the Political Order in Modern Britain*, London: Croom Helm.

Darlington, Ralph 2006, 'In Defence of the *Alter*factual in Historical Analysis', *Socialist History*, 29: 76–88.

————— 2008, *Syndicalism and the Transition to Communism*, Aldershot: Ashgate.

————— 2014, 'The Rank-and-File and the Trade Union Bureaucracy', *International Socialism*, 142.

Darlington, Ralph and Martin Upchurch 2012, 'A Reappraisal of the Rank-and-File versus Bureaucracy Debate', *Capital and Class*, 36: 77–95.

Davies, D.K. 1991, 'The Influence of Syndicalism and Industrial Unionism in the South Wales Coalfield, 1898–1921: A Study in Ideology and Practice', Unpublished PhD Thesis, Cardiff: University of Wales.

Davies, P. 1978, 'The Making of A.J. Cook', *Llafur*, 2(3).

Desmerais, Ralph 1975, 'Lloyd George and the Development of the British Government's Strike-breaking Organisation', *International Review of Social History*, 20(1): 1–15.

Dewar, George 1919, 'The Great Home Problem of 1919', *The Nineteenth Century and After*, 85: 205–25.

de Ste. Croix, Geoffrey 1983, *Class Struggle in the Ancient Greek World*, London: Duckworth.

————— 1984, 'Class in Marx's Conception of History, Ancient and Modern', *New Left Review*, 146(July – August).

Dunn, John 1989, *Modern Revolutions*, Cambridge: Cambridge University Press.

Edwards, Ness 1926, *The History of the South Wales Miners*, London: Labour Publishing Co.

Egan, David 1975, 'The Swansea Conference of the British Council of Soldiers' and Workers' Delegates, July 1917: Reactions to the Russian Revolution of February 1917, and the Anti-War Movement in South Wales', *Llafur*, vol. 1, no. 4.

————— 1986, 'Noah Ablett, 1883–1935', *Llafur*, 4(3).

————— 1988, 'Wales at Work', in *Wales 1880–1914*, edited by Trevor Herbert, Gareth Elwyn Jones, Cardiff: University of Wales Press.

————— 1996, 'A Cult of Their Own: Syndicalism and The Miners' Next Step', in *Miners, Unions and Politics, 1910–47*, edited by Alan Campbell, Nina Fishman and David Howell, Aldershot: Scolar Press.

Englander, David 1991, 'The National Union of Ex-Servicemen and the Labour Movement, 1918–1920', *History*, 76(246), February.

Evans, Eric Wyn 1959, *Mabon: William Abraham 1842–1922: A Study in Trade Union Leadership*, Cardiff: University of Wales Press.

Evans, Neil 1992, 'The War after War: Radical Conflict and the Post-War Crisis in Britain, 1919–25', Unpublished paper presented to the Northern Marxist Historians' Group.

Evans, Raymond 1987, *Loyalty and Disloyalty: Social Conflict on the Queensland Home Front, 1914–18*, Sydney: Allen and Unwin.

Ferguson, Niall (ed.) 1997, 'Introduction', in *Virtual History*, London: Picador.

Fernbach, David 1974, 'Introduction', in *Karl Marx: The First International and After*, London: Penguin.

Foster, John 1991, 'Strike Action and Working Class Politics on Clydeside, 1914–1919', *International Review of Social History*, 35(1): 33–70.

———— 2004, 'Prologue: What Kind of Crisis? What Kind of Ruling Class?', in *Industrial Politics and the 1926 Mining Lockout*, edited by John McIlroy, Alan Campbell and Keith Gildart, Cardiff: University of Wales Press.

Fox, Alan 1985, *History and Heritage: The Social Origins of the British Industrial Relations System*, London: Allen & Unwin.

Francis, D. 1974, 'Dai Dan Evans, 1898–1974', *Llafur*, 1(3).

Francis, Hywell 1973, 'The Anthracite Strike and the Disturbances of 1925', *Llafur*, 1(3).

Francis, Hywell and Dai Smith 1980, *The Fed: A History of the South Wales Miners in the Twentieth Century*, London: Lawrence and Wishart.

Garside, William 1971, *The Durham Miners, 1919–60*, London: Allen and Unwin.

Gilbert, David 1992, *Class, Community and Collective Action: Social Change in Two British Coalfields, 1850–1926*, Oxford: Clarendon Press.

———— 1996, 'The Landscape of Spencerism: Mining Politics in the Nottinghamshire Coalfield, 1910–47', in *Miners, Unions and Politics, 1910–47*, edited by Alan Campbell, Nina Fishman and David Howell, Aldershot: Scolar Press.

Gollan, Robin 1963, *The Coalminers of New South Wales*, Cambridge: Cambridge University Press.

Gleason, Arthur 1920, *What the Workers Want: A Study of British Labour*, London: G. Allen and Unwin

Gould, Gerald 1920, *The Coming Revolution in Great Britain*, London: Collins.

Gould, Stephen Jay 1989, *Wonderful Life: The Burgess Shale and the Nature of History*, London: Penguin.

Gregory, Roy 1968, *The Miners in British Politics, 1906–14*, Oxford: Oxford University Press.

Griffin, Alan 1955, *The Miners of Nottinghamshire: A History of the Nottinghamshire Miners' Association, Vol. 1*, Nottingham: National Union of Mineworkers.

————— 1962, *The Miners of Nottinghamshire, Vol. 2, 1914–44*, London: Allen and Unwin, 1962.

Griffiths, R. 1983, *S.O. Davies: A Socialist Faith*, Dyfed: Gomer.

Griffiths, Trevor 1996, 'Work, Class and Community: Social Identities and Political Change in the Lancashire Coalfield, 1910–1939', in *Miners, Unions and Politics, 1910–47*, edited by Alan Campbell, Nina Fishman and David Howell, Aldershot: Scolar Press.

Grigg, John 1978, *Lloyd George: The People's Champion*, London: Methuen.

Gwyther, C.E. 1967, 'Methodism and Syndicalism in the Rhondda Valley, 1906–1926', Unpublished PhD Thesis, Sheffield University.

Hallas, Duncan 1980, 'Trade Unionists and Revolution: A Response to Richard Hyman', *International Socialism*, 2(8): 80–4.

————— 1985, *The Comintern*, London: Bookmarks.

Harley, J. 1916, 'The Conscription of Industry', *The Contemporary Review*, 109(May).

Harman, Chris 1991, 'The State and Capital Today', *International Socialism*, 2(51): 3–54.

Harrison, Royden 1971, 'The War Emergency Workers' National Committee, 1914–1920', in *Essays in Labour History, 1886–1923*, edited by Asa Briggs and John Saville, London: Macmillan.

————— (ed.) 1978, *Independent Collier: The Coal Miner as Archetypal Proletarian Reconsidered*, London: Harvester.

Hawthorn, Geoffrey 1991, *Plausible Worlds*, Cambridge: Cambridge University Press.

Hay, W.F. 1919, *Industrial Democracy for the Miners: A Plan for the Democratic Control of the Mining Industry*, Glasgow: South Wales Socialist Society.

Henderson, Arthur 1918, *The Aims of Labour*, London: Headley Bros.

Herbert, T. and G.E. Jones (eds.) 1988a, *Wales 1880–1914*, Cardiff: University of Wales Press.

————— 1988b, *Wales Between the Wars*, Cardiff: University of Wales Press.

Hill, Christopher 1975, *The World Turned Upside Down*, London: Penguin.

Hinton, James 1973, *The First Shop Stewards Movement*, London: Allen and Unwin.

————— 1976, 'A Reply to Jean Monds', *New Left Review*, I/97(May – June).

————— 1983, *Labour and Socialism*, Wilmslow: Wheatsheaf.

Hobsbawm, Eric 1973, *The Age of Revolution*, London: Cardinal.

————— 1984, *Worlds of Labour*, London: George Weidenfield and Nicholson.

————— 1994, *Revolutionaries*, London: Phoenix.

Hodges, Frank n.d., *My Adventures as a Labour Leader*, London: George Newnes.

————— 1920, *Nationalisation of the Mines*, London: New Era Series.

Holton, Robert 1976, *British Syndicalism 1900–14: Myths and Realities*, London: Pluto.

————— 1985, 'Revolutionary Syndicalism and the British Labour Movement', in *The Development of Trade Unionism in Great Britain and Germany, 1880–1914*, edited by Wolfgang Mommsen and Han-Gerhard Husung, London: Allen and Unwin.

Hopkin, D. 1974, 'Patriots and Pacifists in Wales, 1914–1918: The Case of Captain Lionel Lindsay and the Reverend T.E. Nicholas', *Llafur*, 1(3).

———— 1977, 'The Merthyr Pioneer, 1911–22', *Llafur*, 3(4).

Horner, Arthur 1960, *Incorrigible Rebel*, London: MacGibbon and Kee.

Howell, David 1983, *British Workers and the Independent Labour Party 1888–1906*, Manchester: Manchester University Press.

———— 1986, *A Lost Left: Three Studies in Socialism and Nationalism*, Manchester: Manchester University Press.

———— 1996, 'All or Now't: The Politics of the MFGB' in *Miners, Unions and Politics, 1910–47*, edited by Alan Campbell, Nina Fishman and David Howell, Aldershot: Scolar Press.

Hutt, Alan 1937, *The Post-War History of the British Working Class*, London: Gollancz.

Hyman, Richard 1980, 'British Trade Unionism: Post-War Trends and Future Prospects', *International Socialism*, 2(8): 64–79.

———— 1989, *Strikes*, London: Macmillan.

———— 2011, 'Will the Real Richard Hyman Please Stand Up?', *Capital and Class*, 36(1): 151–64.

James, David, Tony Jowitt and Keith Laybourn (eds.) 1992, *The Centennial History of the Independent Labour Party*, Krumlin: Ryburn Academic.

Jenkins, Roy 1991, *A Life at the Centre*, London: Macmillan.

Jevons, H. 1915, *The British Coal Trade*, London: Kegan and Paul.

Johnson, Carol 1980, 'The Problem of Reformism and Marx's Theory of Fetishism', *New Left Review*, I/119.

Johnson, Paul 1968, *Land Fit for Heroes*, Chicago, IL: University of Chicago Press.

Kelly, John 1988, *Trade Unions and Socialist Politics*, London: Verso.

Kendall, Walter 1969, *The Revolutionary Movement in Britain, 1900–21: The Origins of British Communism*, London: Weidenfeld and Nicolson.

Kershaw, Ian 2007, *Fateful Choices*, London: Penguin.

Kirby, M.W. 1977, *The British Coalmining Industry, 1870–1946: A Political and Economic History*, London: Macmillan.

Kliman, Andrew 2007, *Reclaiming Marx's Capital*, Lanham: Lexington.

Knee, Fred 1912, 'Socialism vs. Syndicalism', *The British Socialist*, 1(6).

Knox, W. 1984, *Scottish Labour Leaders*, Edinburgh.

Kouvelakis, Stathis 2007, 'Lenin as Reader of Hegel', in *Lenin Reloaded: Towards a Politics of Truth*, edited by Sebastian Budgen et al., Durham, NC: Duke University Press.

Lann, M. 1984, 'Syndicalist Teachers: The Rhondda Strike of 1919', *Llafur*, 4(1).

Lawrence, D.H. 1988, *Aaron's Rod*, Cambridge, Cambridge University Press.

Lawson, J. 1941, *The Man in the Cloth Cap*, London: Methuen.

Leeson, R.A. 1973, *Strike: A Live History, 1887–1971*, London: Allen and Unwin.

Lenin, V.I. 1961, *Collected Works*, Vol. 38, Moscow: Progress Publishers.

———— 1962, *Collected Works*, Vol. 13, Moscow: Progress Publishers.

———— 1993, *Left-Wing Communism: An Infantile Disorder*, London: Bookmarks.

Lewis, R. 1976, 'The South Wales Miners and the Ruskin College Strike of 1909', *Llafur*, 2(1).

———— 1980, 'Leaders and Teachers: The Origins and Development of the Workers' Education Movement in South Wales, 1906–40', Unpublished PhD Thesis, Swansea: University of Wales.

Long, P. 1982, 'Abe Moffat, the Fife Miners and the United Mineworkers of Scotland', transcribed interview, *Scottish Labour History Society Journal*, 17.

Lowe, D. 1923, *From Pit to Parliament: The Story of the Early Life of Keir Hardie*, London: Labour Publishing Co.

Lowe, R. 1978, 'The Failure of the Consensus in Britain: The National Industrial Conference, 1919–21', *Historical Journal*, 21(3): 649–75.

Luxemburg, Rosa 1986, *The Mass Strike*, London: Bookmarks.

MacDonald, Ramsay 1919, *Parliament and Revolution*, Manchester.

MacDougall, Ian 1978, *Essays in Scottish Labour History: A Tribute to W.H. Marwick*, Edinburgh: Donald.

———— (ed.) 1981, *Militant Miners*, Edinburgh: Polygon.

MacDougall, J. 1927, 'The Scottish Coalminer', *The Nineteenth Century and After*, 102: 762–81.

MacIntyre, Alasdair 2008, 'Trotsky in Exile', in *Alasdair Macintyre's Engagement with Marxism*, edited by Paul Blackledge and Neil Davidson, Leiden: Brill.

Macintyre, Stuart 1980, *Little Moscows*, London: Croom Helm.

Mackenzie, Norman and Jeanne Mackenzie (eds.) 1984, *The Diaries of Beatrice Webb, Vol. 3, 1905–24*, London: Virago.

Mann, Tom 1913, *From Single Tax to Syndicalism*, Walthamstow: G. Bowman.

Mantoux, Paul 1964, *Paris Peace Conference, 1919: Proceedings of the Council of Four*, Geneva: Droz.

Marquand, David 1977, *Ramsay MacDonald*, London: Cape.

Marwick, Arthur 1974, *War and Social Change in the Twentieth Century*, London: Macmillan.

Marx, Karl 1975, 'Theses on Feuerbach', in *Early Writings*, London: Penguin.

Marx, Karl and Friedrich Engels 1975, *Collected Works*, Vol. 12, London: Lawrence and Wishart.

Mayer, Arno J. 1968, *Politics and the Diplomacy of Peacemaking: Containment and Counter-Revolution at Versailles, 1918–19*, London: Weidenfeld and Nicolson.

McFarlane, L.J. 1966, *The British Communist Party: Its Origins and Development Until 1929*, London: McGibbon and Kee.

———— 1967, 'Hands of Russia: British Labour and the Russo-Polish War, 1920', *Past and Present*, 38(1): 3–19.

McIlroy, John 2009a [2004], 'South Wales', in *Industrial Politics and the 1926 Mining Lockout: The Struggle for Dignity*, edited by John McIlroy, Alan Campbell and Keith Gildart, Cardiff: University of Wales Press.

McIlroy, John 2009b [2004], 'Nottinghamshire', in *Industrial Politics and the 1926 Mining Lockout: The Struggle for Dignity*, edited by John McIlroy, Alan Campbell and Keith Gildart, Cardiff: University of Wales Press.

McIlroy, John and Alan Campbell 2009a [2004], 'Introduction to the Paperback Edition: Debating the 1926 Mining Lockout', in *Industrial Politics and the 1926 Mining Lockout: The Struggle for Dignity*, edited by John McIlroy, Alan Campbell and Keith Gildart, Cardiff: University of Wales Press.

———— 2009b [2004], 'Fighting the Legions of Hell', in *Industrial Politics and the 1926 Mining Lockout: The Struggle for Dignity*, edited by John McIlroy, Alan Campbell and Keith Gildart, Cardiff: University of Wales Press.

McIlroy, John, Alan Campbell and Keith Gildart (eds.) 2009 [2004], *Industrial Politics and the 1926 Mining Lockout: The Struggle for Dignity*, Cardiff: University of Wales Press.

McKibbin Ross 1974, *The Evolution of the Labour Party, 1910–24*, Oxford: Oxford University Press.

McLean, Iain 1983, *The Legend of Red Clydeside*, Edinburgh: John Donald.

Melling, Joseph 1983, *Rent Strikes: Peoples' Struggles for Housing in West Scotland, 1890–1916*, Edinburgh: Polygon.

———— 1991, 'Whatever Happened to Red Clydeside? Industrial Conflict and the Politics of Skill in the First World War', *International Review of Social History*, 35(1): 3–32.

Mellor, William 1920, *Direct Action*, London: New Era Series.

Michels, Robert, 1962, *Political Parties*, London: Collier Press.

Middlemas Keith (ed.) 1969, *Tom Jones's Whitehall Diary, Vol. 1, 1916–25*, Oxford: Oxford University Press.

———— 1979, *Politics in Industrial Society: The Experience of the British System since 1911*, London: Andre Deutsch.

Miliband, Ralph 1964, *Parliamentary Socialism: A Study in the Politics of Labour*, London: Merlin Press.

———— 1977, *Marxism and Politics*, Oxford: Oxford University Press.

Milton, Nan 1978, *John MacLean: In the Rapids of Revolution: Essays, Articles and Letters, 1902–1923*, London: Alison and Busby.

Mitchell, David J. 1970, *1919: Red Mirage*, London: Jonathan Cape.

Molyneux, John 1986, *Marxism and the Party*, London: Bookmarks.

Mommsen, Wolfgang and Han Gerhard Husung (eds.) 1985, *The Development of Trade Unionism in Britain and Germany*, London: Allen and Unwin.

Monds, Jean 1976, 'Workers' Control and the Historians: A New Economism', *New Left Review*, 1/97.

Morgan, Kenneth O. 1973, 'The New Liberalism and the Challenge of Labour: The Welsh Experience, 1885–1929', *Welsh History Review*, 6(3): 288–312.

———— 1975, 'Socialism and Syndicalism: The Welsh Miners' Debate, 1912', *Bulletin of the Society for the Study of Labour History*, 30: 22–37.

———— 1979, *Consensus and Disunity: The Lloyd George Coalition Government, 1918–22*, Oxford: Clarendon Press.

———— 1988, 'Welsh Politics 1918–39', in *Wales Between the Wars*, edited by Trevor Herbert and Gareth Elwyn Jones, Cardiff: University of Wales Press.

———— 1991, *Wales in British Politics, 1868–1922*, Cardiff: University of Wales Press.

Morton, A.L. and George Tate 1979, *The British Labour Movement*, London: Lawrence and Wishart.

Morris, Richard 1979, 'Mr Justice Higgins Scuppered: The 1919 Seamen's Strike', *Labour History*, 37: 52–62.

Mowat, Charles Loch 1968, *Britain Between the Wars 1918–1940*, London: Methuen.

———— 1971, 'Ramsay MacDonald and the Labour Party', in *Essays in Labour History, 1886–1923*, edited by Asa Briggs and John Saville, London: Macmillan.

Murphy, J.T. 1972, *Preparing for Power*, London: Pluto.

Neville, Robert G. 1974, 'The Yorkshire Miners 1881–1926: A Study in Labour and Social History', Unpublished PhD Thesis, University of Leeds.

———— 1976, 'The Yorkshire Miners and the 1893 Lockout: The Featherstone "Massacre"', *International Review of Social History*, 21(3): 337–57.

O'Brien, Anthony M. 1985, 'Patriotism on Trial: The Strike of the South Wales Miners, July 1915', *Welsh History Review*, 12: 74–104.

Page Arnot, Robert 1919a, *Facts from the Coal Commission*, London: Miners' Federation of Great Britain.

———— 1919b, *Further Facts from the Coal Commission*, London: Miners' Federation of Great Britain.

———— 1953, *The Miners: A History of the Miners' Federation of Great Britain, Vol. 2: Years of Struggle*, London: Allen and Unwin.

———— 1955, *A History of the Scottish Miners from the Earliest Times*, London: Allen and Unwin.

———— 1967, *South Wales Miners: A History of the South Wales Miners' Federation, Vol. 1, 1898–1914*, London: Allen and Unwin.

———— 1974, *South Wales Miners: A History of the South Wales Miners' Federation, Vol. 2, 1914–26*, London: Allen and Unwin.

Postgate, Raymond 1951, *The Life of George Lansbury*, London: Longmans.

Pribicevic, Branco 1957, 'The Demand for Workers' Control in the Railway, Coalmining and Engineering Industries, 1910–22', Unpublished DPhil Thesis, Oxford University.

———— 1959, *The Shop Stewards' Movement and Workers' Control, 1910–22*, London: Blackwell.

Reid, Alistair 1985a, 'Dilution, Trade Unionism and the State in Britain during World War One', in *Shopfloor Bargaining and the State: Historical and Comparative Perspectives*, edited by Steven Tolliday and Jonathan Zeitlin, Cambridge: Cambridge University Press.

———— 1985b, 'The Divisions of Labour and Politics in Britain, 1880–1920', in *The Development of Trade Unionism in Great Britain and Germany, 1880–1914*, edited by Wolfgang Mommsen and Hans-Gerhard Husung, London: Allen and Unwin.

———— 1986, 'Glasgow Socialism', *Social History*, 2(1): 89–97.

———— 1987, 'Class and Organization', *The Historical Journal*, 30(1): 225–38.

Reid, Fred 1978, 'Alexander MacDonald and the Crisis of the Independent Collier, 1872–1874', in *Independent Collier: The Coal Miner as Archetypal Proletarian Reconsidered*, edited by Royden Harrison, London: Harvester.

Riddell, G. 1933, *Riddell's Intimate Diary of the Peace Conference and After, 1918–23*, London: Gollancz.

Ripley, B. and J. McHugh 1989, *John MacLean*, Manchester: Manchester University Press.

Roberts, Andrew (ed.) 2004, *What Might Have Been*, London: Weidenfeld and Nicolson.

Romero, Patricia W. 1990, *Sylvia Pankhurst: Portrait of a Radical*, New Haven, CT: Yale University Press.

Rosenberg, Chanie 1987, *1919: Britain on the Brink of Revolution*, London: Bookmarks.

Rothstein, Andrew 1980, *The Soldiers' Strikes of 1919*, London: Macmillan.

Rubinstein, David 1978, 'The Independent Labour Party and the Yorkshire Miners: the Barnsley By-Election of 1897', *International Review of Social History*, 23(1): 102–34.

Saville, John 1966, 'Notes on Ideology and the Miners before World War One', *Bulletin of the Society for the Study of Labour History*, 23: 25–7.

———— 1988, *The Labour Movement in Britain: A Commentary*, London: Faber.

Schorske, Carl 1983, *German Social Democracy, 1905–1917*, Boston, MA: Harvard University Press.

Smillie, Robert 1924, *My Life for Labour*, London: Mills and Boon.

Smith, Dai 1973, 'What Does History Know of Nail Biting?', *Llafur*, 1(2).

———— 1988, 'From Riots to Revolt: Tonypandy and The Miners' Next Step', in *Wales Between the Wars*, edited by Trevor Herbert and Gareth Elwyn Jones, Cardiff: University of Wales Press.

———— 1988, 'Wales at Work', in *Wales Between the Wars*, edited by Trevor Herbert and Gareth Elwyn Jones, Cardiff: University of Wales Press.

———— 1993, *Aneurin Bevan and the World of South Wales*, Cardiff: University of Wales Press.

Smith, Joan and Harry McShane 1978, *No Mean Fighter*, London: Pluto.

Smith, Steve 1981, 'Craft Consciousness, Class Consciousness: Petrograd 1917', *History Workshop Journal*, 11(1): 33–58.

Snowden, Philip 1913, *Socialism and Syndicalism*, Baltimore.

Stead, Peter 1973a, 'The Welsh Working Class', *Llafur*, 1(2).

———— 1973b, 'Working Class Leadership in South Wales, 1900–20', *Welsh History Review*, 6.

Supple, Barry 1984, 'No Bloody Revolutions but for Obstinate reactions? British Coalowners in their Context, 1919–1920', in *Enterprise and History*, edited by D.C. Coleman and P. Matthias, Cambridge: Cambridge University Press.

———— 1987, *The History of the British Coal Industry, Vol. 4, 1913–1946*, Oxford: Clarendon Press.

Tawney, R.H. 1943, 'The Abolition of Economic Controls, 1918–21', *Economic History Review*, 13(1–2): 1–30.

Taylor, Andrew 1984, '"The Pulse of One Fraternity": Non-Unionism in the Yorkshire Coalfield, 1931–38', *Bulletin of the Society for the Study of Labour History*, 49: 46–56.

———— 1992, 'Trailed on the Tail of a Comet: The Yorkshire Miners and the ILP, 1885–1908', in *The Centennial History of the Independent Labour Party*, edited by David James, Tony Jowitt and Keith Laybourn, Krumlin: Ryburn Academic.

———— 1996, 'The Politics of Labourism in the Yorkshire Coalfield, 1926–1945', in *Miners, Unions and Politics, 1910–47*, edited by Alan Campbell, Nina Fishman and David Howell, Aldershot: Scolar Press.

Taylor, A.J.P. 1965, *English History 1914–45*, Oxford: Clarendon Press.

Thomas, J. 1922, 'The South Wales Coalfield during Government Control, 1914–21', Unpublished MA Thesis, University of Wales.

Thompson, E.P. 1968, *The Making of the English Working Class*, London: Pelican.

———— 1978, *The Poverty of Theory and Other Essays*, London: Merlin.

Tilly, Charles 1993, *European Revolutions, 1492–1992*, Oxford: Blackwell.

Tolliday, Steven and Jonathan Zeitlin 1985, *Shop Floor Bargaining and the State: Historical and Comparative Perspectives*, Cambridge: Cambridge University Press.

Unofficial Reform Committee 1974, 'The Miners' Next Step', reprinted in *Democracy in the Mines: Some Documents of the Controversy on Mines Nationalisation up to the time of the Sankey Commission*, edited by Ken Coates, Nottingham: Spokesman Books.

Waites, Bernard 1976, 'The Effect of the First World War on Class and Status in England, 1910–20', *Journal of Contemporary History*, 11: 27–48.

———— 1987, *A Class Society at War: Britain During World War One*, Leamington Spa: Berg.

Waller, Robert 1983, *The Dukeries Transformed: The Social and Political Development of a Twentieth Century Coalfield*, Oxford: Clarendon.

Webb, Beatrice and Sidney Webb 1907, *History of Trade Unionism*, London: Longman.

———— 1920, *Industrial Democracy*, London: Longman.

Weeks, John 1981, *Capital and Exploitation*, Princeton, NJ: Princeton University Press.

White, S. 1974, 'Labour's Council of Action, 1920', *Journal of Contemporary History*, 9(4): 99–122.

———— 1979, *Britain and the Bolshevik Revolution: A Study in the Politics of Diplomacy, 1920–24*, London: Macmillan.

Williams, Chris 1996, 'The Hope of the British Proletariat: The South Wales Miners, 1910–1947', in *Miners, Unions and Politics, 1910–47*, edited by Alan Campbell, Nina Fishman and David Howell, Aldershot: Scolar Press.

Williams, C.R. 1962, 'The Welsh Religious Revival, 1904–5', *British Journal of Sociology*, 13: 242–59.

Williams, J.E. 1962, *The Derbyshire Miners*, London: Allen and Unwin.

Williams, L.J. 1973, 'The Road to Tonypandy', *Llafur*, 1(2).

Williams, Robert 1921, *The New Labour Outlook*, London: New Era Series.

Wilshire, Gaylord 1912, 'Syndicalism: What it is', *The British Socialist*, 1(5).

Wilson, Trevor 1966, *The Downfall of the Liberal Party 1914–1935*, London: Collins.

———— (ed.) 1970, *The Political Diaries of C.P. Scott, 1911–1928*, London: Collins.

Winter, J.M. 1974, *Socialism and the Challenge of War: Ideas and Politics in Britain 1912–18*, London: Routledge and Kegan Paul.

———— 1985, 'Trade Unions and the Labour Party in Britain', in *The Development of Trade Unionism in Great Britain and Germany, 1880–1914* edited by Wolfgang Mommsen and Hans-Gerhard Husung, London: Allen and Unwin.

Woodhouse, M.G. 1969, 'Rank and File Movements among the Miners of South Wales, 1910–26', Unpublished PhD Thesis, Oxford University.

———— 1978, 'Mines for the Nation or Mines for the Miners? *Alternative perspectives on industrial democracy*, 1919–21', Llafur, 2(3).

Wrigley, Christopher 1976, *David Lloyd George and the British Labour Movement*, London: Harvester.

———— 1979, *1919: The Critical Year*, West Midlands Society for the Study of Labour History.

———— 1990, *David Lloyd George and the Challenge of Labour: The Post-War Coalition, 1918–22*, London: Harvester Wheatsheaf.

———— 1993, 'The State and the Challenge of Labour in Britain 1917–1920', in *Challenges of Labour*, London: Routledge.

———— 2000, 'Counter-Revolution and the "Failure" of Revolution in Interwar Europe', in *Revolutions and the Revolutionary Tradition in the West: 1560–1991*, edited by David Parker, London: Routledge.

Index